Zbigniew W. Raś and Alicja A. Wieczorkowska (Eds.)

Advances in Music Information Retrieval

Studies in Computational Intelligence, Volume 274

Editor-in-Chief

Prof. Janusz Kacprzyk
Systems Research Institute
Polish Academy of Sciences
ul. Newelska 6
01-447 Warsaw
Poland
E-mail: kacprzyk@ibspan.waw.pl

Further volumes of this series can be found on our
homepage: springer.com

Vol. 256. Patricia Melin, Janusz Kacprzyk, and
Witold Pedrycz (Eds.)
*Bio-inspired Hybrid Intelligent Systems for Image Analysis
and Pattern Recognition,* 2009
ISBN 978-3-642-04515-8

Vol. 257. Oscar Castillo, Witold Pedrycz, and
Janusz Kacprzyk (Eds.)
*Evolutionary Design of Intelligent Systems in Modeling,
Simulation and Control,* 2009
ISBN 978-3-642-04513-4

Vol. 258. Leonardo Franco, David A. Elizondo, and
José M. Jerez (Eds.)
Constructive Neural Networks, 2009
ISBN 978-3-642-04511-0

Vol. 259. Kasthurirangan Gopalakrishnan, Halil Ceylan, and
Nii O. Attoh-Okine (Eds.)
*Intelligent and Soft Computing in Infrastructure Systems
Engineering,* 2009
ISBN 978-3-642-04585-1

Vol. 260. Edward Szczerbicki and Ngoc Thanh Nguyen (Eds.)
Smart Information and Knowledge Management, 2009
ISBN 978-3-642-04583-7

Vol. 261. Nadia Nedjah, Leandro dos Santos Coelho, and
Luiza de Macedo de Mourelle (Eds.)
Multi-Objective Swarm Intelligent Systems, 2009
ISBN 978-3-642-05164-7

Vol. 262. Jacek Koronacki, Zbigniew W. Ras,
Slawomir T. Wierzchon, and Janusz Kacprzyk (Eds.)
Advances in Machine Learning I, 2009
ISBN 978-3-642-05176-0

Vol. 263. Jacek Koronacki, Zbigniew W. Ras,
Slawomir T. Wierzchon, and Janusz Kacprzyk (Eds.)
Advances in Machine Learning II, 2009
ISBN 978-3-642-05178-4

Vol. 264. Olivier Sigaud and Jan Peters (Eds.)
From Motor Learning to Interaction Learning in Robots, 2009
ISBN 978-3-642-05180-7

Vol. 265. Zbigniew W. Ras and Li-Shiang Tsay (Eds.)
Advances in Intelligent Information Systems, 2009
ISBN 978-3-642-05182-1

Vol. 266. Akitoshi Hanazawa, Tsutom Miki,
and Keiichi Horio (Eds.)
Brain Inspired Information Technology, 2009
ISBN 978-3-642-04024-5

Vol. 267. Ivan Zelinka, Sergej Celikovský, Hendrik Richter,
and Guanrong Chen (Eds.)
Evolutionary Algorithms and Chaotic Systems, 2009
ISBN 978-3-642-10706-1

Vol. 268. Johann M.Ph. Schumann and Yan Liu (Eds.)
*Applications of Neural Networks in High
Assurance Systems,* 2009
ISBN 978-3-642-10689-7

Vol. 269. Francisco Fernández de de Vega and
Erick Cantú-Paz (Eds.)
Parallel and Distributed Computational Intelligence, 2009
ISBN 978-3-642-10674-3

Vol. 270. Zong Woo Geem
Recent Advances In Harmony Search Algorithm, 2009
ISBN 978-3-642-04316-1

Vol. 271. Janusz Kacprzyk, Frederick E. Petry, and Adnan
Yazici (Eds.)
*Uncertainty Approaches for Spatial Data Modeling and
Processing,* 2009
ISBN 978-3-642-10662-0

Vol. 272. Carlos A. Coello Coello, Clarisse Dhaenens, and
Laetitia Jourdan (Eds.)
Advances in Multi-Objective Nature Inspired Computing,
2009
ISBN 978-3-642-11217-1

Vol. 273. Fatos Xhafa, Santi Caballé, Ajith Abraham,
Thanasis Daradoumis, and Angel Alejandro Juan Perez
(Eds.)
*Computational Intelligence for Technology Enhanced
Learning,* 2010
ISBN 978-3-642-11223-2

Vol. 274. Zbigniew W. Raś and Alicja A. Wieczorkowska (Eds.)
Advances in Music Information Retrieval, 2010
ISBN 978-3-642-11673-5

Zbigniew W. Raś and Alicja A. Wieczorkowska (Eds.)

Advances in Music
Information Retrieval

 Springer

Zbigniew W. Raś
Dept. of Computer Science
University of North Carolina
Charlotte, NC 28223, USA
and
Polish-Japanese Institute
of Information Technology
Department of Intelligent Systems
02-008 Warsaw
Poland
E-mail: ras@uncc.edu

Alicja A. Wieczorkowska
Polish-Japanese Institute
of Information Technology
Multimedia Department
02-008 Warsaw
Poland
E-mail: alicja@pjwstk.edu.pl

ISBN 978-3-642-11673-5   e-ISBN 978-3-642-11674-2

DOI 10.1007/978-3-642-11674-2

Studies in Computational Intelligence   ISSN 1860-949X

Library of Congress Control Number: 2009943434

Typeset & Cover Design: Scientific Publishing Services Pvt. Ltd., Chennai, India.

Printed in acid-free paper

9 8 7 6 5 4 3 2 1

springer.com

Preface

Sound waves propagate through various media, and allow communication or entertainment for us, humans. Music we hear or create can be perceived in such aspects as rhythm, melody, harmony, timbre, or mood. All these elements of music can be of interest for users of music information retrieval systems. Since vast music repositories are available for everyone in everyday use (both in private collections, and in the Internet), it is desirable and becomes necessary to browse music collections by contents. Therefore, music information retrieval can be potentially of interest for every user of computers and the Internet. There is a lot of research performed in music information retrieval domain, and the outcomes, as well as trends in this research, are certainly worth popularizing. This idea motivated us to prepare the book on *Advances in Music Information Retrieval*.

Music is present in our lives in various ways. We enjoy listening to the music, sometimes singing, maybe playing, or even creating our own music. In general, musical activities can be divided into four categories: *Performing, Composing, Improvising, Listening and Music Perception*. From the technical point of view, *Performing* music means taking a high level representation, and transforming it into an acoustic waveform, in front of listeners. The interpretation of a musical score can differ from performer to performer and it depends on the artist's understanding of the structure and meaning of a piece of music. Emotional expression allows a performance to portray certain moods or emotions. The classical view of *musical composition* is that of a composer creating a piece of music which is represented using some representation format designed for this purpose. For example, a widespread representation format is common music notation (sheet music). This format uses specially devised symbols to represent information on the music such as the pitch and the duration of notes. This representation reflects composer's wishes, to be followed by performers. *Improvisation* is an impenetrable skill based very much on instantaneous intuition, inspiration and insight. *Listening and Music Perception* is highly intertwined with both emotions and context. Not surprisingly, many of the users' information seeking actions aim at retrieving music songs based on these perceptual dimensions - moods and themes, expressing how people feel about a piece of music or which situations they associate it with. In order to successfully support music retrieval along these dimensions, powerful methods are needed.

The websites providing music services usually support text-based searching and/or category based browsing only. For content-based search or feature-based filtering systems, one important problem is to describe music by its parameters or features, so that search engines or information filtering agents can use them to measure the similarity of the target and the candidates. MPEG-7 is an international standard, which describes the multimedia content data to allow universal indexing, retrieval, filtering, control, and other activities supported by rich metadata. However, the metadata about the multimedia content itself are still insufficient, because many features of multimedia content are quite perceptual and user-dependent. For example, emotional features are very important for multimedia retrieval, but they are hard to be described by a universal model since different users may have different emotional responses to the same multimedia content. In the last few years a wealth of research effort has been invested to analyze multimedia content, including music. Many techniques for music information retrieval related to harmony, chord progression, timbre, rhythm, and tempo have been proposed.

This book covers some of the above and closely related topics. It is divided into four sections: *MIR Methods and Platforms, Harmony, Music Similarity, and Content Based Identification and Retrieval.* Glossary of basic terms is given at the end of the book, to familiarize readers with vocabulary referring to music information retrieval.

The first section of this book contains five contributions in the area of *Music Information Retrieval: Indexing, Representations, and Platforms.*

The first chapter is written by *Rainer Typke* and *Agatha Walczak-Typke*. It gives an overview of some existing approaches to building an indexing structure that makes efficient retrieval possible even if the underlying dissimilarity measure is not a metric. Authors show that by tailoring an indexing structure to the non-metric distance measure at hand, it can be possible to guarantee that no false negatives are introduced by indexing. They also give an overview of some more generally applicable statistical methods, as well as embeddings into metric spaces.

The second chapter titled *"Clustering Driven Cascade Classifiers for Multi-Indexing of Polyphonic Music by Instruments"* concerns the problem of automatic indexing of polyphonic music by instruments and their categories. A large database of music instrument sounds was built by authors and used for training a number of classifiers. It is shown that automatic indexing systems for polyphonic music driven by a cascade classifier outperforms standard flat classifiers. Clustering analysis is used to build a new hierarchical schema for classifying music instruments. Authors show that a cascade classifier based on this new hierarchical schema outperforms related classifiers based on Hornbostel-Sachs tree. Presented methods for automatic indexing of polyphonic music do not use sound separation algorithms.

In the third chapter titled *"Representations of Music in Ranking Rhythmic Hypotheses"*, Jarosław Wójcik and Bożena Kostek examine the problem of finding the rhythmic structure of music in a symbolic format. Authors present a new method for retrieving a hypermetric structure of rhythm of a musical piece consisting of rhythmic motives, phrases, and sentences. On the basis of the retrieved hypermetric structure, they propose a system capable of creating automatic drum accompaniment to a given melody supporting the composition.

The fourth chapter is written by *Tetsuro Kitahara*. Author describes various mid-level representations of music. In the early 2000s, it was common to use low-level features such as spectral and cepstral features extracted directly from polyphonic audio signals. Various researchers have pointed out the importance of higher-level, musically meaningful representations and have been engaged in discovering such new music representations. The presented review is a brief survey of the latest results of these attempts.

The last chapter of the first section is written by *J. Stephen Downie, Andreas F. Ehmann, Mert Bay,* and *M. Cameron Jones.* Since the Music Information Retrieval Evaluation eXchange (MIREX) began in 2005, it has fostered great advancements not only in many specific areas of MIR, but also in our general understanding of how MIR systems and algorithms are to be evaluated. Authors outline some of the major highlights of the past four years of MIREX evaluations, including its organizing principles, the selection of evaluation metrics, and the evolution of evaluation tasks.

The second part of the book contains three contributions in the area of *Harmony*.

Current chord analysis techniques often disregard specific note information in favor of a chord color or, in other words, pitch class profile technique. Pitch class profiles cannot disambiguate between inversions or voicing of a single chord, nor can they identify where in the musical range a chord may have been played. By enumerating possible and common chord voicings, it is possible to improve standard pitch class profile techniques by identifying the chord voicing. In the first chapter written by *David Gerhard* and *Xinglin Zhang*, authors demonstrate these techniques in a chord detection system they developed. The system makes use of voicing constraints to increase accuracy of chord and chord sequence identification.

In the second chapter written by *Daniele Radicioni* and *Roberto Esposito*, authors present BREVE, a system for performing chord recognition. The system relies on a conditional model, where domain knowledge is encoded in the form of Boolean features. They cast the tonal harmony analysis problem to a sequential one, and adopt a Supervised Sequential Learning (SSL) approach. The implemented system is validated on a corpus of J.S. Bach's chorales.

In the third chapter titled *"Analysis of Chord Progression Data"* authors study the problem of comparing a chord progression against others using chord N-grams. Various approaches for simplifying chord names are compared. Prominent Jazz composers are compared against each other using the distribution of N-grams obtained from their compositions. The authors also explore the use of chord N-grams as the query to retrieve chord progressions by the same composer.

The third part of the book contains five contributions in the area of *Content-based Identification and Retrieval of Musical Information*.

The first chapter is written by *Riccardo Miotto, Nicola Montecchio* and *Nicola Orio*. Authors describe a methodology for the statistical modeling of music works. Starting from either the representation of the symbolic score or the audio recording of a performance, a hidden Markov model is built to represent the corresponding music work. The model can be used to identify unknown recordings and to align them with the corresponding score.

The second chapter titled *"Harmonic and Percussive Sound Separation and its Application to MIR-related Tasks"* presents a simple and fast method to separate a monaural audio signal into harmonic and percussive components. The convergence is guaranteed in the method and it can be implemented in real time. The usefulness is shown in application to automatic chord recognition and rhythm pattern extraction.

In the third chapter written by *Piotr Wrzeciono* and *Krzysztof Marasek*, authors describe properties of violin modes, related search methods, and a violinist's reaction for their presence in the energy spectrum. This description also contains a mathematical model of violin sound, and chromatic scales played on this instrument. Presented classifier is used to evaluate the jurors' appraisals in the tone quality category. It takes into account both subjective and objective parameters. The objective parameters are the instrument's modes – frequency and the mutual energy factor. The subjective qualities are the consonances or dissonances between mode frequencies.

In the fourth chapter titled *"Emotion Based MIDI Files Retrieval System"*, authors present a cooperative query answering system driven by classifiers for automatic indexing of music by emotions. It includes visualization module which is used to boost emotions envoked by music. The training database is multi-labeled and describes music files on different granularity levels using collection of harmonic and rhythmic attributes, and a hierarchical structure of emotions based on Thayer's model. Personalization aspect of the query answering system related to subjectivity in emotion perception is also considered.

In the fifth chapter titled *"On Search for Emotion in Hindusthani Vocal Music"*, authors present the experiments used to extract meaningful emotional sequences of sounds from ragas, and to test what emotions were evoked by these sequences. The listening tests were performed on two groups of listeners: Hindustani listeners, and Western listeners not familiar with Hindustani music. For both groups, authors investigated what emotions were evoked, for the audio segments used in listening test, and for sequences of notes of minimal length, specific for each raga.

The fourth part of the book contains four contributions in the area of *Music Similarity*.

Its first chapter is written by *Joan Serra, Emilia Gomez*, and *Perfecto Herrera*. It comprehensively summarizes the work done in cover song identification while encompassing the background related to this area of research. The most promising strategies are reviewed and qualitatively compared under a common framework, and their evaluation methodologies are critically assessed.

The second chapter is written by *Rudolf Mayer* and *Andreas Rauber*. Authors show the influence of so called text statistic features on song similarity. Musical similarity can be defined on textual analysis of certain parts-of-speech (POS) characteristics. Analogously to the common beats-per-minute (BPM) descriptor in audio analysis, they introduce the words-per-minute (WPM) measure to identify similar songs. The rationale behind WPM is that it can capture the 'density' of a song and its rhythmic sound in terms of similarity in audio and lyrics characteristics.

The third chapter is written by *Marcus T. Pearce, Daniel Mullensiefen*, and *Geraint A. Wiggins*. Authors examine the problem of melodic segmentation in music information retrieval. They review a number of existing algorithms before introducing a new method called IDyOM (Information Dynamics of Music) based on unsupervised statistical learning. The proposed model exploits a putative relationship between predictive modeling and grouping boundaries and, in contrast to rule-based models, is sensitive to stylistic differences between music corpora. The performance of the models is compared in segmenting a large collection of German folk songs and the four best-performing models (including IDyOM) are combined into a hybrid model that outperforms each of its component models.

The last chapter is written by *Shyamala Doraisamy* and *Shahram Golzari*. Authors investigate the Artificial Immune Recognition System (AIRS) as a classifier for musical genres from differing cultures. Musical data of two cultures were used – Traditional Malay Music (TMM) and Latin Music (LM). The performance of AIRS for the classification of these genres is compared with performances using several commonly used classifiers.

We wish to express our thanks to all the authors who contributed the above seventeen chapters to this book.

October 2009   Z.W. Raś
                                                                          A.A. Wieczorkowska

Contents

Part III: Content-Based Identification and Retrieval of Musical Information

Part IV: Music Similarity

Part I
Music Information Retrieval: Indexing, Representations, and Platforms

Indexing Techniques for Non-metric Music Dissimilarity Measures

Rainer Typke and Agatha Walczak-Typke

Abstract. Many dissimilarity measures suitable for music retrieval do not satisfy all properties of a metric. This rules out the use of many established indexing structures, most of which rely on metricity. In this chapter, we give an overview of some existing approaches to building an indexing structure that makes efficient retrieval possible even if the underlying dissimilarity measure is not a metric.

For symmetric prametrics with metric subspaces, a tunneling technique allows one to search a non-metric space efficiently without false negatives. We give a detailed example for this case. In a query-by-example scenario, if queries are already part of a collection, and the triangle inequality is violated, one can enforce it in subsets of the collection by adding a small constant to the distance measure (Linear Constant Embedding). By embedding a non-metric distance function into a metric space in a way that preserves the ordering induced by the function on any query, one can make indexing methods applicable that usually only work for metrics (TriGen). Also, we present several probabilistic methods, including distance based hashing (DBH), clustering (DynDex), and a tree structure with pointers to near neighbours (SASH).

1 Non-metric Distance Measures of Musical Interest

1.1 Metrics and Partial Matching

A distance function $d : X \longrightarrow \mathbb{R}$ is a *metric* if it satisfies the following conditions for every $x, y, z \in X$:

Rainer Typke
Austrian Research Institute for Artificial Intelligence (OFAI)
e-mail: rainer.typke@ofai.at

Agatha Walczak-Typke
Kurt Gödel Research Center, University of Vienna
e-mail: agatha@logic.univie.ac.at

Z.W. Raś and A.A. Wieczorkowska (Eds.): Adv. in Music Inform. Retrieval, SCI 274, pp. 3–17.
springerlink.com   © Springer-Verlag Berlin Heidelberg 2010

1. $d(x,y) \geq 0$ (non-negativity)
2. $d(x,y) = 0$ if and only if $x = y$ (identity of indiscernibles)
3. $d(x,y) = d(y,x)$ (symmetry)
4. $d(x,z) \leq d(x,y) + d(y,z)$ (triangle inequality)

Some distance measures (such as the basic Levenshtein distance [10]) applicable for music retrieval are metrics. However, these metric distances are usually used for matching complete queries to complete database items.

If one alters the distance to allow for partial matching, the distance may fail one or more of the conditions of a metric. By *partial matching* we mean that a distance measure determines two objects to be close to each other if one object is similar to either parts of the other or the complete other object.

More generally, some distance measures which capture human notions of similarity well will fail one or more conditions of a metric. The authors of [14] point out psychological studies and experimental evidence that self similarity is not constant (some things are more "similar to themselves" than others), and that in addition, human similarity assessment can be asymmetrical. Experimental evidence specific to human perceptions of musical similarity [12] seem to support this view.

1.2 Commonly Applied Distance Measure Examples

1.2.1 Edit Distances

Edit distances are a class of distances that measure the distance between two strings of characters by the number of operations required to transform one of them into the other. The various distance measures in this class vary by the type of editing operations allowed, and the cost-weights of the various operations.

The *Hamming distance* between two strings of equal length is the number of positions in which characters differ. In other terms, the Hamming distance allows the operation of substitution, with each use of this operation having cost 1. In a space composed of strings of a fixed length n, the Hamming distance is a metric. The Hamming distance can be viewed as the simplest edit distance.

The *Levenshtein distance* [10] (often called *editing distance*) allows the operations of insertion, deletion, and substitution of a single character. If each operation has the same cost-weight, the Levenshtein distance is a metric. *Generalized Levenshtein distances* allow different cost-weights for different replacement, insertion, or deletion operations. The metricity of a generalized Levenshtein distance depends on the choice of cost-weights.

A different generalization of Levenshtein distance is the *Damerau-Levenshtein distance*. This distance allows all the operations allowed in Levenshtein distance, with the addition of the ability to transpose two characters.

For music retrieval, editing distances are suitable for comparing melodic contours, rhythms, or chords. In the past they have also been used for information theory, coding theory and cryptography (Hamming distance), spell checkers

(Levenshtein distance), natural language processing, DNA and protein comparisons (Damerau-Levenshtein distance).

1.2.2 Earth Mover's Distances

We describe the class of Earth Mover's distances (EMD) in more detail here because we will show an exact indexing solution for a variant of the EMD in Section 2.2.1. Also, it illustrates a common reason for non-metricity: partial matching, and matching only similar subparts of two compared objects.

The EMD measures the distance between weighted point sets. Intuitively speaking, a weighted point set a_i can be imagined as an array of piles of dirt each equal to w_i units, situated at position x_i. The role of the supplier is arbitrarily assigned to one array and that of the receiver to the other one, and the arrays of piles are made to look as similar as possible by shifting dirt from piles in the supplier array to piles in the receiver array. The EMD then measures the minimum amount of work needed to make two arrays of piles as similar as possible in this way. See [4] for a more detailed description of the EMD. We now define the EMD formally:

Fix a *ground distance* d on $\mathbb{R}^k$. The ground distance can, but need not be, a metric.

Let $A = \{a_1, a_2, \ldots, a_m\}$, $B = \{b_1, b_2, \ldots, b_n\}$ be *weighted point sets* such that $a_i = \{(x_i, w_i)\}, i = 1, \ldots, m$, $b_j = \{(y_j, v_j)\}, j = 1, \ldots, n$, where $x_i, y_j \in \mathbb{R}^k$ with $w_i, v_j \in \mathbb{R}^+ \cup \{0\}$ being the respective weights.

Let $W_A = \sum_{i=1}^{m} w_i$ be the total weight of set A; the total weight W_B of the set B is defined analogously.

Let $d_{ij} = d(x_i, y_j)$ denote the ground distance between individual coordinates in A and B, without regard to their weight.

A *flow matrix* $F = (f_{ij})$ between A and B is an $m \times n$ matrix of non-negative real numbers, such that for each $1 \leqslant i \leqslant m$, $\sum_{j=1}^{n} f_{ij} \leqslant w_i$, and for each $1 \leqslant j \leqslant n$, $\sum_{i=1}^{m} f_{ij} \leqslant v_j$. Furthermore, we require that $\sum_i \sum_j f_{ij} = \min(W_A, W_B)$. Denote by $\mathscr{F}$ the collection of all possible flow matrices between A and B.

The *Earth Mover's Distance*, $\text{EMD}_d(A, B)$, between A and B is defined as

$$\text{EMD}_d(A, B) = \frac{\min_{F \in \mathscr{F}} \sum_{i=1}^{m} \sum_{j=1}^{n} f_{ij} d_{ij}}{\min(W_A, W_B)}.$$

For the remainder of this chapter, we will assume that the ground distance for the EMD is the Euclidean metric l_2. With this assumption in mind, we drop the subscript referring to the ground distance from our formulas, writing $\text{EMD}(A, B)$ instead of $\text{EMD}_{l_2}(A, B)$.

The EMD is a useful measure for music similarity, as was demonstrated at the annual MIREX comparison of music retrieval algorithms in 2006 (http://www.music-ir.org/mirex/2006/index.php/Symbolic_ Melodic_Similarity_Results). Useful properties of the EMD include its continuity, its ability to support partial matching, and its robustness against distortions of tempo and pitch when measuring melodic similarity for symbolically encoded music. For doing so, one can represent every note by a point in the two-dimensional space of onset time and pitch. The weights of points can be used to encode the importance of notes. See [18] for details.

The EMD is in general not a metric. Specifically, the triangle inequality does not hold, and while the EMD of a point to itself is 0, there can exist distinct points that are also EMD 0 from one another. While there is no universally accepted terminology in the mathematical literature for weak distance measures, there is some precedent for calling weak distance measures with properties like the EMD *symmetric prametrics* [13].

It should be emphasized that the EMD does behave as a metric if one restricts the domain of the EMD to point sets having a given weight, assuming that the ground distance is a metric [4]. One can take advantage of this property when working with an EMD which is a prametric by decomposing the space of possible point sets into subspaces each containing only point sets having a given weight. We will refer to such subspaces as *metric subspaces* of the EMD space.

1.3   Other Commonly Used Measures

There are other dissimilarity measures which are suitable for music information retrieval applications, many of which are not metric. Examples include statistical modeling of melodic and rhythmic content with Hidden Markov Models (HMM), where the distance measure is the probability of an HMM (which represents a melody in the database) generating the query [16]; the presence of identical or slightly modified n-grams [5] (metricity here depends on how one treats deviations in n-grams); and many others. The focus of this chapter, however, is not to give an overview of dissimilarity measures used in Music Information Retrieval, but an overview of how one can use non-metric measures efficiently.

2   Strategies for Non-metrics

In Table 1, we will give an overview of the indexing methods for non-metrics that will be described in this section.

In our presentation below, we divide these methods into four broad categories: metric embedding, metrizable partition, other partition, and other. *Metric embedding* approaches embed the non-metric space into a metric space with as little distortion of distance as possible. Then, some known indexing method which relies on metricity is applied. We follow Skopal [17] and call such an indexing method a *metric access method (MAM)*. *Metrizable partition* methods partition the search space into pieces which are themselves either metric spaces, or can be easily altered to allow the use of a MAM. *Other partition* methods partition the search space into pieces but do not make any assumptions about metric-like behavior of the distance on the pieces. Finally, *other* methods are ones that do not fall into any of the above three categories.

Within each category, there is a tradeoff between generality on one side and accuracy and efficiency on the other. If one exploits specific properties of a particular distance measure, as we do for the combined tunneling and vantage indexing method described in Section 2.2.1, one is more likely to be able to achieve a recall of 100 % and few false positives than if one only assumes symmetry and the availability of a pairwise distance matrix.

Note that some methods become feasible only if one restricts queries to objects that are already in the database – Chen's "Local Constant Embedding" (see Section 2.2.2) is an example.

2.1 Metric Embedding Methods

2.1.1 BoostMap: Embedding into a Metric Space

Athitsos et al. [1] propose to combine "line projection functions" for embedding data items into a multidimensional space. At first, only a distance matrix is given. The line projection functions have the form:

$$F_{x_1,x_2}(x) = \frac{d(x,x_1)^2 + d(x_1,x_2)^2 - d(x,x_2)^2}{2d(x_1,x_2)}$$

If two items x_1, x_2 from the database are given, such a function can be used to project any other item onto a line, and thus embed it in a one-dimensional space, such that proximity to x_1 and x_2 is preserved if the triangle inequality holds. If it does not hold, in practice, such embedding functions may still preserve proximity most of the time.

One can view each line projection function as a weak classifier that reports, with a relatively high error rate, for any triplet (q, x_1, x_2), whether q is closer to x_1 or to x_2. By combining a suitable group of different line projection functions (functions that differ in their x_1 and/or x_2), one can obtain a multidimensional embedding that behaves as a strong classifier. A suitable group can be identified with the AdaBoost framework [15].

The AdaBoost-inspired training algorithm delivers an embedding and a weighted Manhattan distance. The latter is usually not a metric because some of the weights can be negative, but it can probably be evaluated more efficiently than the original function.

For an excellent survey and analysis of more embedding methods (SparseMap, FastMap, MetricMap), see [8].

2.1.2 TriGen: Embedding a Semi-metric into a Metric Space

Introduction and Assumptions

TriGen, introduced by Skopal [17], is a method for embedding a semi-metric space into a metric space in which an appropriate metric-access method for similarity search can then be used. Generally, in similarity search, dataset objects are ordered according to a single query object, and then the most similar, based on distance, are chosen. TriGen finds an function that embeds the original similarity measure into a metric, but does not alter similarity orderings.

In this method, the initial similarity measure is treated as a black box, and none of its particular topological or geometric properties are utilized. Only one main assumption is made on a similarity measure: it is assumed to have bounded value. For the purposes of the method, the measure is assumed to be a *semi-metric*, that is, a distance measure that fails the triangle inequality, but otherwise has the same properties of non-negativity, symmetry, and identity of indiscernibles as a metric.

In case a given similarity measure fails some more of the properties of a metric, the measure is slightly and easily modified. Non-negativity can be guaranteed by a shift of values. Identity of indiscernibles can be guaranteed by requiring every two non-identical objects be at least ε distant, for ε some positive lower bound. Symmetry is enforced by taking the distance between two points to be the minimum of their asymmetric distances. None of these modifications change similarity orderings.

Enforcing the triangle inequality

Enforcement of the triangle inequality is then the only property left, and one that requires much more effort.

We reintroduce some of Skopal's notation: A *triangular triplet* is a triplet of non-negative real numbers $(a, b, c), a, b, c \geqslant 0$, such that $a + b \geqslant c$ $b + c \geqslant a$, and $a + c \geqslant b$. A distance measure d *generates a triangular triplet* if there are objects $o_i, o_j, o_k \in \mathbb{S}$ in the search space $\mathbb{S}$ such that $(\mathrm{d}(o_i, o_j), \mathrm{d}(o_j, o_k), \mathrm{d}(o_i, o_k))$ is a triangular triplet. Of course, a metric only generates triangular triplets, and if a measure only generates triangular triplets, then it satisfies the triangular inequality.

Given a similarity measure d, we call $\mathrm{d}^f(o_i, o_j) = f(\mathrm{d}(o_i, o_j))$ a *similarity-preserving modification* of d (or *SP-modification*) if f is a strictly increasing bounded function for which $f(0) = 0$. Here, the function f is referred to as an *SP-modifier*.

We define a similarity ordering $SO_{\mathrm{d}(q)}$ for d with respect to a query q as $(o_i, o_j) \in SO_{\mathrm{d}(q)} \leftrightarrow \mathrm{d}(q, o_i) < \mathrm{d}(q, o_j)$, for $o_i, o_j \in \mathbb{S}$ objects in the search space. Then, SO_{d} is the space of similarity ordering for a given similarity measure d.

The goal is to find an SP-modifier, called *metric-preserving*, that maps triangular triplets to triangular triplets and so preserves the triangle inequality. Skopal shows that an SP-modfier that preserves the triangular inequality must be subadditive, that is, $f(x) + f(y) \geqslant f(x + y)$, for all x, y. He also shows that any concave SP-modifier is metric-preserving.

Indeed, the strictly concave SP-modifiers are good candidates for embedding functions, as shown by Theorem 1 of [17]:

Theorem 1. *Given a semi-metric d, there always exists a strictly concave SP-modifier f such that the SP-modification d^f is a metric.*

The proof of this theorem demonstrates that the more concave a function is, the more triplets become triangular.

Having too concave a function can be problematic. Too concave a function increases the intrinsic dimensionality of the search space with respect to the modified distance. This in turn lowers efficiency of search. Thus, the problem becomes one of finding a function that is concave enough to produce the correct number of triangular triplets, but convex enough to give good efficiency. The TriGen algorithm is an optimization algorithm that finds such a convex function.

2.2   Metrizable Partition Methods

2.2.1   Tunneling and Vantage Indexing

Manhattan EMD

The instance of the EMD described in this subsection is of particular interest for searching rhythmic patterns. We call this EMD the *Manhattan EMD*.

Rhythmic patterns can naturally be represented as sequences of onset times. Each onset is represented with a one-dimensional point of weight 1. The EMD between point sets which represent onset sequences can be used to identify rhythmic patterns which contain a subset or superset of the query, that is, many onsets with similar relative positions. Both queries and results would be onset sequences, and the information need would be onset sequences which contain the query, are contained by the query, or contain sequences of onsets whose relative positions deviate as little as possible from those of the query.

To render irrelevant a musical segment's tempo and location within a piece of music, we scale every segment's numeric representations to a fixed segment duration (say, 60) and translate them so that they start at position 0.

When comparing normalized sequences of onsets containing the same number of onsets, the Manhattan EMD equals the sum of the absolute differences between corresponding onsets a_i and b_i in onset sequences $A = a_1 \ldots a_n$ and $B = b_1 \ldots b_n$, divided by the number of onsets:

$\mathrm{EMD}(A,B) = \frac{\sum_{i=1}^{n} |a_i - b_i|}{n}$. Thus, if we restrict ourselves to point sets of a certain given length, the Manhattan EMD (with l_2 as ground distance) is a metric and is equal to the l_1 norm (also known as "Manhattan norm"). For unequal numbers of onsets, however, the Manhattan EMD violates the triangle inequality, and a distance of zero between two sequences of onsets does not imply that they are identical. In this case, the EMD calculation cannot be simplified like this, but has to be done like in the general case.

Since every normalized segment starts with 0 and ends with 60, we omit these two numbers and view n-onset segments as points in an $(n-2)$-dimensional space. All segments lie in the subset of $\mathbb{R}^{n-2}$ where the coordinates are strictly increasing.

Vantage indexing for metric subspaces

Vantage indexing (as described, for instance, by Vleugels and Veltkamp [20]) is an approach to the retrieval of objects from a large database which are similar to a given query. The search is restricted to items whose pre-calculated distances to a small set of prechosen *vantage objects* are similar to the query's distances to the same vantage objects.

If one works with the l_1 norm, a ball (the set of all points whose distance lies within a certain radius around a point of interest) has the shape of a cross-polytope. A one-dimensional cross-polytope is a line segment, a two-dimensional cross-polytope is a square, for three dimentions, an octahedron, and so forth.

In [19], we show how, by optimally placing vantage objects, we can achieve 100 % recall and 100 % precision for retrieving the contents of balls in the shape of a cross-polytope: for a j-dimensional space, we need at least 2^{j-1} vantage objects because a cross-polytope has 2^j facets (a *facet* is a $(j-1)$-dimensional structure). We place those vantage objects in the corners of the space that is inhabited by database objects. Since our database objects have a known maximum coordinate (they are onset sequences with onsets between 0 and a known upper limit), it is known that only a limited area of the space can hold objects.

Partial matching by tunneling between metric subspaces

When searching sequences of onsets that were detected in audio files, there are two problems: the detection is not always completely reliable, producing both false negatives and false positives, and often there will be multiple voices with onsets, while a user might want to search for rhythmic patterns which occur only in one voice. Therefore, for successfully searching sequences of onsets from audio data, one should be able to ignore a certain number of onsets and still find matching subsequences or supersequences. Also in [19], we show how one can use tunnels between metric subspaces to achieve partial matching while still benefiting from the optimum vantage indexing within the subspaces of sequences with equal numbers of onsets.

The EMD provides partial matching as described above for point sets whose weight sums are unequal. Unfortunately, the EMD does not obey the triangle inequality in such a case. This makes it impractical to directly apply vantage indexing since there would be no way of controlling the number of false negatives. Also, the locality sensitive hashing method, which also relies on the triangle inequality, becomes unusable.

To find near neighbours according to the non-metric EMD, we pre-calculate "tunnels" between metric subspaces. Those tunnels link items to their nearest neighbours in other subspaces. Finding all near neighbours of a given query then involves searching the metric subspace where the query resides, as well as following, for every near neighbour of the same dimensionality, the tunnel to its nearest neighbours in other metric subspaces. This yields almost the same result as an exhaustive linear search, but requires only logarithmic complexity.

To be more precise: it is still possible to avoid any false negatives (by sufficiently enlarging the search radius), but false positives can occur.

The database shown in Figure 1 contains 6 point sets $P_1, \ldots, P_6$. Three, $P_1, \ldots, P_3$, are two-dimensional, the others, one-dimensional. The query Q is one-dimensional. The area of interest within the search radius r around Q is marked grey.

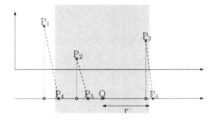

Fig. 1 False negatives and false positives resulting from tunneling, and how to avoid them

False positives: It is conceivable that the projection of a higher-dimensional object onto Q's subspace lies just outside the search radius, but its nearest neighbour in Q's subspace happens to lie within the search radius. An example is P_1, whose projection onto the subspace (shown as a circle) has a nearest neighbour beyond the border of the grey area.

False negatives: It is also possible that while the projection of a higher-dimensional object onto Q's subspace lies inside the search radius, the closest object in Q's subspace lies outside the search radius. In this case, illustrated with P_3 and P_6, the higher-dimensional object will not be retrieved. In the extreme case that there is no single object inside the search radius in Q's subspace, no higher-dimensional objects whatsoever will be retrieved.

Controlling false negatives and false positives. To avoid all false negatives and limit the badness of false positives to a threshold e, one can add the projections as additional "ghost points" to the database if their nearest neighbour in the subspace is further away than $e/2$, and extend the search radius by $e/2$.

The distance of false positives to the query will be at most e higher than desired because in the worst case, the nearest neighbour of the projection will lie on the line connecting the projection with the query. The nearest neighbour can be up to $r + e/2$ away from the query, while the projection can be another $e/2$ away from the nearest neighbour, leading to a total maximum distance of $r + e$ for false positives.

Such additional ghost points would be added as part of the task of building the index, and so would not slow down the search process. It would also not increase the computational complexity of building the index – the only price is some extra space for storing ghost points wherever some point from higher dimensions gets projected into a sparsely populated area in a subspace. There is a tradeoff between the required additional space and the maximum possible distance error for false positives.

2.2.2 Local Constant Embedding: Creating Subspaces Where the Triangle Inequality Holds

Chen and Liang [3] suggest to partition the database and enforce the triangle inequality within each partition by creating a new distance measure. This new distance measure is created from the original one by adding a constant c. This constant varies between the partitions and is kept as small as possible. Note that despite the authors' claim of the contrary, the new distance measure still generally is not a metric – even if it obeys the triangle inequality, it will violate the identity of indiscernibles if $c > 0$, that is, $d'(x,y)$ will always be greater than zero, even if $x = y$, because $d'(x,y) \geq c$.

To build the index, all possible combinations of three items from the database are inspected for violations of the triangle inequality. Unfortunately, there are quite a few triplets: a database with n elements contains $\binom{n}{3} = \frac{n!}{6(n-3)!}$ of them. For each triplet, the smallest possible constant can be determined quite easily – if the distance measure obeys the triangle inequality for the triplet, it is zero, and otherwise it equals the severity of the violation. For instance, if $d(a,b) = 1, d(b,c) = 2$, and $d(a,c) = 7$, the new distance measure could be chosen as $d'(x,y) = d(x,y) + 4$, because the direct route from a to c is 4 larger than the indirect route via b.

Once the severities of triangle violations are known for all triplets, the items are grouped into partitions such that as many items as possible belong to a group with an added constant that is as small as possible.

The index then is searched by searching each partition and merging the results. Each partition can be searched with a method that relies on the triangle inequality, for instance by using vantage objects, without introducing any false negatives. However, the added constant will lead to false positives. In the worst case, the constant can be big enough to prevent any pruning effect.

Such an index can only be searched efficiently for near neighbours of items that are already in the database. After all, the knowledge that the triangle inequality is not violated is only given for the existing partitions. Unknown queries might introduce violations of the triangle inequality to any partition of the database. Chen and Liang suggest a "dynamic" method for supporting unknown queries, but the necessity to look at all possible triplets essentially remains. For a new query, one has to consider all triplets involving the query, that is, all $\binom{n}{2}$ pairs from the database combined with the query.

2.3 Other Partition Methods

2.3.1 DynDex: Clustering

Goh, Li, and Chang [7] propose a cluster-based method for indexing with non-metric distance functions. The underlying assumption here is that near neighbours of a query are probably in the cluster that is closest to the query, thus once a good clustering has been determined, one can always limit the search to a small part of the database by determining the closest cluster or clusters and searching only these clusters or this cluster exhaustively.

As a preparation, all items in the database are clustered using an algorithm which only needs a distance matrix and the desired number of clusters as input, such as CLARANS [11]. For each cluster, a small set of representative items is calculated, for example one single, very central item.

To find the nearest neighbours of a query, the representatives of every cluster are compared to the query, and the closest cluster (the cluster belonging to the representative that is closest to the query) is then exhaustively searched for near neighbours of the query.

To improve recall, one can search more than one cluster. Also, to reduce the impact of errors that were introduced by the clustering, one can use a bagging technique. A "bag" is a set of clusters. Different sets of clusters are generated by using different seed objects for the clustering algorithm, and then all bags are searched in the same way; finally, the search results are merged.

DynDex works well in practice for some data, but it does not guarantee a recall of 100 %. It does guarantee that some pruning takes place – the fewer clusters one searches exhaustively, the larger the pruning effect. However, the part of the database that needs to be considered grows linearly with the size of the database, since more items in the database will lead to either larger clusters or the need for more clusters.

2.3.2 Distance-Based Hashing

Athitsos et al. [2] propose an indexing method for arbitrary distance measures, called Distance-Based Hashing (DBH), that is inspired by locality-sensitive hashing (LSH) [6]. For LSH, one needs hash functions that are locality-sensitive, that is, likely to hash items which are close to each other into the same bucket. LSH is therefore only applicable if one has locality-sensitive hash functions for the given distance measure. DBH, on the other hand, generally uses "line projection" functions (the same functions that are used for BoostMap) to construct hash functions. Parameters t_1 and t_2 for a discretized version of a line projection function are chosen such that its value is 0 for about half the database items (it is 0 for items that are projected into the interval $[t_1, t_2]$), and 1 for the rest of the items. k of these discretized functions are combined to be used as a k-bit hash function. Several such hash functions are used to hash each database item into several buckets. The query object is then compared to each database object that is hashed into the same bucket as the query by at least one of the used hash functions.

Since DBH does not assume anything about the distance measure, and the hash functions are not necessarily locality-sensitive, one cannot say much about the number of false negatives and false positives without sampling the database and counting them. There is no guarantee for the absence of either false negatives or false positives, but in practice the method seems to offer a good tradeoff between pruning the database and finding almost all near neighbours for some databases. How well it works depends on the data and the distance measure.

2.4 Other Method

2.4.1 Spatial Approximation Sample Hierarchy

Houle and Sakuma [9] propose a multi-level structure of random samples, where data items are connected to their approximate near neighbours.

On every level but the top level, each data item is connected to some limited number of near neighbours on a higher level ("parents"). On every level except for the lowest one, each data item is also connected to a limited number of near

Table 1 Indexing methods for non-metrics

METHOD	ACCURACY		APPLICABILITY	
	No false negatives (100% recall)?	No false positives (100% precision)?	Handles lack of triangle inequality	Query need not be in collection
Tunneling for Manhattan EMD [19], Section 2.2.1	✔	Only in metric subspaces	✔	✔
LCE [3], see Section 2.2.2	✔	✘	✔	✘
TriGen [17], see Section 2.1.2	Preserves the order of most similar items. Recall depends on the chosen metric access method.	Depends on the chosen metric access method.	✔	✔
DynDex [7], see Section 2.3.1	✘	✘	✔	✔
BoostMap [1], see Section 2.1.1	✘	✘	✔	✔
DBH [2], see Section 2.3.2	✘	✘	✔	✔
SASH [9], see Section 2.4.1	✘	✘	✔	✔

neighbours on lower levels ("children"). When the index structure is built, already existing connections are used to limit the number of nodes that need to be considered for adding another node to the index structure.

Candidates for near neighbours of a query are found in a way very similar to what would be done if the query were to be added to the index structure. That way, only a limited part of the database is considered.

This index method relies on a pairwise distance measure, but makes no other assumptions. It does not guarantee a perfect recall. Its performance varies greatly depending on the kind of data indexed.

3 Overview

In this section, we give a brief overview of the basic ideas, the efficiency, and – in Table 1 – the accuracy and applicability of each of the methods that were presented in this chapter.

- **Tunneling for Manhattan EMD [19], Section 2.2.1:** Search metric subspaces with vantage objects, use pointers to near neighbours to search across subspaces. **Efficiency:** Logarithmic if the number of crossed dimensions is bounded.
- **Linear constant Embedding (LCE) [3], see Section 2.2.2:** Enforce the triangle inequality in separate subsets of the database by adding a small constant to each distance. **Efficiency:** Might be worse than an exhaustive search – if the added constants are large enough, the pruning effect can be zero. Building an index is extremely expensive (involves looking at each of the $\binom{n}{3} = \frac{n!}{6(n-3)!}$ groups of 3 elements from the database).
- **TriGen [17], see Section 2.1.2:** For a bounded non-metric, enforce the triangle inequality using a convex function which embeds the non-metric into a metric space. Then use an access method developed for metric spaces. TriGen is the algorithm that indentifies an optimal convex embedding function. **Efficiency:** depends on the particular non-metric in question, and on the metric access method subsequently applied. Can be very efficient thanks to metricity.
- **DynDex [7], see Section 2.3.1:** Clustering, searching clusters whose representatives are close to the query. **Efficiency:** Linear and always better than an exhaustive search (query is compared to representatives of all clusters, and a fixed number of clusters are searched exhaustively).
- **BoostMap [1], see Section 2.1.1:** Embed all items into a metric space such that the similarity rankings are approximately preserved. **Efficiency:** Efficient search in a metric space is a solved problem.
- **DBH [2], see Section 2.3.2:** Hash each item into l buckets using line projection functions. Search all buckets into which the query gets hashed. **Efficiency:** It depends on the data – usually better than linear.
- **SASH [9], see Section 2.4.1**: A hierarchical graph with connections between approximate near neighbours. **Efficiency:** Never worse than an exhaustive search, usually better. Otherwise no guarantee.

4 Conclusions

Efficient near neighbour search for non-metric distance measures is important for many MIR applications. Many distance measures which reflect human notions of dissimilarity well are not metrics. One common way of losing metricity is partial matching. Music collections which are of interest in real life often contain millions of items, which makes indexing techniques necessary which allow for faster than linear searching.

There are various approaches to the problem of indexing data for the purpose of efficient near neighbour search where the distance measure is non-metric. We have looked at methods which enforce the triangle inequality for all data [17] or at least within partitions of the database [3], at hashing [2], clustering [7], and linking near neighbours [9]. We also presented a specialized indexing method for a non-metric variant of the Earth Mover's Distance.

One cannot achieve perfect recall or precision with the known most generally applicable methods – methods which do not assume anything besides the existence of a pairwise distance matrix and which work for queries that are not known at the time the index is built. TriGen comes closest to this goal among the generally applicable methods we surveyed, but preserving order is not necessarily sufficient for avoiding false negatives and/or false positives for r-near neighbours since the distance estimate gets distorted even if the ordering is preserved. Perfect recall is attainable if one drops the requirement of supporting queries that are unknown at the time the index is built ([3]). However, for guarantees about recall and/or precision even if the query is unknown when the index is built, one needs to tailor the indexing method to the underlying distance measure. If one does that, one can achieve high efficiency and perfect recall even for some non-metric distance measures.

References

1. Athitsos, V., Alon, J., Sclaroff, S., Kollios, G.: Boostmap: A method for efficient approximate similarity rankings. In: IEEE Computer Society Conference on Computer Vision and Pattern Recognition, vol. 2, pp. 268–275 (2004)
2. Athitsos, V., Potamias, M., Papapetrou, P., Kollios, G.: Nearest neighbor retrieval using distance-based hashing. In: ICDE, pp. 327–336. IEEE, Los Alamitos (2008)
3. Chen, L., Lian, X.: Efficient similarity search in nonmetric spaces with local constant embedding. IEEE Transactions on Knowledge and Data Engineering 20(3), 321–336 (2008)
4. Cohen, S.: Finding Color and Shape Patterns in Images. Ph.D. thesis. Stanford University (1999)
5. Stephen Downie, J.: Music retrieval as text retrieval (poster abstract): simple yet effective. In: SIGIR 1999: Proceedings of the 22nd annual international ACM SIGIR conference on Research and development in information retrieval, pp. 297–298. ACM, New York (1999)
6. Gionis, A., Indyk, P., Motwani, R.: Similarity search in high dimensions via hashing, pp. 518–529 (1999)
7. Goh, K.-s., Li, B., Chang, E.: Dyndex: a dynamic and non-metric space indexer. In: ACM Multimedia, pp. 466–475 (2002)
8. Hjaltason, G.R., Samet, H.: Properties of embedding methods for similarity searching in metric spaces. IEEE Trans. Pattern Anal. Mach. Intell. 25(5), 530–549 (2003)
9. Houle, M.E., Sakuma, J.: Fast approximate similarity search in extremely high-dimensional data sets. In: International Conference on Data Engineering, pp. 619–630 (2005)
10. Levenshtein, V.I.: Binary codes capable of correcting deletions, insertions, and reversals. Doklady Akademii Nauk SSSR 163(4), 845–848 (1965)
11. Ng, R.T., Han, J.: Efficient and effective clustering methods for spatial data mining. In: VLDB, pp. 144–155 (1994)
12. Pardo, B., Shifrin, J., Birmingham, W.: Name that tune: a pilot study in finding a melody from a sung query. J. Am. Soc. Inf. Sci. Technol. 55(4), 283–300 (2004)

13. Pontryagin, L.S. (ed.): General topology. I. Encyclopaedia of Mathematical Sciences, vol. 17. Springer, Berlin (1990); Basic concepts and constructions. Dimension theory, A translation of Sovremennye problemy matematiki. Fundamentalnye napravleniya, Tom 17, Akad. Nauk SSSR, Vsesoyuz. Inst. Nauchn. i Tekhn. Inform., Moscow (1988) [MR0942943 (89m:54001)], Translation by D. B. O'Shea, Translation edited by A. V. Arkhangel′ skiĭ and L. S. Pontryagin

14. Santini, S., Jain, R.: Similarity measures. IEEE Trans. Pattern Anal. Mach. Intell. 21(9), 871–883 (1999)

15. Schapire, R.E., Singer, Y.: Improved boosting algorithms using confidence-rated predictions. Machine Learning 37(3), 297–336 (1999)

16. Shifrin, J.B., Birmingham, W.P.: Effectiveness of hmm-based retrieval on large databases. In: International Conference on Music Information Retrieval, pp. 33–39 (2003)

17. Skopal, T.: On fast non-metric similarity search by metric access methods. In: Ioannidis, Y., Scholl, M.H., Schmidt, J.W., Matthes, F., Hatzopoulos, M., Böhm, K., Kemper, A., Grust, T., Böhm, C. (eds.) EDBT 2006. LNCS, vol. 3896, pp. 718–736. Springer, Heidelberg (2006)

18. Typke, R.: Music Retrieval based on Melodic Similarity. Doctoral Thesis, Universiteit Utrecht (2007), http://rainer.typke.org/books.html

19. Typke, R., Walczak-Typke, A.: A tunneling-vantage indexing method for non-metrics. In: 9th International Conference on Music Information Retrieval, pp. 683–688 (2008)

20. Vleugels, J., Veltkamp, R.C.: Efficient image retrieval through vantage objects. Pattern Recognition 35(1), 69–80 (2002)

Clustering Driven Cascade Classifiers for Multi-indexing of Polyphonic Music by Instruments

Wenxin Jiang, Zbigniew W. Raś, and Alicja A. Wieczorkowska

Abstract. Recognition and separation of sounds played by various instruments is very useful in labeling audio files with semantic information. Numerous approaches on acoustic feature extraction have already been proposed for timbre recognition. Unfortunately, none of these monophonic timbre estimation algorithms can be successfully applied to polyphonic sounds, which are more usual cases in the real music world. This has stimulated the research on a hierarchically structured cascade classification system under the inspiration of the human perceptual process. This cascade classification system makes first estimate on the higher level of the decision attribute, which stands for the musical instrument family. Then, the further estimation is done within that specific family range. However, the traditional hierarchical structures were constructed in human semantics, which are meaningful from human perspective but not appropriate for the cascade system. We introduce the new hierarchical instrument schema according to the clustering results of the acoustic features. This new schema better describes the similarity among different instruments or among different playing techniques of the same instrument. The classification results show a higher accuracy of cascade system with the new schema compared to the traditional schemas.

1 Introduction

Different classifiers have been used in musical instrument estimation domain, usually for a small number of instruments [1], [7], [12]. Still, it is a non-trivial problem

Wenxin Jiang
University of North Carolina, Dept. of Computer Science, Charlotte, NC 28223, USA
e-mail: wjiang3@uncc.edu

Zbigniew W. Raś
University of North Carolina, Dept. of Computer Science, Charlotte, NC 28223, USA &
Polish-Japanese Institute of Information Technology, Koszykowa 86, 02-008 Warsaw, Poland
e-mail: ras@uncc.edu

Alicja A. Wieczorkowska
Polish-Japanese Institute of Information Technology, Koszykowa 86, 02-008 Warsaw, Poland
e-mail: alicja@pjwstk.edu.pl

Z.W. Raś and A.A. Wieczorkowska (Eds.): Adv. in Music Inform. Retrieval, SCI 274, pp. 19–38.
springerlink.com   © Springer-Verlag Berlin Heidelberg 2010

to choose the one with the optimal performance in terms of estimation accuracy for most western orchestral instruments. It is common to try different classifiers on the same training database which contains features extracted from audio files and select the classifier which yields the highest accuracy for the training database. The selected classifier is used for the timbre estimation on analyzed music sounds. There are also boosting systems [3], [2] consisting of a set of weak classifiers and iteratively adding them to a final strong classifier. Boosting systems usually achieve a better estimation model by training each given classifier on a different set of samples from the training database, which uses the same number of features (attributes). In other words, boosting system works under assumption that there is a (big) difference between different groups of subsets of the training database, so different classifiers are trained on the corresponding subset based on their expertise. However, due to the homogeneous characteristics across all the data samples in a training database, musical data usually cannot take full advantage of such panel of learners because none of the given classifiers would get a majority weight. Thus the improvement cannot be achieved by such a combination of different classifiers. Also, in many cases, the speed of classification is also an important issue.

To achieve the applicable classification time while preserving high classification accuracy, we introduce the cascade classifier which may further improve the instruments' recognition of the *MIR system*.

Cascade classifiers have been investigated in the domain of handwritten digit recognition. Thabtah [18] used filter-and-refine processes and combined it with k-Nearest Neighbor (KNN) classifier to give the rough but fast classification with lower dimensionality of features at filter step and to rematch the objects marked by the previous filter with higher accuracy by increasing dimensionality of features. Also, Lienhart [8] used CART trees as base classifiers to build a boosted cascade of simple feature classifiers to achieve rapid object detection. It is possible to construct a simple instrument family classifier with a low recognition error, which is called a classification pre-filter. When one musical frame is labeled by a specific family, the training samples in other families can be immediately discarded, and further classification is then performed within small subsets, which could be identified by a stronger classifier through adding more features or even calculating the complete spectrum. Since the number of training samples is reduced, the computational complexity is reduced while the recognition rate still remains high.

2   Hierarchical Structure of Musical Instrument Sound Classification

According to the experience regarding human recognition of musical instruments, it is usually easier for one to tell the difference between violin and piano than violin and viola. This is because violin and piano belong to different instrument families and thus have quite different timbre qualities. Violin and viola fall into the same instrument family which indicates they share quite similar timbre quality. If we build the classifiers both on the family level and the instrument level, then the polyphonic music sound is first classified at the instrument family level. After a specific instrument family label is assigned to the analyzed sound by the classifier, it can be further classified at the instrument level by another classifier which is built on the training data of that specific instrument family. Since there is a smaller number of possible instruments in this family, the classifier trained on this family has the appropriate expertise for the classification of the instruments within it.

Before we discuss how to build classifiers on different levels, let us first have a look at the hierarchical structure of the western instruments. Erich von Hornbostel and Curt Sachs published an extensive scheme for musical instrument classification in 1914. Their scheme is widely used today, and is most often known as the Hornbostel-Sachs system. Figure 1 shows a part of the Hornbostel/Sachs instrument classification tree. The Hornbostel-Sachs system includes aerophones (wind instruments), chordophones (string instruments), idiophones (made of solid, non-stretchable, resonant material), and membranophones (mainly drums). Idiophones and membranophones are called percussion. Additional groups include electrophones, i.e. instruments where the acoustical vibrations are produced by electric or electronic means (electric guitars, keyboards, synthesizers), complex mechanical instruments (including pianos, organs, and other mechanical music makers), and special instruments (include bullroarers, but they can be classified as free aerophones). Each category can be further subdivided into groups, subgroups etc. and finally into instruments. Idiophones' subcategories include instruments classified as: struck (e.g. gongs), struck together (by concussion - e.g. claves, clappers, castanets, finger cymbals), scrapped, rubbed, stamped, shaken (e.g. rattles), and plucked (e.g. Jew's harp). Membranophones include the following subgroups: cylindrical drum, conical drum, barrel drum, hourglass drum, goblet drum, footed drum, long drum, kettle or pot drum, frame drum (e.g. tambourine), friction drum, and mirliton/kazoo. Chordophones' subcategories include: zithers, lutes plucked (e.g. mandolins, guitars, ukuleles), lutes bowed (e.g. viols - fretted neck, fiddles, violin, viola, cello, double bass, and hurdy-gurdy - no frets), and harp. Aerophones are classified into the following subgroups: free aerophone (e.g. bullroarers), end-blown flute, side-blown flute, nose flute, globular flute (e.g. ocarina), multiple flutes, panpipes, whistle mouthpiece (e.g. recorder), air chamber (e.g. accordion); single reed instruments (such as clarinet, saxophones), double reed (such as oboe, bassoon) and lip vibrated (trumpet or horn) - instruments classified according to the mouthpiece used to set air in motion to produce sound. Some of aerophones subcategories are also called woodwinds (single reed, double reed, flutes) or brass (lip vibrated), but this criterion is not based on the material the instrument is made of, but rather on the method of sound production. In woodwinds, the change of pitch is mainly obtained by the change of the length of the column of the vibrating air. Additionally, over-blow is applied to obtain second, third or fourth harmonic to become the fundamental. In brass instruments, over-blows are very easy because of wide bell and narrow pipe, and therefore over-blows are the main method of pitch changing.

Sounds can be also classified according to the articulation, i.e. the method the instrument is played. According to articulation, sounds can be basically classified in the following 3 ways: (1) sustained or non-sustained sounds, (2) muted or not

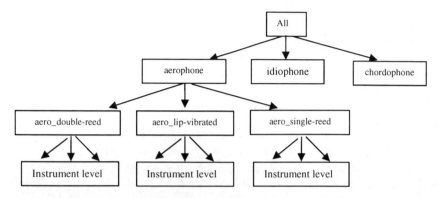

Fig. 1 A part of the Hornbostel-Sachs hierarchical tree

muted sounds, (3) vibrated and not vibrated sounds. This classification may be difficult, since the vibration may not appear in the entire sound; some changes may be visible, but no clear vibration. Also, brass is sometimes played with moving the mute in and out of the bell. According to MPEG7 classification [9], there are four classes of musical instrument sounds: (1) Harmonic, sustained, coherent sounds - well detailed in MPEG7, (2) Non-harmonic, sustained, coherent sounds, (3) Percussive, non-sustained sounds - well detailed in MPEG7, (4) Non-coherent, sustained sounds.

Musical instruments may produce sound of definite or indefinite pitch. Still, most of musical instrument sounds of definite pitch have some noises/continuity in their spectra. In our experiments, we do not include membranophones because the instruments of this family usually do not produce the harmonic sound, so they need special techniques to be identified. This chapter focuses on the instruments producing basically harmonic sounds.

Figure 2 shows another tree structure of instrument sound classification, in which sounds are grouped according to the way how the musical instruments are played.

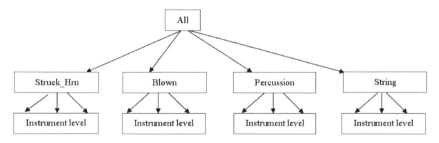

Fig. 2 A part of the play method (articulation) hierarchical tree

3   MIR Framework Based on Cascade Classification System

In this section, we describe cascade classification strategy investigated for musical instruments estimation [15]. Based on how the hierarchical instrument family structures, we have implemented the cascade classification system for the polyphonic sound estimation, as shown in the Figure 3.

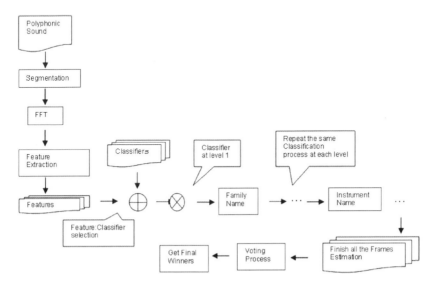

Fig. 3 Timbre estimation with classifier and feature selection

Let $S = \{F, C, D\}$ be the multiple-classifier timbre estimation system, where $D = \{d_1, \ldots, d_n\}$ is the set of all possible decision values of musical instruments in S, $F = \{f_1, \ldots, f_m\}$ is the collection of all available feature sets which could be extracted from the input signal and then used by the classifiers to identify the target frame. $C = \{C_1, \ldots, C_w\}$ is the set of classifiers built on feature sets F after they are extracted from the standard instrument sounds and saved as the feature training database. Let $X = \{x_1, \ldots, x_t\}$ be the set of segmented frames from the input audio sound. There are multiple processes of classification for each frame x_t. First at the root of the hierarchical tree, then down to its lower level, we have pairs (C_z, f_y), $1 \leq z \leq w$, $1 \leq y \leq m$, where (C_z, f_y) means "use classifier C_z on the feature set f_y", to perform classification at each level and get the estimation confidence, i.e. the probability of the classification result given by classifier (related to the similarity between the analyzed frame and reference frames), $conf(x_i, \alpha) = C_z(f_y)$, where α is the specific node of the tree. The result should satisfy two constraints: $conf(x_i, \alpha) \geq \lambda_1$ and $sup(x_i, \alpha) \geq \lambda_2$, where λ_1 is the minimum confidence for the correct classification and λ_2 is the threshold for minimal support (λ_1 and λ_2 are given by the user). Support is defined as the number of matched frames of a particular instrument from the reference database (during classification process, the algorithm tries matching frames to all available frames representing all instruments in the database, and the most similar ones are returned as the matched frames). Confidence is the ratio of support over the total number of matched frames.

After classifications are finished at all the levels, we get the final instrument estimations $\{d_p\}$ for the frame x_i (multiple instrument estimations are given by classifier for each frame), where $d_p \in D$, and the overall confidence for each instrument estimation is calculated by multiplying the confidence obtained previously at each classification level $conf(x_i, d_p) = \prod_{\alpha=1}^{v} conf(x_i, \alpha)$, where v is the total number of ancestors for node x_i in the path of hierarchical tree. After all the individual frames are estimated by the classification system, a smoothing process is performed within a smoothing window. It is done by calculating the average confidence for each possible instrument within the window $\overline{conf(d_p)} = \sum_{q=1}^{s} conf(d_p)_q/s$ where s is the size of a smoothing window. The smoothing window is an indexing granule, which is the smallest segment of the signal on which the estimation of instruments is yielded by the indexing system - the indexing system yields the list of instruments for each smoothing window instead of each frame; the smoothing window size usually is 1-2 seconds, whereas the analyzing frame is 0.12 second, which would be rather short for such labeling, and too short for the users to estimate instruments by themselves.

Next, the output of the final results is further controlled by the threshold λ_3. If the mean value of the confidence within smoothing window $\overline{conf(d_p)} > \lambda_3$, where λ_3 is given by the user, the instrument candidate is kept, otherwise it is discarded as the noise signal which only occurs in a very short time period. According to the indexing resolution requirement, smoothing window can be adjusted to the desired size.

The advantage of this process is that it uses the information of music context to further adjust the results from the frame-wise estimation phase. The system will perform timbre estimation for the polyphonic sound with high accuracy while still preserving the applicable analyzing speed by choosing the best feature and classifier for the classification process at each level based on the knowledge derived from the training database.

4 Hierarchical Structure Based on Clustering Analysis

Clustering is the classification of data objects into similarity groups (clusters) according to a defined distance measure. It is widely used as one of the important techniques of machine learning and pattern recognition in such fields as biology, genomics and image analysis. However, it has not been well investigated in the music domain, since the category information of musical instruments has already been defined by musicians as the two hierarchical structures demonstrated in the previous section. These structures group the musical instrument sounds according to their semantic similarity which is concluded from the human experience. However, the instruments that are assigned to the same family or subfamily by these hierarchical structures often sound quite different from another. On the other hand, instruments that have similar timbre qualities can be assigned to very different groups by these hierarchical structures. Thus, the inconsistency between the timbre quality and the family information causes the incorrect timbre estimation, given by the cascade classification system based on Hornbostel-Sachs instrument classification, used in our previous research [5].

For instance, the trombone belongs to the aerophone family, but the system often classifies it as the chordophone instrument, such as violin. In order to take the full advantage of the cascade classification strategy, we have built a new hierarchical structure of musical instruments by the matching learning technique.

Cluster analysis is commonly used to search for groups in data. It is most effective when the groups are not known a priori. We use the cluster analysis methods to re-organize the instrument groups according to the similarity of timbre relevant features among the instruments.

4.1 Clustering Analysis Methods

There exist many clustering algorithms. Basically, all the clustering algorithms can be divided into two categories: partitional clustering and hierarchical clustering. Partitional clustering algorithms determine all clusters at once without hierarchical merging or dividing process. K-means clustering is most common method in this category [11]; K is the empirical parameter. Basically, it randomly assigns instances to K clusters. Next, new centroid for each of the K clusters and the distance of all items to these K centroids are calculated. Items are re-assigned to the closest centroid and the whole process is repeated until cluster assignments are stable. Hierarchical clustering generates a hierarchical structure of clusters which may be represented in a structure called dendrogram. The root of the dendrogram consists of a single cluster containing all the instances, and the leaves correspond to individual instances. Hierarchical clustering can be further divided into two types according to whether the tree structure is constructed by following agglomerative or divisive approach. Agglomerative approach works in the bottom-up manner, it recursively merges smaller clusters into larger ones till some stoping condition is reached. Algorithms based on divisive (or top-down) approach begin with the whole set and then recursively split this set into smaller ones till some stoping condition is reached.

We have chosen the hierarchical clustering method to learn the new hierarchical schema for music instruments, since it fits our scenario well. There are many options to compute the distance between two clusters. The most common methods are the following [19]:

- Single linkage (nearest neighbor). In this method, the distance between two clusters is determined by the distance of the two closest objects (nearest neighbors) in different clusters. This rule will string objects together to form the clusters, and the resulting ones tend to represent long "chains".
- Complete linkage (furthest neighbor). In this method, the distances between clusters are determined by the greatest distance between any two objects in different clusters (the "furthest neighbors"). This method usually performs quite well in cases when the objects actually form naturally distinct "clumps." If the clusters tend to be of a "chain" type, then this method is inappropriate.
- Unweighted pair-group method using arithmetic averages (UPGMA). In this method, the distance between two clusters is calculated as the average distance

between all pairs of objects in two different clusters. This method is also very efficient when the objects form natural distinct "clumps" and it performs equally well with "chain" type clusters.

- Weighted pair-group method using arithmetic averages (WPGMA). This method is identical to the UPGMA method, except that in the computations, the size of the respective clusters is used as a weight. Thus, this method should be used when the cluster sizes are suspected to be greatly uneven [17].
- Unweighted pair-group method using the centroid average (UPGMC). The centroid of a cluster is the average point in the multidimensional space, calculated as the mean value (for each dimension separately). In a sense, it is the center of gravity for the respective cluster. In this method, the distance between two clusters is determined as the difference between centroids.
- Weighted pair-group method using the centroid average (WPGMC). This method is identical to the previous one, except that weighting is introduced into the computation. When there are considerable differences in cluster sizes, this method is preferable to the previous one.
- Ward's method. This method is distinct from all other methods because it uses an analysis of variance approach to evaluate the distances between clusters. In short, this method attempts to minimize the Sum of Squares of any two hypothetical clusters that can be formed at each step. In general, this method is good at finding compact, spherical clusters. However, it tends to create clusters of small size.

To complete the above definitions of a distance measure between two clusters, we also have to define the distance between their instances or centroids. Here are some most common distance measures between two objects:

1. Euclidean: Usual square distance between the two vectors. Disadvantages: not scale invariant, not for negative correlations
$$d_{xy} = \sqrt{\sum (x_i - y_i)^2}$$
2. Manhattan: Absolute distance between the two vectors.
$$d_{xy} = \sum |x_i - y_i|$$
3. Maximum: Maximum distance between any two components of x and y
$$d_{xy} = max|x_i - y_i|$$
4. Canberra: Canberra distance examines the sum of series of a fraction differences between coordinates of a pair of objects. Each term of fraction difference has value between 0 and 1. If one of coordinates is zero, the term corresponding to this coordinate become unity regardless the other value, thus the distance will not be affected; if both coordinates are zero, then the term is defined as zero.
$$d_{xy} = \sum \frac{|x_i - y_i|}{|x_i| + |y_i|}$$
5. Pearson correlation coefficient (PCC) is a correlation-based distance. It measures the degree of association between two variables.
$$\rho_{xy} = \frac{[cov(X,Y)]^2}{var(X)var(Y)} \ , \ d_{xy} = 1 - \rho_{xy}$$ where $cov(X,Y)$ is the covariance of the two variables, $var(X)$ and $var(Y)$ - the variance of each variable.

6. Spearman's rank correlation coefficient is another correlation based distance.

$$\rho_{xy} = 1 - \frac{6\sum d_i^2}{n(n^2-1)} , d_{xy} = 1 - \rho_{xy}$$

where $d_i = x_i - y_i$ is the difference between the ranks of corresponding values x_i and y_i, and n is the number of values in each data set (same for both sets). Rank is calculated in the following way: 1 is assigned to the smallest element of each data, 2 to the second smallest element, and so on; the average ranking is calculated if there is a tie among different elements.

It is critical to choose an appropriate distance measure for objects in a musical domain because different measures may produce different shapes of clusters which represent different schema of instrument family. Different features also require the appropriate measures to be chosen in order to give better description of feature variation. The inappropriate measure could distort the characteristics of timbre which may cause the incorrect clustering.

4.2 Evaluation of Different Clustering Algorithms for Different Features

As we can see, each clustering method has its own different advantage and disadvantage over others. It is a nontrivial task to decide which one is the most appropriate method for generating the hierarchical instrument classification structure. Not only the specific cluster linkage method needs to be decided in the hierarchical clustering algorithms, but also the good distance measurement has to be chosen in order to generate the good schema that represents the actual relationships among those instruments. We designed quite intensive experiments with the "cluster" package in R system [14]. The R package provides two hierarchical clustering algorithms: hclust (agglomerative hierarchical clustering), and diana (divisive hierarchical clustering). Table 1 shows all the clustering methods that we tested. We evaluated six different distance measurements (Euclidean, Manhattan, Maximum, Canberra, Pearson correlation coefficient, and Spearman's rank correlation coefficient) for each algorithm. For the agglomerative type of clustering (hclust), we also evaluated seven different cluster linkages that are available in this package: Ward, single (single linkage), complete (complete linkage), average (UPGMA), mcquitty (WPGMC), median(WPGMA), and centroid (UPGMC).

We have chosen the middle C pitch group which contains 46 different musical sound objects. We have extracted three different feature sets (MFCC [10], spectral flatness coefficients [9], and harmonic peaks [9]) from those sound objects. Each feature set produces one dataset for clustering. Some sound objects belong to the same instrument. For example, "ctrumpet" and "ctrumpet harmonStemOut" are objects produced by the same instrument: trumpet. We have preserved these particular object labels in our feature database without merging them as the same label because they could have very different timbre quality which the conventional hierarchical structure ignores. We have tried to discover the unknown musical instrument

Table 1 All distance measures and linkage methods tested for agglomerative and divisive clustering

Clustering algorithm	Cluster Linkage	Distance Measure
hclust (agglomerative)	average	6 distance metrics
	centroid	6 distance metrics
	complete	6 distance metrics
	mcquitty	6 distance metrics
	median	6 distance metrics
	single	6 distance metrics
	ward	6 distance metrics
diana (divisive)	N/A	6 distance metrics

group information solely by the unsupervised machine learning algorithm, instead of applying any human guidance. Each sound object was segmented into multiple 0.12s frames and each frame was stored as an instance in the testing dataset. Since the segmentation is performed with overlap of 2/3 of the frame, there were totally 2884 frames from the 46 objects in each of the three feature datasets.

When our algorithm finishes the clustering job, a particular cluster ID is assigned to each frame. Theoretically, one may expect the same cluster ID to be assigned to all the frames of the same instrument sound object. However, the frames from the same sound object are not uniform and have variations in their feature patterns as the time evolves. Therefore, clustering algorithms do not perfectly identify them as the same cluster. Instead, some frames are assigned into other groups where majority of the frames come from other instrument sounds. As a result, multiple (different) cluster IDs are assigned to the frames of the same instrument object.

Our goal is to cluster the different instruments into the groups according to the similarity of timbre relevant features. Therefore, one important step of the evaluation is to check if a clustering algorithm is able to cluster most frames of an individual instrument sound into one group. In other words, a clustering algorithm should be able to differentiate most of the frames of one instrument sound from the others. It is evaluated by calculating the accuracy of a cluster ID assignment. We use the following example to illustrate this evaluation process. A hierarchical cluster tree T_m is produced by a clustering algorithm A_m. There are totally n instrument sound objects in the dataset (n=46). The clustering package provides function *cutree* to cut T_m into n clusters. One of these clusters is assigned to each frame. Table 2 shows a contingency table (x_{ij} represent numbers) derived from the clustering results after the *cutree* is applied. It is a $n \times n$ matrix, where x_{ij} is the number of frames of *instrument$_i$* that are labeled by *cluster$_j$*, and $x_{ij} \geq 0$.

In order to calculate the accuracy of the cluster assignment, we need to decide which cluster ID corresponds to which instrument object. If cluster k is assigned to *instrument$_i$*, x_{ik} is the number of correct assignments for *instrument$_i$*, the accuracy of the clustering for *instrument$_i$* is $\beta_i = x_{ik}/(\sum_{j=1}^{n} x_{ij})$.

Table 2 Format of the contingency table derived from clustering result

	$|Cluster_1|$	$\cdots$	$|Cluster_j|$	$\cdots$	$|Cluster_n|$
$Instrument_1$	x_{11}	$\cdots$	x_{1j}	$\cdots$	x_{1n}
$\cdots$	$\cdots$	$\cdots$	$\cdots$	$\cdots$	$\cdots$
$Instrument_i$	x_{i1}	$\cdots$	x_{ij}	$\cdots$	x_{in}
$\cdots$	$\cdots$	$\cdots$	$\cdots$	$\cdots$	$\cdots$
$Instrument_n$	x_{n1}	$\cdots$	x_{nj}	$\cdots$	x_{nn}

During clustering process, each frame of the sound object is clustered into one particular group, and the group ID (i.e. instrument) is assigned to this frame. For each row, the maximum value is found among n columns, and next the column corresponding to the position of this maximum becomes the class label for frames represented in this row. However, it may happen that the maximum value is found in the same column also for other rows, and then the same group ID is linked to two different sound objects, which means these two different instrument sounds could not be distinguished by this particular clustering scheme. Clearly, we would like to avoid such an ambiguity. On the other hand, we have to cluster many frames of one sound object into a single group. Therefore, we would need permutations to calculate the theoretic best solution for the whole table, but such a large number of computations cannot be performed.

The overall accuracy for the clustering algorithm A_m is the average accuracy of all the instruments $\overline{\beta} = (\sum_{i=1}^{n} \beta_i)/n$. To find the maximum $\overline{\beta}$ among all possible cluster assignments to instruments, we should permute this matrix in order to find the maximum accuracy for the whole matrix (for each row of matrix, there are multiple values that could be selected among n columns), but it is not applicable to perform such a large number of calculations. This is why we have chosen maximum x_{ij} in each row to approximate the optimal $\overline{\beta}$.

Since it is possible to assign the same cluster to multiple instruments, we have taken the number of clusters as well as accuracy into account. The final measurement to evaluate the performance of clustering is $score_m = \overline{\beta} \cdot w$, where w is the number of clusters, $w \leq n$. This measure reflects how well the algorithm clusters the frames from the same instrument object into the same cluster. It also reflects the ability of algorithm to separate instrument objects from each other.

In the experiments, we used two hierarchical clustering algorithms, hclust and diana. Table 3 presents 15 results which yielded the highest score among 126 experiments based on hclust algorithm.

From the results, the Ward linkage outperforms other methods and it yields the best performance when Pearson distance measure is used on the flatness coefficients feature dataset.

Table 4 shows the results from diana algorithm. In this algorithm, Euclidean yields the highest score on the mfcc feature dataset.

During the clustering process, we cut the hierarchical clustering result at a certain level, when obtaining groups which could represent instrument objects. If most of

Table 3 Evaluation result of `hclust` algorithm

Feature	method	metric	$\bar{\beta}$	w	score
Flatness Coefficients	ward	pearson	87.3%	37	32.30
Flatness Coefficients	ward	euclidean	85.8%	37	31.74
Flatness Coefficients	ward	manhattan	85.6%	36	30.83
mfcc	ward	kendall	81.0%	36	29.18
mfcc	ward	pearson	83.0%	35	29.05
Flatness Coefficients	ward	kendall	82.9%	35	29.03
mfcc	ward	euclidean	80.5%	35	28.17
mfcc	ward	manhattan	80.1%	35	28.04
mfcc	ward	spearman	81.3%	34	27.63
Flatness Coefficients	ward	spearman	83.7%	33	27.62
Flatness Coefficients	ward	maximum	86.1%	32	27.56
mfcc	ward	maximum	79.8%	34	27.12
Flatness Coefficients	mcquitty	euclidean	88.9%	33	26.67
mfcc	ward	average	87.3%	30	26.20

Table 4 Evaluation result of `diana` algorithm

Feature	metric	$\bar{\beta}$	w	score
Flatness Coefficients	euclidean	77.3%	24	18.55
Flatness Coefficients	kendall	75.7%	23	17.40
Flatness Coefficients	manhattan	76.8%	25	19.20
Flatness Coefficients	maximum	80.3%	23	18.47
Flatness Coefficients	pearson	79.9%	26	20.77
mfcc	euclidean	78.5%	29	22.78
mfcc	kendall	77.2%	27	20.84
mfcc	manhattan	77.7%	26	20.21
mfcc	pearson	83.4%	25	20.86
mfcc	spearman	81.2%	24	19.48

the frames from the same instrument object are clustered into one group, then this algorithm is selected to generate the hierarchical tree.

When we compare the two algorithms (`hclust` and `diana`), `hclust` yields better clustering results than `diana`. Therefore, we chose agglomerative clustering algorithm to generate the hierarchical schema for musical instruments, using Ward as the linkage method, Pearson distance measure as the distance metric, and Flatness Coefficients as the feature dataset to perform clustering analysis.

5 New Hierarchical Tree

Figure 4 shows the dendrogram result generated by the hierarchical clustering algorithm we chose (i.e. agglomerative clustering), as mentioned in Section 4.2.

From this new hierarchical classification, we discover some instrument relationships which are not represented in the traditional schemas.

A musical instrument can produce sounds with quite different timbre qualities when different playing techniques are applied. One of the common techniques is muting. A mute is a device fitted to a musical instrument to alter the sound produced. It usually reduces the volume of the sound as well as affects the timbre. There are several different mute types for different instruments. The most common type used with the brass is the straight mute - a hollow, cone-shaped mute that fits into the bell of the instrument. This results in a more metallic, sometimes nasal sound, and when played at loud volumes can result in a very piercing note. The second common brass mute is the cup mute. Cup mutes are similar to straight mutes, but attached to the end of the mute's cone is a large lip that forms a cup over the bell. The result is removal of the upper and lower frequencies and a rounder, more muffled tone. In the case of string instruments of the violin family, the mute takes the form of a comb-shaped device attached to the bridge of the instrument, dampening vibrations and resulting in a "softer" sound.

In the hierarchical structure shown in Figure 4, "trumpet" and "ctrumpet harmonStemOut" represent two different sounds produced by the trumpet. "ctrumpet harmonStemOut" is produced when a particular mute is applied, called Harmon mute (different from the common straight or cup mutes). It is a hollow, bulbous metal device placed in the bell of the trumpet. All air is forced through the middle of the mute. This gives the mute a nasal quality. Protruding at the end of the device, there is a detachable stem extending through the centre of the mute. The stem can be removed completely or can be inserted to varying degrees. Name of this instrument sound object shows whether the stem is extended or completely removed, which darken the original piercing, strident timbre quality.

From the spectra of various sound objects (Figure 5), we can clearly observe big differences between them. The spectra also show that "Bach trumpet" has more similar spectral pattern to "trumpet". The relationships between C trumpet, C trumpet muted (Harmon, stem out) and Bach trumpet are accurately represented in the new hierarchical schema. Figure 4 shows that "ctrumpet" and "bachtrumpet" are clustered into the same group. "ctrumpet harmonStemOut" is clustered in one single group instead of merging with "ctrumpet" since it has a very unique spectral pattern. The new schema also discovers the relationships among "French horn", "French horn muted" and "bassoon". Instead of clustering two "French horn" sounds in one group as the conventional schema does, bassoon is considered as the sibling of the regular French horn. "French horn muted" is clustered in another different group together with "English Horn" and "Oboe" (the extent of the difference between groups is measured by the distance between the nodes in the hierarchical tree).

According to this result, the new schema is more accurate than the traditional schema, because it represents the actual similarity of timbre qualities of musical instruments. Not only it better describes the differences between instruments, but it also distinguishes the sounds produced by the same instrument that have quite different timbre qualities due to different playing techniques.

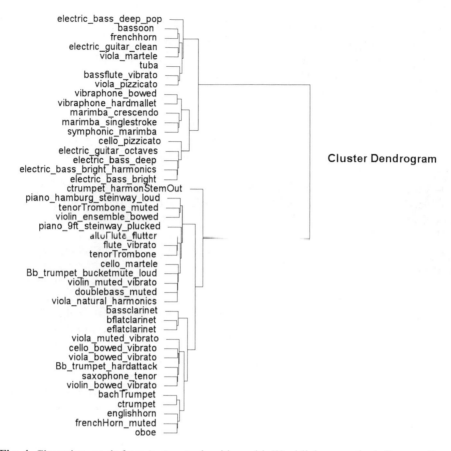

Fig. 4 Clustering result from hclust algorithm with Ward linkage method, Pearson distance measure, and Flatness Coefficients used as the feature set

6 Experiments and Evaluation

In order to evaluate the new schema, we developed the cascade classification system based on the multi-label classification method and tested it with the new schema, as well as with the two previous conventional hierarchical schemas: Hornbostel-Sachs and Playing Method. The system used MS SQLSERVER2005 database system to store training dataset and MS SQLSERVER analysis server as the data mining server to build decision tree and process the classification request.

Training data: The audio files used in this research consist of stereo musical pieces from the McGill University Master Samples (MUMS, [13]). Each file has two channels: left channel and right channel, in .au (or .snd) format. These audio data files are treated as mono-channel, where only left channel is taken into consideration, since successful methods for the left channel can also be applied to the right channel, or

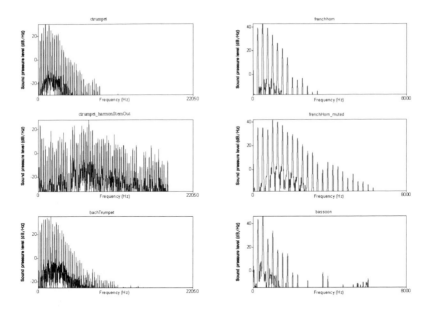

Fig. 5 Spectrum comparison of different instrument objects. On the left hand side: C Trumpet, C Trumpet muted (Harmon, Stem out) and Bach Trumpet; on the right hand side: French horn, French horn muted, bassoon

any channel if more channels are available. 2917 single instrument sound files were used, representing 45 different instruments.

Each sound stands for one note played by a specific instrument. Many instruments can produce different timbres when they are played using different techniques. Therefore, sounds of various pitch and articulation were investigated for each of these 45 instruments.

Power spectrum and 33 spectral flatness coefficients were extracted from each frame of these single instrument sounds, according to the equations described by the MPEG-7 standard [9]. The frame size was 120 ms and the overlap between two adjacent frames was 80ms, to reduce the information loss caused by windowing function (therefore, the hop size was 40ms). The total number of frames for the entire feature database reaches to about one million, since each sound is analyzed in many frames. For instance, the instrument sound which only lasts three seconds is segmented into 75 overlapped frames. The classifier is trained by the obtained feature database.

Testing data: 308 mixed sounds were synthesized by randomly choosing two single instrument sounds from 2917 training data files. Spectral flatness coefficients were extracted from the frames of mixes, in order to perform instrument family estimation on the higher level of the hierarchical tree. After reaching the bottom level of the hierarchical tree, we used the power spectrum from the frames representing mixes, in order to match against the reference spectral database. Since the spectrum matching is performed in a small subgroup, the computation complexity is reduced. The same analyzing frame size and hop size were used for the mixes as in the case of training data.

Table 5 Comparison between non-cascade classification and cascade classification with different hierarchical schemas

Experiment	classification method	Description	Recall	Precision	F-Score
1	Non-Cascade	Feature-based	64.3%	44.8%	51.4%
2	Non-Cascade	Spectrum-Match	79.4%	50.8%	60.7%
3	Cascade	Hornbostel-Sachs	75.0%	43.5%	53.4%
4	Cascade	play method	77.8%	53.6%	62.4%
5	Cascade	machine Learned	87.5%	62.3%	69.5%

The average recall, precision and F-score of all the 308 sounds estimations were calculated to evaluate each method. The definitions of recall and precision are shown in Figure 6. I_1 is the number of actual instruments playing in the analyzed sound. I_2 is the number of instruments estimated by the system. I_3 is the number of correct estimations.

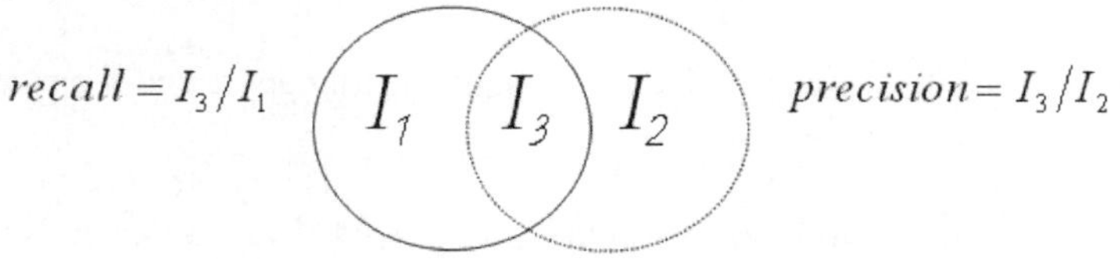

$$recall = I_3/I_1 \qquad I_1 \quad I_3 \quad I_2 \qquad precision = I_3/I_2$$

Fig. 6 Precision and Recall

Recall is the measurement to evaluate the recognition rate and precision is to evaluate the recognition accuracy. The F-score is often used in the field of information retrieval for measuring search, document classification, and query classification performance. It is the harmonic mean of precision and recall. F-score is calculated as

$$\text{F-score} = \frac{2 \times precision \times recall}{precision + recall}$$

Since timbre estimation was performed for indexing segments (smoothing window), containing multiple frames, as described in Section 3, the measures mentioned above were calculated for indexing segments of size 1 second.

K-Nearest Neighbor (KNN) [6] was used as the classifier, with $k = 3$. As shown in Table 5, in Experiment 1 we applied the multiple label classification [5] based on features representing spectral flatness coefficients only. In Experiment 2 we used the power spectrum matching method, instead of features [4]. In Experiment 3 and

Experiment 4 we used two traditional hierarchical structures (Hornbostel-Sachs and play method) in order to perform cascade classification based on both the power spectrum and spectral flatness coefficients. In Experiment 5 we applied the new hierarchical structure as the basis of the cascade system. The indexed window size for all the experiments is one second, and the output of total number of estimations for each indexed window is controlled by confidence threshold $\lambda = 0.4$, which is the minimum average confidence of instrument candidates.

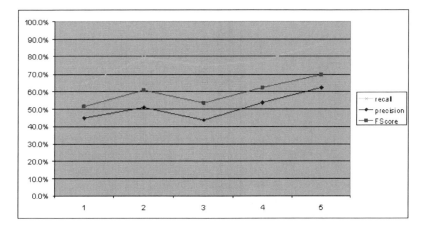

Fig. 7 Comparison between non-cascade classification and cascade classification with different hierarchical schemas

Figure 7 shows that generally the cascade classification improves the recall compared to the non-cascade methods. The non-cascade classification based on spectrum-match (Experiment 2) shows higher recall than the cascade classification approaches based on the traditional hierarchical schema (Experiment 3 and Experiment 4). However, the cascade classification based on the new schema learned by the clustering analysis (Experiment 5) outperforms the non-cascade classification. It increases the recall by 8 percent points, precision by 12 percent points and general F-score by 9 percent points. This shows that the new schema yields significant improvement in comparison to the other two traditional schemas. Also, since the hierarchical tree has more levels, the size of the subset on the bottom level is reduced to a very small size, which significantly reduces the cost of spectrum matching.

We evaluated the classification system by the mixed sounds which contain two single instrument sounds. In the real world recordings, there could be more than two instruments playing simultaneously, especially in the orchestra music. Therefore, we also created 49 polyphonic sounds by randomly selecting three different single instrument sounds and mixing them together. Next, we tested those three-instrument mixes, using various classification methods (Table 6).

As we can see from Table 6, the lowest precision and recall is obtained for the algorithm based on sound separation, i.e. separating sounds of mixes and then

Table 6 Classification results of 3-instrument mixtures with different algorithms

Experiment	Classifier	Method	Recall	Precision	F-Score
1	Non-Cascade	single-label based on sound separation	31.48%	43.06%	36.37%
2	Non-Cascade	feature-based multi-label classification	69.44%	58.64%	63.59%
3	Non-Cascade	spectrum-match multi-label classification	85.51%	55.04%	66.97%
4	Cascade (Hornbostel-Sachs)	multi-label classification	64.49%	63.10%	63.79%
5	Cascade (playmethod)	multi-label classification	66.67%	55.25%	60.43%
6	Cascade (machine learned)	multi-label classification	63.77%	69.67%	66.59%

performing instrument estimation on separated sounds. This is because there is no much information left in the sound mix for the further classification of the third instrument after two signal subtractions corresponding to the first two instrument estimations are made. The cascade method based on multi-label classification again yields high recall and precision of results.

This experiment shows the robustness and effectiveness of the algorithm for the polyphonic sounds which contain more than two timbres. As the dendrogram in Figure 4 shows, the new schema has more hierarchical levels and looks more complex and obscure to users. However, we only use it as the internal structure for the cascade classification process, and we do not use it in the query interface. Therefore, when the user submits a query to *QAS* defined in the user's semantic structure (e.g. searching instrument sounds which are close in Hornbostel-Sachs classification, or with respect to the play method), system translates it to the internal schema (based on clustering). After the estimation is done, the answer is converted back to the user semantics. The user does not need to know the difference between French horn and French horn muted since only French horn is returned by the system as the final estimation result. The internal hierarchical representation of musical instrument sound classification is used as an auxiliary tool, assisting answering user's queries.

7 Conclusion

In this chapter we have discussed the timbre estimation based on hierarchical classification. In order to deal with polyphonic sounds, multi-label classifiers were used, with classification based on spectrum matching and also based on feature vectors extracted from spectra. Given the fact that spectrum matching in a large training database is much more expensive than feature based classification, the cascade classifier was introduced to give a good solution for achieving both high recognition rate and high efficiency. Cascade classification system needs to acquire knowledge how to choose the appropriate classifier and features at each level of hierarchical tree. The experiments have been conducted to discover such knowledge based on the training database. As a result, we introduced a new hierarchical structure for the cascade classification system based on the obtained hierarchical clustering. Compared to the traditional schemas which are manually designed by the musicians, the new schema better represents the relationships between musical instrument sounds in terms of their timbre similarity, since the hierarchical structures are directly derived from the acoustic features based on their similarity matrix. This new hierarchical schema shows better results in the cascade classification of musical instrument sounds.

Acknowledgments

This material is based in part upon work supported by the National Science Foundation under Grant Number *IIS-0414815*. Any opinions, findings, and conclusions or recommendations expressed in this material are those of the author(s) and do not necessarily reflect the views of the National Science Foundation.

This work was also partially supported by the Research Center of Polish-Japanese Institute of Information Technology, supported by the Polish National Committee for Scientific Research (KBN).

References

1. Brown, J.C., Houix, O., McAdams, S.: Feature dependence in the automatic identification of musical wind instruments. J. Acoust. Soc. of America 109, 1064–1072 (2001)
2. Freund, Y., Schapire, R.E.: A decision-theoretic generalization of on-line learning and an application to boosting. Journal of Computer and System Sciences 55(1), 119–139 (1997)
3. Freund, Y.: Boosting a weak learning algorithm by majority. In: 3rd Annual Workshop on Computational Learning Theory (1990)
4. Jiang, W., Wieczorkowska, A., Ras, Z.: Music Instrument Estimation in Polyphonic Sound Based on Short-Term Spectrum Match. In: Hassanien, A.-E., et al. (eds.) Foundations of Computational Intelligence. Studies in Computational Intelligence, vol. 202, pp. 259–273. Springer, Heidelberg (2009)
5. Jiang, W., Cohen, A., Ras, Z.: Polyphonic music information retrieval based on multi-label cascade classification system. In: Ras, Z.W., Ribarsky, W. (eds.) Advances in Information & Intelligent Systems. Studies in Computational Intelligence. Springer, Heidelberg (2009)
6. Kaminskyj, I.: Multi-feature Musical Instrument Sound Classifier. Mikropolyphonie WWW Journal (6) (2001),
   http://farben.latrobe.edu.au/mikropol/articles.html
7. Kostek, B., Czyzewski, A.: Representing Musical Instrument Sounds for Their Automatic Classification. J. Audio Eng. Soc. 49(9), 768–785 (2001)
8. Lienhart, R., Kuranov, A., Pisarevsky, V.: Empirical Analysis of Detection Cascades of Boosted Classifiers for Rapid Object Detection. Pattern Recognition Journal, 297–304 (2003)

9. Lindsay, A.T., Herre, J.: MPEG-7 and MPEG-7 Audio-An Overview. J. Audio Eng. Soc. 49, 589–594 (2001)
10. Logan, B.: Frequency Cepstral Coefficients for Music Modeling. In: 1st Ann. Int. Symposium On Music Information Retrieval (2000)
11. MacQueen, J.B.: Some Methods for classification and Analysis of Multivariate Observations. In: Proceedings of 5th Berkeley Symposium on Mathematical Statistics and Probability, vol. 1, pp. 281–297. University of California Press (1967)
12. Martin, K.D., Kim, Y.E.: Musical Instrument Identification: A Pattern-Recognition Approach. In: 136th Meeting of the Acoustical Soc. of America, Norfolk, VA 2pMU9 (1998)
13. Opolko, F., Wapnick, J.: MUMS - McGill University Master Samples. CD's (1987)
14. R Development Core Team. R: Language and Environment for Statistical Computing. R Foundation for Statistical Computing, Vienna, Austria (2005) ISBN 3-900051-07-0, http://www.R-project.org
15. Ras, Z., Dardzinska, A., Jiang, W.: Cascade Classifiers for Hierarchical Decision Systems. In: Proceedings of IIS 2008 Conference, in Challenging Problems of Science. Intelligent Information Systems, vol. XVI, pp. 171–180. Academic Publishing House EXIT, Warsaw (2008)
16. Ras, Z., Wieczorkowska, A.: Indexing audio databases with musical information. In: SCI 2001, Orlando, Florida, vol. 10, pp. 279–285 (2001)
17. Sneath, P.H.A., Sokal, R.R.: Numerical Taxonomy. Freeman, San Francisco (1973)
18. Thabtah, F.A., Cowling, P., Peng, Y.: Multiple Labels Associative Classification. Knowledge and Information Systems 9(1), 109–129 (2006)
19. The Statistics Homepage: Cluster Analysis, http://statsoft.com/textbook/stathome.html

Representations of Music in Ranking Rhythmic Hypotheses

Jaroslaw Wojcik and Bozena Kostek

Abstract. The chapter presents first the main issues related to music information retrieval (MIR) domain. Within this domain, there exists a variety of approaches to musical instrument recognition, musical phrase classification, melody classification (e.g. query-by-humming systems), rhythm retrieval, retrieval of high-level- musical features such as looking for emotions in music or differences in expressiveness, music search based on listeners' preferences, etc. The objective of this study is to propose a method for retrieval of *hypermetric* rhythm on the basis of melody. A stream of sounds in MIDI format is introduced at the system input. On the basis of a musical content the method retrieves a *hypermetric* structure of rhythm of a musical piece consisting of rhythmic motives, phrases, and sentences. On the basis of the *hypermetric* structure retrieved, a system capable of creating automatic drum accompaniment to a given melody supporting the composition is proposed. A method does not use any information about rhythm (time signature), which is often included in MIDI information. Neither rhythmic tracks nor harmonic information are used in this method. The only information analyzed is a melody, which may be monophonic as well as polyphonic. The analysis starts after the entire piece has been played. Recurrence of melodic and rhythmic patterns and the rhythmic salience of sounds are combined to create an algorithm that finds the metric structure of rhythm in a given melody.

1 Introduction

Music Information Retrieval (MIR) is a multi-discipline area. In this Chapter, some aspects related to MIR are shortly reviewed in the context of possible and actual applications within this domain with the main stress on rhythm retrieval. To

Jaroslaw Wojcik . Bozena Kostek
Multimedia Systems Department, Electronics, Telecommunications and Informatics Faculty
Gdansk University of Technology, Poland
e-mail: bozenka@sound.eti.pg.gda.pl

Z.W. Raś and A.A. Wieczorkowska (Eds.): Adv. in Music Inform. Retrieval, SCI 274, pp. 39–64.

categorize MIR, a list of topics may be recalled after ISMIR (*International Conferences on Music Information Retrieval*) [12]:

- MIR systems (content-based querying, instrument/ genre/ style/ mood classification, recommendation/ play-list generation, fingerprinting/ DRM (Digital Rights Managements), transcription/ annotation, text/ web mining, OMR (Optical music recognition), database systems/ indexing, etc.).
- Human issues (user interfaces, user models, emotion,aesthetics, perception, cognition, social, legal and ethical issues, etc.).
- Data and metadata (audio, MIDI, score, libraries and collections, etc.).
- Musical knowledge (melody and motives, harmony, chords and tonality, rhythm, beat, tempo and form, timbre, instrumentation and voice, genre, style and mood, performance, composition, etc.).

Prospective applications related to MIR envision the management of virtually unlimited quantities of music information contained in vast Internet repositories, however such an ambitious aim is not yet fulfilled.

MIR research area lies within scientists' interest because of wide possibilities to use retrieval methods in practical applications. Such applications let composers check effortlessly whether a similar melody has already been created. They also can serve as easy-to-use tools to seek certain music given only the humming of a melody and thus making any other knowledge about the piece, e.g. its author, unnecessary. Computer programs with MIR software installed offer facilities in both music composing and performing.

Music analysis is clearly a hierarchical approach, since music itself is hierarchically structured. This implies that MIR, being a hierarchical domain, should be analyzed from this point of view. There are numerous papers published about the application of the hierarchical approach to music recognition [8][23][31][35].

It should also be emphasized that the diversity of music, musical styles, genres, and instruments, as well as the variety of performers and performance techniques implies the multiplicity of MIR-based systems, and at the same time numerous classification methods [16][17][35][37].

Rhythm is one of the elements of musical style, which may be valuable in retrieval. The rhythmic structure together with patterns retrieved carry information about the genre of a piece. The important feature of a rhythm finding model is whether it accepts audio or symbolic input [9]. The difference between those types of models is fundamental – in case of audio input a method to extract sounds from an acoustic signal is employed first, then the quantization or metric analysis can be performed. The process to convert audio data into symbolic representation of music is called *audio transcription*. It should be remembered that audio transcription methods were engineered to extract pitches, onsets and durations of sounds from raw audio data. The same transcription methods can be employed in rhythm retrieval systems if it is assumed to find the musical rhythm in raw audio recordings. Current transcription methods work well for signal containing melody contour only, the results get worse if the piece is polyphonic (an example of research by Dovey can be studied in this context [4]) multi-instrumental, contains drum tracks (a paper by Ryynanen & Klapuri is a good example of such research - [28]) non-instrumental tracks, human singing voice or if the signal is distorted. After the transcription stage, the symbolic representation is available, thus high level analysis of musical content, retrieval or classification methods can be performed.

Beat tracking is part of automatic music transcription systems, and is also used in music information retrieval, metadata generating and accompaniment [5]. A good example of a beat tracking system is the one proposed by Goto [8], recognizing the hierarchical structure of rhythm up to a measure level. The system processes popular music sampled from compact discs in real time, and the method deals with audio performances also if they contain drum tracks. The knowledge used to accomplish this includes onset times, changes in harmony and drum patterns. It is to remind that notes are defined by *note onset time, pitch, and duration*. The note onset is the beginning of the note. The note onset detection aims to find the start of musical events. In the real-life performances the onsets of sounds are not equally spaced, because of performer's inaccuracies in playing. The process of rounding the inter-onset interval (interval between onsets) and durations of sounds to the time grid is named *quantization*. The quantization of onsets of sounds is a process modifying the onsets to be placed in the particular locations of bars, namely in the multiplies of the shortest rhythmic values in a song, the differences between two subsequent onsets after quantization are thus equal to the natural multiple of an atomic period. Sound durations also are a subject of quantization – they should belong to a finite set of possible durations. In this case, rhythm quantization is the operation aiming at qualify the sound as a note of particular rhythmic value, i.e. one-eights, quarter-notes, half-notes or their multiplies. A paper by Desain [3] provides a concise overview of past approaches to the quantization of time intervals.

Possible applications of *metric rhythm* retrieval method, i.e. to find the entire hierarchic structure of related rhythmic levels, are: music recommendation, plagiarism detection, automatic synchronization of music with other elements of multimedia applications, support in creating musical scores that base on MIDI instruments played melodies, automatic drum accompaniment or retrieval based on musical genre.

The Chapter is organized as follows: Section 2 reviews computational rhythm retrieval-related studies. Then, in Section 3 the experimental setup is presented and as well as the layout of all experiments performed within this study is shown. Also, the method proposed by the Authors is visualized as a block diagram. Experiments 1-3 have been already presented by the authors in other publications that is why only their experimental outcome is recalled, since it is used as the basis in Experiments 4 and 5, the most important part of this Chapter. To this end, some notions such as creating and ranking rhythm hypotheses were introduced in the consecutive Sections. Especially important is the representation of music used in the context of ranking rhythmic hypotheses. Experiment 6 is devoted to subjective tests to validate the authors' theoretical and experimental approach to rhythm retrieval. Detailed results of experiments and Concluding remarks are contained in Sections 4-5.

2 Related Works

The scope of the studies concerning rhythm is very wide and, among other issues, involves the quantization process of the onsets and durations of notes, the extraction of rhythm events from audio recordings, and the search for meter of compositions. A good review of automatic rhythm description systems is brought by Gouyon and Dixon [9], and Schuller et al. [30].

The task of computational rhythm retrieval is complex; it consists of a few stages. The simplified approach to this task may be reduced to retrieving the sequence of onset times and/or durations of sounds from the musical data – this process is called quantization or rhythm parsing. In some other approaches, the time signature is retrieved on the basis of a musical content. For clarifying, the time signature appears at the beginning of a piece of music, it is the symbol that tells the meter of the piece [13] It looks like a fraction written without a horizontal line between numerator and denominator. The lower number is the main rhythmic value appearing in the piece, and the upper number determines how many rhythmic values are stored within one measure. In the class of computational rhythm retrieval methods, usually the period of time is found, which divides the stream of sounds into repeating fragments. This task may additionally be combined with phenomenal accent retrieval in such a way, that the phase of phenomenal accentuations in a piece is found. If the algorithmically discovered accentuations line up with human feet taping to the melody, it may be concluded that the rhythmic level and a period equal to the meter have been found correctly. The next complication is to retrieve metric rhythm. Existing metric rhythm research usually focuses on retrieving low rhythmic levels – usually to the level of a measure, those methods are enough to emulate human perception of a local rhythm. According to McAuley & Semple [24] trained musicians perceive more levels, though. High-level perception is required from drum players, thus computational approach needs to retrieve the so-called *hypermetric* structure of a piece. If it reaches high rhythmic levels such as phrases, sentences and periods, automatic drum accompaniment applications can be developed. That is the motivation behind the authors' approach to the rhythm retrieval, namely creating the drum accompaniment automatically.

Most of the methods in related research retrieve either single rhythmic levels or a rhythmic structure, of which highest rhythmic level reaches the meter. First we will try to describe the term *rhythmic/hypermetric* structure more thoroughly. Metric structure includes meter, tempo, and all rhythmic aspects which produce temporal regularity or structure, against which the foreground details or durational patterns are projected. Hypermeter is large-scale meter (as opposed to surface-level meter) created by hypermeasures which consist of hyperbeats [13].

Lester proposed: "Meter is ... an organization of pulses that are of functionally equivalent duration. For a meter, and, by extension, a hypermeter, to exist, there must be a stream of pulses to be organized". Hypermeter, if it is to be analogous to meter, must concern itself with groupings of equivalent pulses, not with the pairing of structural events. Meter is thus an aspect of grouping, or partitioning, which is in turn a vital aspect of rhythm [11].

ADVANCES IN MUSIC INFORMATION RETRIEVAL; ED. BY
ZBIGNIEW W. RAS.

BERLIN: SPRINGER, 2010   Cloth   410 P.
SER: STUDIES IN COMPUTATIONAL INTELLIGENCE;
V. 274.
ED: U. NORTH CAROLINA, CHARLOTTE. NEW COLLECTION.

LCCN 2009-943434

ISBN 3642116736   LIB PO# SLIP ORDERS   ML74.7

LIST	169.00
DISC	17.0%
NET	140.27

6207 UNIV OF TEXAS/SAN ANTONIO

DATE 9/01/10 MUS.APR   6108-09

YBP - CONTOOCOOK, N.H. 03229

SUBJ: 1. MUSIC---COMPUTER NETWORK RESOURCES.
      2. INFORMATION RETREIVAL.

CLASS ML74.7   DEWEY# 780.26   LEVEL ADV-AC

PLEASE CHECK (✓) REASON FOR RETURN. THANK YOU.

- ☐ 1 Reject--no reason
- ☐ 2 Duplicate from another source
- ☐ 3 Damaged/defective
- ☐ 4 No to series (specify if subseries)
- ☐ 5 Out of scope/subject excluded
- ☐ 6 Sufficient coverage
- ☐ 7 Not on press list
- ☐ 8 Too expensive
- ☐ 9 Billing/shipping error

- ☐ 10 Wrong content level
- ☐ 11 Ordered in error
- ☐ 12 Poor quality
- ☐ 13 Specialized interest
- ☐ 14 Peripheral subject
- ☐ 15 Duplicated by YBP
- ☐ 16 Reprint
- ☐ 17 Too popular/low level
- ☐ 18 Narrow geographic area

Your reasons for returning this book will help us to assess your profile and recommend potential revisions to your approval plan.

Comments:

Init:

For Kramer, distribution of accents is crucial to the perception of meter. Musical events cause accents, and the recurrence of accents at specific timepoints creates meter. He identifies three types of accent.

1) Stress accent: performance/notational conventions, e.g. dynamics;
2) Rhythmic accent: a point of stability (e.g. a cadence; probably also agogic);
3) Metric accent: a point of initiation.

His, third category is a statement of a simple musical reality: accents often coincide with the notated beat. Where the accents coincide regularly with the notated downbeat, he draws the analogy of measure beat. At this point, *hypermetric* structures are obtained [7].

There exist two significant works concerning retrieval of the hypermetric structure of rhythm with rhythmic levels exceeding meter. The first approach, being at the same time one of the first methods of creating and ranking hypermetric hypotheses, was proposed by Rosenthal [26][27], but performance of the method does not exceed 65%. The method was implemented in the "*Machine Rhythm*" system. It takes unquantized musical data presented in a symbolic format as inputs. The model is able to cope with slight changes in tempo. Temperley & Sleator [32], authors of the second approach, the so-called preference-rule method, admit, that the rhythmic levels above the meter are not retrieved correctly with their method. Therefore one of the aims of this study was to propose an approach overcoming limitations of existing methods.

Rhythm retrieval models proposed by various researchers are usually based on the same assumptions coming from the domain of music theory, but the ways they operate and music representations might be different. The theory constituting the foundations of most computational models of metric rhythm retrieval is the Generative Theory of Tonal Music (GTTM) by Lerdahl & Jackendoff [22]. Main ideas postulated in GTTM, used to construct rhythm finding models, concern phenomenal accent and hierarchy of rhythmic structure. Phenomenal accent depends on such features of a piece as: local amplitude stresses, duration of sounds, frequencies of sounds, locations of sound onsets and changes in dynamics and harmony. The hierarchy of rhythmic structure concerns either low rhythmic levels or so-called *groupings* corresponding to phrases, sentences or periods. Rhythm finding systems very often rank the hypotheses, using the sound salience function. In fact, Lerdahl & Jackendoff, the authors of GTTM [22] claim that the physical attributes of sounds such as pitch, duration and amplitude influence the rhythmical salience of sounds. A number of research studies are based on this theory. An approach proposed by Rosenthal citero ranks higher hypotheses in which long sounds are placed in accented positions. Similar approaches were presented by Povel & Essens citepo, Allen & Dannenberg [1] and Parncutt [25]. In a multiple-agent approach by Dixon [6] two salience functions are proposed, combining duration, pitch and amplitude. First of them is a linear combination of physical attributes – Dixon calls it an *additive function* (Eq. 1) the other – *multiplicative function* can be calculated with Eq. 2. Both Equations use $p[p_{min}, p_{max}]$ values satisfying conditions of Eq. 3.

$$s_{add}(d,p,v) = c_1 \cdot d + c_2 \cdot p[p_{\min}, p_{\max}] + c_3 \cdot v \tag{1}$$

$$s_{mul}(d,p,v) = d \cdot (c_4 - p[p_{\min}, p_{\max}]) \cdot \log(v) \tag{2}$$

$$p[p_{\min}, p_{\max}] = \left\{ \begin{array}{ll} p_{\min}, & p \leq p_{\min} \\ p, & p_{\min} < p < p_{\max} \\ p_{\max} & p_{\max} \leq p \end{array} \right\} \tag{3}$$

In the above Eqs., p_{min}, p_{max} and $c_k, k = 1,2,3,4$ are constants set experimentally, d denotes the duration of sound expressed in seconds, p is MIDI pitch and v is the amplitude of a sound (MIDI velocity). The values of p_{min}, p_{max} and c_1, k=1,2,3,4 were set experimentally and received values: $p_{min} = 48$, $p_{max} = 72$, $c_1 = 300$, $c_2 = -4$, $c_3 = 1$, $c_4 = 84$. Either additive or multiplicative functions count the rhythmical salience of sounds. The real influence of physical attributes on the rhythmical salience was not estimated experimentally. The authors of this Chapter used and verified these functions in their experiments.

As mentioned before, the literature about metric rhythm is vast, but only a few works refer to the hypermetric rhythm retrieval. A prototype system of automatic drum accompaniment, created on the basis of hypermetric structures, had not as yet been proposed by researchers, that is why one of our aims was to build such a system. The working principle of the system engineered will be shown later.

3   Hypermetric Rhythm Retrieval Approach

The block diagram of the experiments that were conducted by the authors is presented in Fig. 1. The experimental setup consisted of the rhythm retrieval and automatic hypothesis creating and ranking processes is visualized in Fig. 2. The method proposed was validated through experiments conducted on a database of national anthems, retrieved from the Internet [39]. Binary MIDI files were converted to the textual form, that enabled musical analysis and MIDI commands insertion. Note onsets were quantized then, and atomic period u was calculated. Rhythmic hypotheses are formulated with a method proposed by the authors, i.e. periods and phases of rhythmic levels are created first, then related rhythmic levels are grouped into families (hypotheses). The entire hypothesis forming stage has already been described by the authors [18] and presented in Wojcik's Ph.D. thesis [39], thus in the following Chapters we will focus on the hypothesis ranking and then the drum accompaniment creation.

3.1   Experiments 1-3

The aim of Experiments 1-3 was to estimate the rhythmic salience of sounds. The salience-based approach employed artificial neural networks, association rules from the data mining domain as well as rough sets [19][20][21]. The main conclusion coming from that research was that duration is the only attribute of sound, that

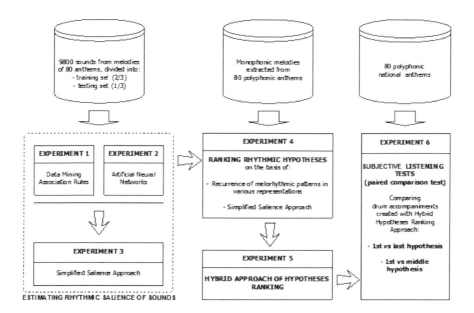

Fig. 1 Experiment layout

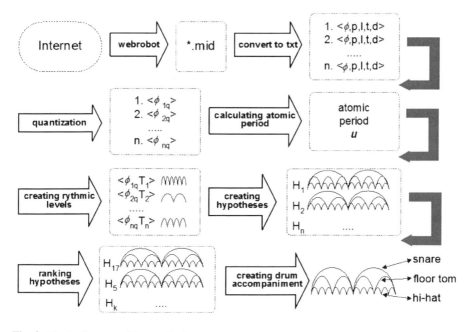

Fig. 2 Block diagram of the method

should be considered in ranking of the rhythmic hypotheses. The outcome of these studies showed also a strong tendency of "long notes to be placed in accented positions" [21].

Thorough analysis of Experiments 1-3 have already been presented by the authors in previous publications [21][38][40], thus they will only be recalled here. For further analysis only their outcome is needed. The aim of the Experiment 1 was to examine whether physical attributes influence the rhythmic salience of sounds and how the stability of results depends on the choice of data for experiments. It also included study on the influence of the number of discretization subranges on the *precision/recall* of the method, and the way in which the rhythmic salience of sound depends on its physical attributes. This was done by using Data Mining Association Rules.

To judge both the ranking functions and at the same time Dixon's additive and multiplicative functions, a *precision/recall* evaluation measure is to be employed. With the same measure, it is possible to choose the most optimal number of discretization subranges. In Information Retrieval (IR) domain precision and recall are calculated with expressions (4) and (5).

$$Precision = N/number\,of\,documents\,in\,answer \qquad (4)$$

$$Recall = N/number\,of\,relevant\,documents\,in\,database \qquad (5)$$

In the above expressions, N denotes the number of relevant documents in the answer. In the evaluation method proposed, a single sound plays the role of a document. Relevant documents are those sounds which are accented. The sounds are sorted descending, according to the value of each ranking function. N highly ranked sounds are placed in the answer. The number of sounds placed in the answer equals to the number of the relevant documents (sounds placed in the accented positions). Denominators of precision and recall get the same value, which results in the equality of precision and recall giving a single measure, allowing for an easy comparison of ranking approaches. This measure will be called *PR (precision/recall)*. The best of the proposed functions or approaches is the one, which gets the highest the highest *PR* of retrieval, calculated according to Eq.6.

$$PR = number\,of\,sounds\,denominated\,as\,accented/number\,of\,accented\,sounds \quad (6)$$

The aim of the Experiment 2 was to apply another computational intelligence technique, namely Artificial Neural Network (ANN) to resolve salience problem. ANNs were fed with musical data to train them to recognize accented sounds on the basis of physical attributes. In particular, it was expected to receive confirmation whether physical attributes influence a tendency of sounds to be located in accented positions. Further, it was to answer how complex is the way the rhythmic salience of sound depends on its physical attributes, and moreover to observe the stability of ANNs answers.

Experiments 1 and 2 showed that the dependency between physical attributes of a sound and its salience is simple – an approach taking only one attribute into account (i.e. duration, pitch, amplitude) ranks accented sounds even slightly better than a method considering a combination of attributes. Therefore the aim of the Experiment 3 was to examine whether the proposed simplified sound duration-based approach to the salience problem performs equally well as the one based on complex Dixon's functions [6].

The results obtained from these experiments (1-3) proved that regardless of what type of melodies were used, the findings were similar. The accuracies of methods retrieving accented sounds using the salience approach, proposed and verified in Experiments 1-3, were compared to the approach proposed in related literature. Dixon's additive and multiplicative functions are the only ones which express quantitatively the value of salience, thus they could be treated as a reference in the study of salience. Therefore computational intelligence-based approach verified experimentally that both intuitive considerations of researchers and functions proposed by Dixon are correct. In this way our approach was verified with other literature findings and we could use the outcomes for the experiments carried out previously.

Further analysis in this study will be limited to the description of Experiments 4, 5 and 6 and to the presentation of the results. Since assumptions in Experiments 1-3 (i.e. salience approach) were different from those in Exp. 4-6, i.e. they were carried out on monophonic melodies, also the context of the previous and consecutive notes was not considered, thus the results are not straightforward comparable. However, as mentioned before, they were very similar to those obtained by other researchers in the rhythm retrieval area.

As explained later, in Experiments 4-5 the context of neighboring notes was taken into account. In addition, Experiments 4 and 5 are conducted on polyphonic versions of national anthems. The proposed method can also process polyphony, i.e. melodic contours are extracted automatically. If two or more sounds appear simultaneously, two approaches are verified – either the upper or the lowest sound of a chord is included into a melodic contour. In Experiment 6, experts could listen to a reference musical piece, which was polyphonic. Experiments 4-5 were conducted on 80 national anthems. In Experiment 6, experts listened to 47 anthems in alphabetical order, starting from the anthem of Afghanistan. The number of anthems had to be limited, because the duration of a single session of subjective listening test should not exceed 20 minutes. Exceeding this time could result in the loss of objectivity of experts' answers, because of tiredness, which may affect the auditory system. Subjective listening tests were conducted according to criteria taking into account the influence of test conditions on human perception, studied by Kostek [14].

3.2 Experiment 4 - Creating and Ranking Rhythmic Hypotheses

In the salience-based approach (Experiments 1-3), the context of neighboring sounds was not considered. However, human perception of rhythm is based on the recurrence of musical patterns, thus a succession of sounds should be taken into account.

Another interesting problem is how to represent music to achieve high performance of hypothesis ranking method. Therefore, in Experiment 4, it was examined what is the influence of the representation of a melody on the correctness of ranking the rhythmic hypotheses. This issue will be discussed later on.

The accuracy of salience approach applied to rank rhythmic hypotheses is compared to accuracies of methods of recurring melodic and rhythmic patterns.

A method to generate rhythmic hypotheses related to a given melody has been proposed. All sets of related rhythmic levels must be found. To find all possible periods, the engineered algorithm searches for the *atomic period* (u) of the piece, which can be found in two steps: subtracting onsets between all pairs of two adjacent sounds in the piece and finding the smallest of differences, which will further be called the atomic period. The assumption is made that musical data are quantized, and there are no ornaments in a piece. Quantized data is devoid of slight time differences between onsets of sounds, which should appear simultaneously or in particular locations in the piece, that is in multiplies of the atomic period. The differences in non-quantized music performed by a professional player are very little, thus they may be not be perceived by human ear. There exist several methods of quantization, they can be easily found in the literature e.g. by Cemgil et al. [2]. The remaining possible periods of the piece are found by multiplying an atomic period by prime numbers – in Western music in most cases those numbers are 2 or 3. Divisions such as 5, 7 happen very rarely, 11 – almost never. That is why in this approach double and triple divisions are considered. Let us call the atomic period a period of layer number zero. By multiplying atomic period u by 2 or 3 two new onsets in the first layer are obtained. Onsets in the second layer are derived by multiplying all first layer onsets by 2 or 3. Creating new layers is finished when the smallest onset in the recently created layer exceeds half duration of the piece.

For each created period *onsets* are counted. The first onset of each period is the onset of the first sound in the piece. Adding subsequent natural numbers of atomic periods to the first onset creates the consecutive onsets. For each period as many onsets are generated as is the period length, expressed in the number of atomic periods. All related rhythmic levels form the hypothesis. The first step in this experiment was to assign a correct hypothesis to each of the melody pieces. The hypothesis is a sequence of rhythmic levels as shown in the example in Fig. 3. The hypothesis is presented there in a graphical form as a series of dots over a score of a piece. The hypothesis is written in square brackets. Each rhythmic level in a hypothesis is a pair of numbers – the first number denotes a period and the second is the onset of a rhythmic level. A rhythmic level with a period equal to the atomic period is the first rhythmic level in each hypothesis. This level is the child of all other rhythmic levels in the hypothesis.

3.2.1 Musical Piece Representations

Computational representations of any of the elements of a musical piece are simplified to the extent allowing optimal processing. However, it cannot lose too much information, since it is supposed to provide high precision and recall of retrieval. In the aspects mentioned above, melody can be represented in a number of the ways. The most suitable representations of non-polyphonic melodies, are as follows:

- melody profile,
- sequence of frequencies,
- sequence of intervals,
- sequence of approximate intervals
- sequence of directions of intervals.

The sequence of directions of intervals is often employed in Music Information Retrieval systems, and it is sufficient enough for successful melody retrieval. In recent works, one may also find examples of studies that use melodic (or fraction of) intervals as the melody representation. However, the authors decided to check the above cited melody representations, only.

Melody Profile

Melody profile contains information in the frequency domain (pitches of sounds) and in the time domain (onsets and durations of sounds). This representation gives information on the rhythm and tonality of a melody. For the example given in Fig. 3, the melody profile would be:

C1 (1/16) C1 (3/16) C1 (1/16)
F1 (1/4) F1 (1/4) G1 (1/4) G1 (1/4)
C2 (11/16) A1 (1/16) F1 (3/16) F1 (1/16)
A1 (3/16)

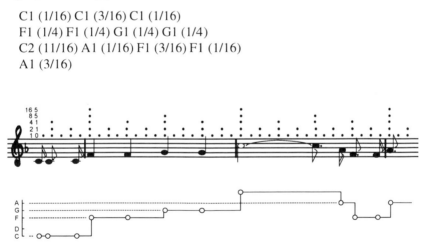

Fig. 3 Correct hypothesis [<1, 0> <2, 1> <4, 1> <8, 5> <16, 5>] and a melody profile to the score of "La Marseillaise"

Numbers shown in Fig. 3 denote the consecutive rhythmic levels and they refer to the correct hypothesis. The correct hypothesis is achieved through adding subsequent rhythmic levels. The first step to build the hypothesis is to add the first rhythmic level <1, 0>, which exists in each hypothesis. Then one of the following levels <2, 0>, <2, 1>, <3, 0>, <3, 1> or <3, 2> is added. After adding each single rhythmic level to the existing hypothesis, authors of this Chapter listened to drum accompaniment, in order to determine the adequateness of the hypothesis, which for the excerpt above is as seen in Fig. 3. The hypothesis is presented n a graphical form as a series of dots over a score of a piece. Dots regard locations of drum fill-ins. The accompaniment is added to the melody by inserting a drum channel, whose number is 10 in the MIDI file. Hi-hat hits are inserted in the locations of rhythmic events associated with the first rhythmic level. The consecutive drum instruments associated with higher rhythmic levels are: bass drum, snare drum, open triangle, splash cymbal, long whistle and a Chinese cymbal, as shown in Table 1.

Table 1 Drum instruments added at a particular rhythmic level

Rhythmic level	Name of the instrument
1	Closed hi-hat
2	Bass drum
3	Snare drum
4	Open triangle
5	Splash cymbal
6	Long whistle
7	Chinese cymbal

Sequence of Frequencies

In comparison to the melody profile, the *sequence of frequencies* representation contains information about the tonality of a piece, but loses rhythmic information – i.e. the duration of all sounds is noted as equal. The graphical representation of *sequence of frequencies* representation to the melody of "La Marseillaise" is presented in Fig. 4.a. The textual representation in this example would be:

C1 C1 C1 F1 F1 G1 G1 C2 A1 F1 F1 A1

Sequence of Intervals

Sequence of intervals representation is a further simplification of music. It contains neither rhythmic nor tonal information. The sizes of intervals are preserved, however. If frequencies were expressed in halftones, and the first sound of a melody was marked by an asterisk '*', the textual *sequence of intervals* drawn in Fig. 4.b would be:

* 0 0 5 0 2 0 5 −3 −4 0 4

Sequence of Approximate Intervals

Music representations described above, i.e. *melody profile, sequence of frequencies, sequence of intervals*, as well as *sequence of directions of intervals* are representations commonly used in MIR research. Wojcik in his P.D. work proposed a new representation of music [39], which is a *sequence of approximate intervals*. This is very similar to the *MelodyContour* descriptor designed in the MPEG-7 standard as a five-level contour [10]. In the *sequence of approximate intervals*, intervals are reduced and classified depending on their size and shift. The idea of approximate intervals is presented in Table 2. The first column in the table, C(1_0), presents the consecutive intervals, expressed in semitones, which represents a *sequence of interval* representation. The second column simplifies intervals to entire tones. Since intervals approximated by entire tones (range of two semitones) can be rounded to the upper semitone or the lower one, two shifts are distinguished in the table, namely C(2_0) and C(2_1). The methods work analogically with approximation by the range of three semitones, in this case there are three possible shifts. The next column in the table would be C(4_0). The values inserted in the table are equal to the lowest interval in each range as seen in the first column.

Table 2 Approximate intervals music representation

C(1_0)	C(2_0)	C(2_1)	C(3_0)	C(3_1)	C(3_2)
-2	-2	-3	-3	-2	-4
-1	-2	-1	-3	-2	-1
0	0	-1	0	-2	-1
1	0	1	0	1	-1
2	2	1	0	1	2
3	2	3	3	1	2

Sequence of Directions of Intervals

The sequence of *directions of intervals* preserves no information of intervals sizes. In this representation, it is only known whether the consecutive sound has lower (d – down), higher (u – up) or equal frequency (s – the same) in comparison to the previous sound. Example illustrated in Fig. 4.c would be represented textually as follows:
    * s s u s u s u d d s u

3.2.2   Creating and Ranking Rhythmic Hypotheses

After the system generated all rhythmic hypotheses according to the approach proposed in this Chapter, then the hypotheses were ranked employing approach proposed by the authors:

    1. sound duration-based hypotheses ranking method,
    2. repeating melodic patterns represented as:
    - sequences of frequencies,
    - sequences of intervals,
    - sequences of approximate intervals,
    - sequences of directions of intervals;

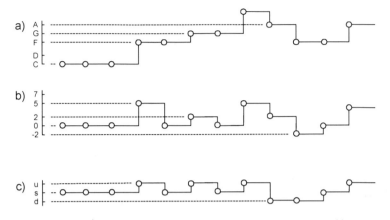

Fig. 4 Representations of a melody: a) sequence of frequencies, b) sequence of intervals, c) sequence of directions of intervals

3. repeating rhythmic patterns.

In each of the above mentioned approaches for all pieces, the accuracy of the method is counted with the aid of Eq. 7, the obtained method accuracies are then averaged. The aim of this experiment is to find the most accurate methods to rank rhythmic hypotheses.

For each ranking method, hypotheses are sorted descending, according to the ranking values, each hypothesis gets a *ranking position RP*, belonging to the range 1..*NoH*, where *NoH* is the number of the hypothesis. The accuracy of the method is equal to the expression given below:

$$Accuracy = 100\% - \frac{RP - 1}{NoH} \qquad (7)$$

Each musical piece in the dataset receives the accuracy calculated. The authors validate engineered methods by averaging the accuracy of a single hypothesis ranking function for all pieces.

Sound Duration-Based Hypothesis Ranking Method

The idea for the hypothesis ranking approach based on sound duration comes directly from the salience research proposed in previous studies by the authors [18]. Since the experiments performed proved experimentally that sounds of long duration are placed in the accented positions, the hypotheses ranking function awarding duration is employed (see Eq. 8). Let the rhythmic levels, expressed as a sequence of equally spaced rhythmic events, described by pairs $<T,\varphi_1>$ be given, where φ_1 is an onset of the first rhythmic event in a piece, and a period T is the time between two adjacent events. T is constant for each rhythmic level. A hypothesis containing the set of rhythmic levels receives the ranking value $HypRANK_{dur}$ calculated with Eq. 8, where NoL is the number of rhythmic levels in the hypothesis, NoN is the number of notes in a piece and *NoRE* is the number of rhythmic events belonging to the rhythmic level.

$$HypRANK_{dur} = \sum_{i=(1,0)}^{NoL} \sum_{k=1}^{NoN} \begin{cases} Note_k \cdot Duration \leftarrow j*T + \phi_1 = Note_k \cdot Onset, j = 1..NoRE \\ 0 \leftarrow otherwise \end{cases}$$

(8)

3.2.3 Recurrence of Melorhythmic Patterns

Recurrence of Melody Patterns

In each of the above mentioned approaches (melodic patterns are represented as a sequence of frequencies, a sequence of intervals and a sequence of directions of intervals), for each hypothesis and for each rhythmic level, a sum of repetitions of each pattern was calculated and assumed to be a hypothesis ranking value. Each pattern is a fragment of a melody which begins at locations calculated with Eq. 9 and ends at locations calculated with Eq. 10, where φ_1 is the onset of the first onset in a song, n is a natural number and T is the period of the rhythmic level. Rhythmic level $<1,0>$ is a descendant of all rhythmic levels in the hypothesis, for which the value of ranking is currently calculated, so the repetitions of this constant element of each sum can be skipped.

$$\varphi_1 + n \cdot T$$

(9)

$$\varphi_1 + (n+1) \cdot T$$

(10)

In the hypothesis ranking methods with the recurrence of melodic patterns, a polyphony problem should also be treated. Two approaches to this problem can be proposed. In the first approach, a sequence of the highest sounds of the chords were treated as a melody, in the second one a sequence of the lowest sounds in the chords were considered to create a melody. For this reason, the hypothesis ranking method for each of the three representations of a melody has two accuracy values calculated, thus all melody pattern representations will receive six average accuracy values.

Recurrence of Rhythmic Patterns

The influence of repeating rhythmic patterns on the ranking of the rhythmic hypotheses is also examined in this experiment. The recurrence of repeating rhythmic patterns was proposed by Wojcik in his doctoral thesis to rank the rhythmic hypotheses [39]. The rhythmic pattern is a binary vector in this approach. Period T, expressed in the number of atomic periods u, is the length of the vector. Values '1' are set at the positions where the note onsets appear, the remaining positions are zeros. In the example from Fig. 3, the first full pattern in the piece on rhythmic level $<4, 1>$ is 1001. For rhythmic level $<8, 5>$, the first full pattern is placed in the first half of the second bar – this pattern is 10001000. The ranking value of each hypothesis with rhythmic patterns is calculated in the same manner as in the case of melody patterns, proposed above – the sum of repetitions of each pattern is calculated for each hypothesis and for each rhythmic level in the hypothesis. Rhythmic levels with short periods belong to more hypotheses than the levels with long periods. Since a single rhythmic level may belong to many hypotheses, it is possible to use the recurrence matrix, proposed in the next Section, to count the number of repetitions only once, and then to use this value as many times as necessary.

Melorhythmic N-gram Recurrence Detection Algorithm

The main aim of the analysis of a piece rhythm is to extract repeating patterns from the stream of notes. The rhythmic structure of a piece is hierarchical – longer rhythmic levels contain shorter ones, which creates a hierarchical structure, that are called *rhythmical families* by Rosenthal [27]. The frequencies of all patterns are calculated in order to be used in the hypothesis ranking stage. Neither the onsets of patterns nor their durations are assumed to be known. Thus, this task can be performed by finding all repeating patterns with all possible durations. If the representation of rhythm is textual, the task might be formulated as follows:

A text string R, representing melodic or rhythmical phrases is given. The length of R is m. The algorithm should calculate the frequency of the appearance of all n-grams, which are the substrings of text string R. The length of the n-gram is n. The matrix $\mathbf{M}[i,j]$ is created $i, j = 1...m$. Each cell of $\mathbf{M}_{i,j}$ is a natural number equal to the number of identical n-grams in string R, which begin at the ith position, and have the length equal to j.

Example:

Let $R = 110011001011$. The matrix $\mathbf{M}$ for such a string is presented in Table 3.

Cells in matrix $\mathbf{M}$ contain values, which indicate how often the n-gram beginning in the column over the particular value appears in the whole string of signs. The length of that n-gram equals to the number of a row (from 1 to 6 in Table 3) in which the particular value lies. Thus, each number 5 in 1st row means that 1-gram '0' appeared five times in the whole string, number 2 in the 5th row means that 5-gram '11001' appeared twice in the whole string. Recurrence of $(k+1)$-gram is not calculated if the k-gram appeared only once in the whole string. Those cells are marked with a hashmark '#'. In each row but "1", there also exists an l-gram, whose last sign is the last note of a piece, thus this pattern cannot become longer, and, as a result, a recurrence of $(l+1)$-gram also is not calculated. Such cells are marked with an asterisk '*'.

The graphical representation of repeating patterns is a tree. A node of such a tree is a structure containing each substring of R and its frequency in string R, separated by two vertical dots $(R:Frq_R)$. A descendant in the tree is a substring, one symbol longer than its ancestor. The proposed representation contains patterns:

– of all possible durations,
– onsets in all possible places in the piece,

Table 3 N-gram recurrence matrix for string 110011001011

n	1	1	0	0	1	1	0	0	1	0	1	1
1	7	7	5	5	7	7	5	5	7	5	7	7
2	3	3	2	3	3	3	2	3	3	3	3	*
3	2	2	2	2	2	2	2	1	1	2	*	*
4	2	2	1	1	2	2	1	#	#	*	*	*
5	2	1	#	#	2	1	#	#	*	*	*	*
6	1	#	#	#	1	#	#	*	*	*	*	*

– repeating at least once in the piece.

With all possible n-grams extracted, it is possible to calculate the frequencies of all patterns in the piece. Pattern frequency information contained in the matrix is used in experiments in hypothesis ranking approaches based on repetitions of patterns represented as sequences of pitches, intervals or directions of intervals. Repeating some of the fragments in the piece is also one of the basic rules of music composition. It can be assumed that long, repeating fragments of a melody are characteristic motives for a particular piece. Those motives are called key melodies by Tseng [33][34]. They can represent binary rhythmic patterns or melodic patterns in any of the above-mentioned representations. The patterns in turn might be treated as indexes or descriptors of a musical piece and can also be retrieved with a pattern recurrence matrix. The onset of the key melody is located in each column having a value 1 in the last row of the matrix (n=6 in the example shown in Table 3), the duration of the key melody is then equal to n-1. In the above example, a key melody is 11001.

Further the repeating patterns in all proposed representations are used to rank hierarchical rhythmic hypotheses. In the frequency domain patterns, represented as the sequence of frequencies, sequence of intervals, sequence of approximate intervals and sequence of directions of intervals, were utilized. In the time domain, binary rhythm patterns were used. Influence of the representation of music on the accuracy of the hypotheses ranking method is also analyzed.

3.3 Experiment 5 - Hybrid Approach

The performance of a few hypotheses ranking methods might be relatively high. Thus, it is worth combining those methods and creating a hybrid approach, which could result in high accuracy of the final method. Let the methods of high accuracies be called *promising methods*. Hybrid approach could be stated as follows: a set of hypotheses ranking methods $\{M_1, M_2, \ldots, M_p\}$, pieces $\{U_1, U_2, \ldots, U_q\}$, and rhythmic hypotheses $\{H_1(U_j), H_2(U_j), \ldots, H_r(U_j)\}$ for each piece U_j are given. Hypothesis H_k is ranked with the method M_i, the position occupied by H_k is $RankPos(M_i)$. For each hypothesis of a piece, a sum of $RankPos(M_i)$ is counted for all promising methods M_i in a set of promising methods $PromMeth$. For each set of promising methods $PromMeth$, hypotheses are sorted ascending, according to the sum of ranking values. Then, the accuracy of the hybrid is counted in an analogous manner to the accuracy of a single method i.e. with the aid of an expression proposed in the previous Section (Eq. 7).

Therefore in Experiment 5, a hybrid approach of ranking rhythmic hypotheses is verified. Accuracy and stability of the results of the hybrid approach and methods, examined in Experiment 4, are also compared. Experiment 5 is conducted with the aid of an application realizing the hypermetric rhythm approach. A system, named DrumAdd, accepting melodies in MIDI format analyzes a given melody and generates drum accompaniment automatically [21][39].

3.4   Experiment 6 – Subjective Listening Tests

Experiment 6 is the final one, in which subjective listening tests were conducted in order to verify whether the proposed ranking hypothesis approach remains in a good agreement with the human perception of rhythm. Subjects listened to the musical pieces with automatically added drum accompaniments.

The aim of subjective tests was to verify, whether the method of ranking hypermetric hypotheses, proposed in this Chapter, agrees with the human perception of rhythm. A group of ten subjects was formed of students, undergraduate students and Ph.D. students of the Multimedia Systems Department of Gdansk University of Technology, most of them having some musical background. The age of experts was between 23 and 34, since it is assumed that listeners in this age group have a good perceptual memory.

Experts listened to the reference pieces, which were polyphonic national anthems, not containing drum tracks. After listening to the reference piece, each expert listened to a pair of melodies of the same anthem with drum accompaniment added automatically. One of accompaniments in a pair was created on the basis of the hypothesis, which was highly evaluated by the hypermetric rhythm retrieval approach proposed in this Chapter. The second version of each anthem contained a drum track, created either on the basis of the last or on the middle rhythmic hypothesis in the ranking list.

Each expert was expected to indicate which accompaniment in a pair is more adequate to the rhythm of the reference piece. Of course, the experts did not know the positions of the hypotheses in the ranking list. The non-parametric paired comparison test [14][15] conducted within this experiment does not require a judgment scale – experts were expected to indicate a better accompaniment in a pair. Each expert evaluated the accompaniments ordered alphabetically, starting from the anthem of Afghanistan. The experiment was conducted on 47 pieces (as already mentioned), the reason for this is that the duration of a single session of subjective listening test should not exceed 20 minutes. The last was Guyana anthem. The anthems of Bangladesh, Bulgaria and Dominica are musical pieces of non-constant meter, so they were not included in the set. From the point of view of the objectification of results, the duration of compared pieces is one of the important factors. Experts could listen to a piece for until they perceived its rhythm well – usually it was enough to listen to few bars. Listeners could decide when to end their listening to the piece or they could start listening to the piece again for deeper analysis of its rhythm.

Two sessions were conducted. In the latter one, each expert evaluated the same accompaniments as in the first session, but in a different order. With tests conducted this way, it is possible to estimate the extent to which experts' votes are consistent. Stability of each expert's judgment is based on the statistical paired test [14] and determines a listener's reliability. The critical value of inconsistent answers should not exceed the value calculated with Eq. 11:

$$0.5 \cdot (n - 1 + x_\alpha \sqrt{n}) \tag{11}$$

In this experiment the critical value equals 28.64, it was calculated for the number of $n=47$ pairs and significance level $\alpha = 5\%$. The value of $x_\alpha = 1.645$ was taken from statistical tables. As presented in Fig. 5, all experts made fewer mistakes than the critical value, so they can be considered as reliable listeners.

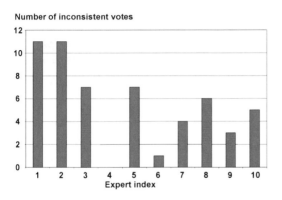

Fig. 5 Experts' reliability

4 Results

For each rhythmic hypothesis, eight ranking positions are calculated – one per each ranking method mentioned below:

- sound duration-based method (salience-based method),
- repeating melodic patterns with melody created of highest sounds of the chords; melody represented as a *sequence of frequencies*,
- repeating melodic patterns with melody created of highest sounds of the chords; melody represented as a *sequence of intervals*,
- repeating melodic patterns with melody created of highest sounds of the chords; melody represented as a *sequence of directions of intervals*,
- repeating melodic patterns with melody created of lowest sounds of the chords; melody represented as a *sequence of frequencies*,

- repeating melodic patterns with melody created of lowest sounds of the chords; melody represented as a *sequence of intervals*,
- repeating melodic patterns with melody created of lowest sounds of the chords; melody represented as a *sequence of directions of intervals*,
- repeating melodic patterns, where a melody is represented as a *sequence of approximate intervals*,
- repeating rhythmic binary patterns.

The results for all hypothesis ranking approaches are shown in Tables 4 and 5. The values in rows B, C and D of Table 4 are accuracy methods, where the melodic contour, automatically extracted from polyphonic pieces, consists of the pitches of the highest sounds in chords, whereas the values in rows O, P and Q are respective accuracies, but the lowest sounds of the chords constitute the melodic contour. Results in the Table 5 present accuracies for melody represented as 'sequence of approximate intervals'. The hypothesis ranking method which received the smallest accuracy is a recurrence of melodic patterns represented as 'sequence of intervals', however its performance is still high (92.74% - row C in Table 4 and in Table 5). The accuracy of the remaining methods exceeded 95%, performances of 10 of 17 methods exceeded 97%. Since the accuracies of all methods in Tables 4 and 5 are high, thus it might be concluded, that all melorhythmic patterns and the duration-based approach are good enough to rank rhythmic hypotheses. For this reason, the authors recommend to contain all methods in the hybrid approach.

Table 4 Average accuracies of ranking hypothesis methods (Part 1 – sound duration-based, upper and low sounds of melodic patterns, rhythmic patterns)

	Ranking method	Average method accuracy [%]	Standard deviation
A.	Salience-based (sound duration-based)	96.12	7.5
B.	Sequence of frequencies, highest sounds	96.33	8.5
C.	Sequence of intervals, highest sounds	92.74	16.7
D.	Sequence of directions of intervals, highest sounds	98.99	4.3
O.	Sequence of frequencies, lowest sounds	97.98	5.8
P.	Sequence of intervals, lowest sounds	97.62	7.5
Q.	Sequence of directions of intervals, lowest sounds	98.95	3.9
R.	Rhythmic patterns	96.49	9.4

Remark 1

The level to which a melody is simplified influences ranking accuracy. This conclusion is deduced from the results presented in Table 5, where performances of recurring melodic patterns, represented as approximate intervals consisting of highest sounds of chords are presented. It may be observed that the larger the size of approximate interval, the larger the accuracy of the method associated with this interval.

Table 5 Average accuracies of ranking hypothesis methods (Part 2 – sequence of approximate intervals, highest sounds of chords)

	Ranking method	Average method accuracy [%]	Standard deviation
C.	C(1_0) (Sequence of intervals)	92.74	16.6
F.	C(2_0)	95.40	11.4
G.	C(2_1)	95.52	10.9
H.	C(3_0)	97.05	7.7
I.	C(3_1)	97.01	7.7
J.	C(3_2)	96.85	8.2
K.	C(4_0)	97.50	6.7
L.	C(4_1)	97.66	6.2
M.	C(4_2)	97.86	5.1
N.	C(4_3)	97.82	5.7

Results from Table 5 are visualized in Fig. 6, where accuracies of approximate intervals are clearly clustered within separable ranges of accuracies. The authors did not observe the influence of shifts of approximate intervals on the accuracy of the method.

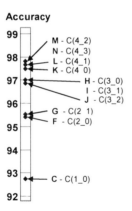

Fig. 6 Accuracy of method based on recurrence of melodic patterns represented as approximate intervals (size 1-4, all possible shifts)

Remark 2

There exists a relation between the performances of the methods and their standard deviations. This can be observed in Fig. 7, which presents the average performance of each approach (value in coordinate 0Y is equal to the accuracy averaged for all anthems) and standard deviation in the first coordinate (0X), calculated from the series of accuracies of all anthems. It can be concluded from Fig. 7 that the higher accuracy, the smaller standard deviation. In other words: results are more stable for methods having good performances.

Fig. 7 presents also the performance and stability of the hybrid approach (experiment 5) in comparison to the performances and stabilities of elementary methods. Hybrid approach achieved accuracy equal to 99.11%, which is the highest value in comparison to all elementary methods. The results are also the most stable – standard deviation achieved the value of 2.2, which is the smallest among all elementary methods. It can be easily noticed that the performance of the hybrid method is located closer to an ideal ranking method than all elementary methods - the coordinates of the ideal ranking approach are (0, 100).

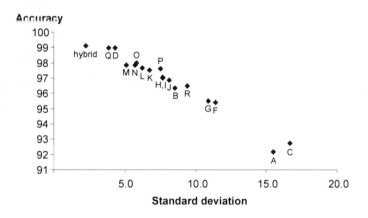

Fig. 7 Relation between accuracies of all hypothesis ranking methods and their standard deviations

In Figs. 8 and 9, the percentage of votes given by listeners for the rhythmic hypotheses which were ranked by hypermetric approach as the first ones in the ranking list (upper plots) and percentage of inconsistent answers (bottom plots) can be seen. Fig. 8 presents the results for the 'last' series, whereas in Fig. 9, the results of the 'middle' series can be observed. Fig. 10 presents the comparison of inconsistencies between 'last' and 'middle' series.

Remark 3

The inconsistency of votes is strongly related to the percentage of votes given on a hypothesis ranked as the highest one. This relation can be easily noticed in Figs. 8 and 9. Listeners were not confident about their answers, when they compared two hypotheses, one of which was not significantly better than the other one. It can thus be concluded that although a single hypothesis is considered as a correct one, there exists a degree of acceptance for the remaining hypotheses.

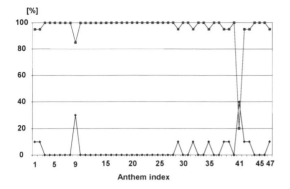

Fig. 8 Inconsistency of votes (bottom plot) and percentage of votes on a highly ranked hypothesis (upper plot), when compared to the last hypothesis

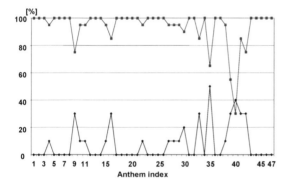

Fig. 9 Inconsistency of votes (bottom plot) and percentage of votes on a highly ranked hypothesis (upper plot), when compared to a middle hypothesis

Remark 4

The average percentage of inconsistent answers in the 'last' series equals 3.62, whereas the average percentage of inconsistent answers in the 'middle' series is equal to 8.09. The percentage of inconsistent answers for each anthem for the 'last' and 'middle' series, which can be seen in Fig. 10, confirms this remark visually. In addition, the average percentage of votes, given on the first hypothesis was equal to 96.91%, when the first and last listed hypotheses were compared, whereas averagely 93.62% of votes were given on the first hypothesis, when it was compared to the middle one. Above results indicate that:

- hypothesis ranked as the highest one is usually recognized by expert as the best one, when compared to either middle or last hypotheses,
- experts were more confident about their votes, while comparing the first hypothesis against the last one rather than against the middle one.

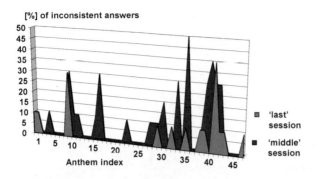

Fig. 10 Percentage of inconsistent answers in the 'last' and 'middle' series

It can thus be concluded that the middle hypothesis is usually more adequate to human rhythm perception than the last one, and the highest ranked hypothesis is more adequate than the middle one. Therefore, hypermetric rhythm retrieval method based on hypotheses ranking approach, proposed by authors, remains in a good agreement with the human perception of rhythm.

5 Conclusions

In this Chapter, the hypermetric rhythm retrieval approach, engineered on the basis of computational intelligence systems, as well as on the recurrence of melorhythmic patterns approach was studied, and in addition its adequacy compared to the human perception of rhythm was verified in subjective tests. Automatic hierarchical rhythm retrieval is analogous to the retrieval of musical words from the stream of sounds. The experiments proved that it is possible to separate musical words automatically using melorhythmic patterns and computational intelligence systems. If the separation of words in musical piece is made correctly, it is used to generate automatic drum accompaniment to the given melody.

Overall, employing an approach to create and then rank rhythm hypotheses has two major advantages over methods that are based on searching for beats in music, namely it reveals that it is possible to extend existing automatic music transcription systems by automatically generating rhythm accompanying a given melody.

Acknowledgments

The research was partially supported by the Polish Ministry of Science and Education within the project No. PBZ-MNiSzW-02/II/2007.

References

1. Allen, P.E., Dannenberg, R.B.: Tracking musical beats in real time. In: Int. Comp. Music Conf., Glasgow, Scotland, pp. 140–143 (1990)
2. Cemgil, A.T., Desain, P., Kappen, B.: Rhythm quantization for transcription. Comp. Music J. 24(2), 60–76 (2000)
3. Desain, P.: A connectionist and a traditional AI quantizer, symbolic versus sub-symbolic models of rhythm perception. Contemporary Music Review 9, 239–254 (1993)
4. Dovey, M.J.: Overview of the OMRAS project: online music retrieval and searching. J. American Society for Information Science and Technology 55(12), 1100–1107 (2004)
5. Dixon, S., Cambouropoulos, E.: Beat tracking with musical knowledge. In: Proc. of the European Conf. on Artificial Intelligence, Amsterdam, pp. 626–630 (2000)
6. Dixon, S.: Automatic Extraction of Tempo and Beat from Expressive Performances. J. of New Music Research, Swets & Zeitlinger 30(1), 39–58 (2001)
7. Ethier, G.: Techniques of Hypermetric Manipulation in Canadian Blues. Canadian Journal for Traditional Music (2001)
8. Goto, M.: An Audio-Based Real-Time Beat Tracking System for Music with or without Drum-Sounds. J. of New Music Research 30, 159–171 (2001)
9. Gouyon, F., Dixon, S.: A review of automatic rhythm description systems. Computer Music Journal 29, 34–54 (2005)
10. http://books.google.com/books?id (information on MPEG-7 standard - MelodyContour)
11. http://cjtm.icaap.org/content/28/28_ethier.html
12. http://www.ismir.net/ (website of ISMIR conferences)
13. http://www.reference.com/browse/wiki/Hypermeter
14. Kostek, B.: Soft Computing in Acoustics. In: Applications of Neural Networks, Fuzzy Logic and Rough Sets to Musical Acoustics. Studies in Fuzziness and Soft Computing. Physica Verlag, Heildelberg (1999)
15. Kostek, B.: Perception-Based Data Processing in Acoustics. In: Applications to Music Information Retrieval and Psychophysiology of Hearing. Cognitive Technologies. Springer, Heidelberg (2005)
16. Kostek, B., Czyzewski, A.: Representing Musical Instrument Sounds for their Automatic Classification. J. Audio Eng. Soc. 49, 768–785 (2001)
17. Kostek, B.: Applying computational intelligence to musical acoustics. Archives of Acoustics 32(3), 617–629 (2007)
18. Kostek, B., Wojcik, J., Holonowicz, P.: Estimation the Rhythmic Salience of Sound with Association Rules and Neural Network. In: Intelligent Information Systems, Gdansk, Poland (2005)
19. Kostek, B., Wojcik, J.: Machine Learning System for Estimation Rhythmic Salience of Sounds. Int. J. of Knowledge-Based and Intelligent Engineering Systems (2005)
20. Kostek, B., Wojcik, J.: Automatic Salience-Based Hypermetric Rhythm Retrieval. In: MUE (Multimedia and Ubiquitous Engineering) International Conference, April 26-28, pp. 1220–1226 (2007)
21. Kostek, B., Wojcik, J., Szczuko, P.: Automatic Rhythm Retrieval from Musical Files. In: Peters, J.F., Skowron, A., Rybiński, H. (eds.) Transactions on Rough Sets IX. LNCS, vol. 5390, pp. 56–75. Springer, Heidelberg (2008)
22. Lerdahl, F., Jackendoff, R.: A Generative Theory of Tonal Music. MIT Press, Cambridge (1983)
23. Lewis, R.A., Cohen, A., Jiang, W., Ras, Z.: Hierarchical Tree for Dissemination of Polyphonic Noise. In: Chan, C.-C., Grzymala-Busse, J.W., Ziarko, W.P. (eds.) RSCTC 2008. LNCS (LNAI), vol. 5306, pp. 448–456. Springer, Heidelberg (2008)

24. McAuley, J.D., Semple, P.: The effect of tempo and musical experience on perceived beat. Australian Journal of Psychology 51(3), 176–187 (1999)
25. Parncutt, R.A.: A perceptual model of pulse salience and metrical accent in musical rhythms. Music Perception 11(4), 409–464 (1994)
26. Rosenthal, D.F.: Emulation of human rhythm perception. Comp. Music J. 16(1), 64–76 (Spring 1992)
27. Rosenthal, D.F.: Machine Rhythm: Computer Emulation of Human Rhythm Perception, PhD thesis, MIT Media Lab, Cambridge, Mass (1992)
28. Ryynanen, M.P., Klapuri, A.: Polyphonic music transcription using note event modeling. In: Workshop on Applications of Signal Processing to Audio and Acoustics, pp. 319–322 (2005)
29. Povel, D.J., Essens, P.: Perception of temporal patterns. Music Perception 2(4), 411–440 (1985)
30. Schuller, B., Eyben, F., Rigoll, G.: Tango or Waltz?: Putting Ballroom Dance Style into Tempo Detection. EURASIP Journal on Audio, Speech, and Music Processing, Article ID 846135 2008 (2008)
31. Takeda, H., Nishimoto, T., Sagayama, S.: Rhythm and Tempo Analysis Toward Automatic Music Transcription. In: IEEE International Conference on Acoustics, Speech and Signal Processing, ICASSP 2007 (2007)
32. Temperley, D., Sleator, D.: Modeling meter and harmony: A preference-rule approach. Comp. Music J. 15(1), 10–27 (1999)
33. Tseng, Y.H.: Multilingual Keyword Extraction for Term Suggestion. In: Proc. of the 21st Annual Int. ACM SIGIR Conf. on Research and Development in Information Retrieval, New York, pp. 377–378 (1998)
34. Tseng, Y.H.: Content-based retrieval for music collections. In: Proc. of SIGIR 1999, 22nd Int. Conf. on Research and Development in Information Retrieval, New York, pp. 176–182 (1999)
35. Wieczorkowska, A.: Learning from Soft-Computing Methods on Abnormalities in Audio Data. In: Chan, C.-C., Grzymala-Busse, J.W., Ziarko, W.P. (eds.) RSCTC 2008. LNCS (LNAI), vol. 5306, pp. 465–474. Springer, Heidelberg (2008)
36. Wieczorkowska, A., Ras, Z.W., Zhang, X., Lewis, R.: Multi-way Hierarchic Classification of Musical Instrument Sounds. In: IEEE CS International Conference on Multimedia and Ubiquitous Engineering (MUE 2007), Seoul, Korea, April 26-28 (2007)
37. Wieczorkowska, A., Ras, Z.W.: Editorial: Music Information Retrieval. J. Intell. Inf. Syst. 21(1), 5–8 (2003)
38. Wojcik, J., Kostek, B.: Intelligent Technologies for Inconsistent Processing. In: Nguyen, N.T. (ed.) Intelligent Methods for Musical Rhythm Finding Systems. Int. Series on Advanced Intelligence, vol. 10, pp. 187–202 (2004)
39. Wojcik, J.: Methods of Forming and Ranking Rhythmic Hypotheses in Musical Pieces, Ph.D. Thesis, Faculty of Electronics, Telecommunications and Informatics, Gdansk University of Technology, Gdansk (2006)
40. Wojcik, J., Kostek, B.: Computational complexity of the algorithm creating hypermetric rhythmic hypotheses. Archives of Acoustics 33(1), 57–63 (2008)

Mid-level Representations of Musical Audio Signals for Music Information Retrieval

Tetsuro Kitahara

Abstract. In this chapter, we introduce mid-level representations of music for content-based music information retrieval (MIR). Although low-level features such as spectral and cepstral features were widely used for audio-based MIR, the necessity for developing more musically meaningful representations has recently been recognized. Here, we review attempts of exploring new representations of music based on this motivation. Such representations are called *mid-level representations* because they have levels of abstraction between those of waveform representations and MIDI-like symbolic representations.

1 Introduction

From a physical point of view, music when recorded in mono is just a vibration represented as one-dimensional time-series data but it obviously has the nature of the multiple dimensionality. Melody, harmony, timbre, and rhythm are important aspects of music. Composers carefully organize the sounds of musical instruments, often sounding simultaneously, to express the ideas of these aspects in their mind.

To handle music data on a computer, we have to represent these aspects in a machine-readable form. In fact, the musical instrument digital interface (MIDI)[1], MusicXML [1], and WEDELMUSIC Format [2] are widely used for representing the melodic aspect on a computer. It is, however, still a challenging problem to transform musical audio signals to such representations.

Tetsuro Kitahara

Graduate School of Science and Tehcnology, Kwansei Gakuin Univeristy, 2-1 Gakuen, Sanda 669-1337, Japan e-mail: t.kitahara@kwansei.ac.jp / CrestMuse Project, CREST, JST, Japan

[1] MIDI has originally been developed and is widely used as an industry-standard protocol that enables electronic musical instruments to communicate. It is, however, also used as a representation of music; in fact, standard MIDI files are one of the most popular file formats for storing music data. In this chapter, we focus on the properties of MIDI only as a representation of music.

Z.W. Raś and A.A. Wieczorkowska (Eds.): Adv. in Music Inform. Retrieval, SCI 274, pp. 65–91.
springerlink.com

In this chapter, we introduce state-of-the-art machine-readable representations of the melodic, harmonic, timbral, and rhythmic aspects of music. In particular, we focus on non-symbolic representations. Because converting audio signals to symbolic representations involves the process of deterministic estimation, a lot of estimation errors are inevitable. Non-symbolic representations, which will be introduced below, can be obtained without deterministic estimation processes and therefore would be robustly obtained. These representations are called *mid-level representations* because they are lower-level than symbolic representations but higher-level than the waveforms themselves.

The rest of this chapter is organized as follows: In Section 2, we discuss an issue in music representation. In Section 3, we briefly review various mid-level representations of music for the melodic, harmonic, timbral, and rhythmic aspects. As an example of the representations introduced in Section 3, we describe an *Instro-gram*, a mid-level representation of instrumentation that we developed, in detail in Section 4. In Section 5 we discuss these representations from different perspectives and finally we conclude the chapter in Section 6.

2   Representations of Music

Representing music—this has been an important subject since early times. When the people did not have audio recording technology, writing music on sheets was the only way for *recording* music. The invention of sheet music (also called musical scores) has enabled us to communicate about music over the temporal and spatial restrictions.

Since the era of computers came, various forms of music representations have been invented. These are classified based on their levels of abstraction. The representations equivalent to the commonly used European modern music notation, for example, belong to the group of the highest level of abstraction. MIDI belongs to the group of the secondary highest level because MIDI sequences are closer to the actual performances. These representations are called *symbolic* representations because music is represented as a sequence of symbols: every note is represented as one symbol (in the music notation) or a pair of Note-On and Note-Off messages (in the MIDI sequences). These representations are indispensable for us to create music on a computer. On the other hand, it is difficult to obtain these representations from audio signals with sufficient accuracy. With the current technology, only a mixture of three or four sound sources can be transcribed [3] even though it has approximately a 20-year history.

A spectrogram (Figure 1) is the most basic representation of audio signals. It is an image, with the time and frequency axes, that shows how the spectral density of a signal varies with time. It can be easily obtained from audio signals with a series of band-pass filters or the short-time Fourier transform (STFT). A spectrogram contains most information about the original signal except the phase, but reading musical semantics from it is not easy. This representation is therefore considered to belong to the group of the lowest level of abstraction.

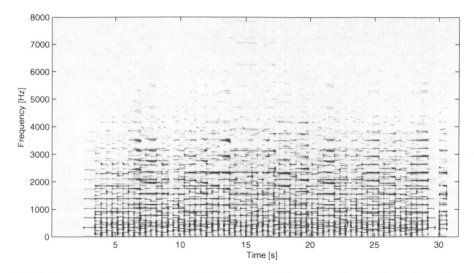

Fig. 1 An example of spectrogram. This is generated from an audio signal of trio music. Although only up-to-three tones simultaneously sound, many simultaneous frequency components (shown like horizontal bars) can be seen because each tone has many overtones.

Mid-level representations of music could be a solution to the problem of the gap between symbolic and low-level representations. This kind of gap is known as the *symbol grounding problem* or *semantic gap*, which is a common problem in many research fields dealing with the real world. Improving the accuracy of symbol grounding is a very important research subject, and indeed, various researchers around the world have been tackling the improvement of symbol grounding (see [3] etc. for details). There is, however, a different approach to the symbol grounding problem. This approach is based on the idea that a symbol system is not always necessary to capture semantics from real-world objects. Music representations with certain levels of abstraction are usable enough for sophisticated MIR even if they are not accurately transformed into symbols. Selecting appropriate abstraction levels and designing representations with the levels are also important research subjects.

The concept of mid-level representation was introduced by Ellis et al. [4] and has been widely accepted as an important concept in the field of computational auditory scene analysis (CASA). They pointed out the following as properties desirable in auditory mid-level representations:

- **Sound source separation.** The representations should decompose sound to a granularity at least as fine as the sources of interest: in the situation that Bill is playing the trombone with TV in the background, pieces that can be labeled as TV noise or trombone.
- **Invertibility.** The original sound should be regenerated from its representation. In particular, the representation should make it possible to regenerate a meaningful part of the sound (e.g., the trombone without the TV noise).

- **Component reduction.** The sound should be represented in a relatively small number of objects corresponding to meaningful structure.
- **Abstract salience of attributes.** The features made explicit by the representations should approach the perceptual attributes of the final result dependent on the tasks or applications.
- **Physiological plausibility.** The researchers on CASA should respect the physiological knowledge and should not pursue hypothesis clearly inconsistent with physiology.

The representations that have all these properties are also useful for MIR, but there can be many representations useful for MIR even if they do not have all these properties. Not limiting *mid-level representations* to those having all these properties, we are dealing here with those having the following two properties:

- They are not symbolic.
- They represent musical semantics as clearly as possible.

Here, the musical semantics include melodic, harmonic, timbral, and rhythmic aspects which are not clearly or separately represented in spectrograms.

3   Examples of Mid-level Music Representations

In this section, we review various mid-level representations of music motivated by the above-mentioned discussion. The representations are sorted by two different dimensions as shown in Figure 2. One dimension refers to which aspect in music is represented. Here we address the melodic, harmonic, timbral, and rhythmic aspects. The other dimension refers to the level of abstraction.

Note, however, that the order of abstraction is not determined between all pairs of representations (even though the figure is drawn as it is possible). This is because the representations represent slightly different content even if they are classified into the same aspect group.

3.1   *Representations of Melodic Aspect*

The most commonly used representation of the melodic aspect is the modern notation originated in European classical music, where notes are placed on a five-line staff. This can represent almost all elements for reproducing music except the timbre, so it is still its important representation. XML versions of this representation, such as MusicXML [1] and WEDELMUSIC Format [2], have therefore been proposed and are used for creating and exchanging sheet music on a computer.

In MIDI sequencers, a piano roll (Figure 5) is also commonly used for representing music, especially as a visualization of MIDI data. A piano roll in MIDI sequencers represents notes as colored bars on a plane with the time and frequency axes. The onset time, pitch, and duration of each note is represented as the horizontal position, vertical position, and horizontal length of the bar, respectively.

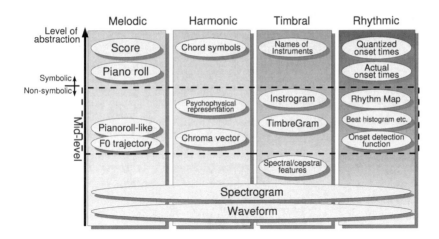

Fig. 2 Overview of mid-level representations reviewed in this chapter. Note that, although the representations are sorted by the levels of abstraction, some orders are not necessarily admissible.

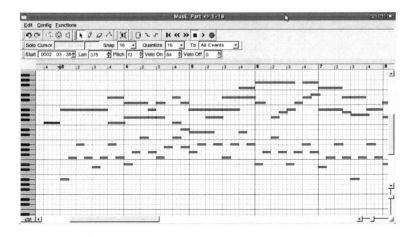

Fig. 3 An example of piano-roll representations

These, however, cannot be considered as mid-level representations; these symbolic representations is difficult to accurately obtain from audio signals. New representations of the melodic aspect have recently been proposed where a melody is represented as a sequence of continuous values.

Continuous Trajectory of Melody's F0

Goto proposed a new method for estimating the F0 of the melody and bass lines contained in real-world CD recordings[2] [5]. He claimed that the result of F0 estimation should not be represented in a symbolic form because musically untrained people understand music without representing its score in their mind. He therefore represented the melody and bass lines as a sequence of continuous quantitative values: $S_m(t) = \{F_m(t), A_m(t)\}$ and $S_b(t) = \{F_b(t), A_b(t)\}$, where $F_i(t)$ and $A_i(t)$ are the F0 and amplitude at time t and "m" and "b" denote the melody and bass lines, respectively.

The basic idea of his F0 estimation method, *PreFEst*, is to estimate the most predominant harmonic structure at every frame after a high-pass (for the melody) or low-pass (for the bass) filter is applied. The melody and bass lines, in general, have the strongest tones in the high- and low-pitch regions, respectively. When the relative dominance of each harmonic structure contained in these filtered signals is calculated, the most predominant harmonic structures in the high- and low-pass-filtered signals can be considered the melody and bass tone, respectively.

To measure the relative dominance of harmonic structures, Goto focused on the analogy of a mixture of harmonic structures to a mixture of probabilistic distributions. A complex probabilistic distribution is often approximated as a mixture (weighted sum) of simpler distributions (typically, Gaussians) using the Expectation-Maximization (EM) algorithm. If a harmonic structure is represented as a parametric model, a complex power spectrum can also be approximated as a weighted sum of the harmonic-structure models using the EM algorithm. The calculated weight for each harmonic-structure model can be considered the relative dominance of the harmonic structure.

In the first version of the PreFEst [6], the harmonic-structure model (he calls it the *tone model*), given the F0 F, is designed as follows:

$$p(x|F) = \alpha \sum_{h=1}^{N} c(h) G(x; F + 1200 \log_2 h, W),$$

$$G(x; m, \sigma) = \frac{1}{\sqrt{2\pi\sigma^2}} \exp\left(-\frac{(x-m)^2}{2\sigma^2}\right),$$

where α is a normalizing factor, $N = 16$, $W = 17$ cent, $c(h) = G(h; 1, 5.5)$, and x is the log frequency. Then, the observed power spectrum after the high- or low-pass filter is approximated as the mixture density
$p(x; \theta^{(t)})$ defined as

$$p(x; \theta^{(t)}) = \int_{Fl}^{Fh} w^{(t)}(F) p(x|F) dF,$$

$$\theta^{(t)} = \{w^{(t)}(F) \mid Fl \leq F \leq Fh\},$$

---

[2] http://staff.aist.go.jp/m.goto/PROJ/f0.html

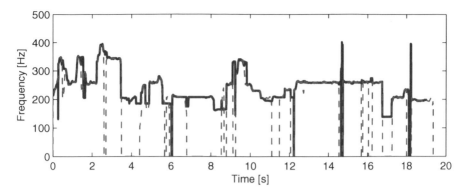

Fig. 4 Continuous trajectory of melody's F0 estimated with the PreFEst. The solid and dashed line represent the estimated F0 trajectory and ground truth, respectively. The result of estimation was provided by Dr. Masataka Goto.

where Fl and Fh denote the lower and upper limits, respectively, of the possible F0 range, and $w^{(t)}$ is the weight of a tone model $p(x|F)$ that satisfies $\int_{Fl}^{Fh} w^{(t)}(F)dF = 1$. Once the weights $w^{(t)}(F)$ are estimated using the EM algorithm, the frequency F maximizing the weights is considered the most predominant and is output as the result of F0 estimation.

An example of melody's representations obtained with the PreFEst is shown in Figure 4. This is the temporal trajectory of the frequency that maximizes the above-mentioned weights at every frame.

Marolt also proposed a mid-level representation that is obtained with a PreFEst-based F0 estimator [7]. The estimated F0s for every frame are linked in time, resulting in a series of pitch tracks called *melodic fragments*. After that, the fragments are clustered based on their timbral similarity.

Pianoroll-like Representation

A pianoroll-like but non-symbolic representation was proposed by Sagayama et al[3] [8]. The basic idea is to suppress the components of overtones in a spectrogram by using the deconvolution of the observed spectrum $v(x)$ with a common harmonic structure pattern $h(x)$. That is, the pianoroll-like representation is obtained at every frame as $u(x) = h^{-1}(x) * v(x)$, where $*$ denotes convolution and x is the log frequency also here. All tones appearing in a signal to be analyzed are assumed to have a common harmonic structure pattern. Although the harmonic structure patterns of tones with different F0s are different in practice even for the same instrument, the specmurt analysis successfully suppresses the overtones to some extent.

A similar representation can be obtained using other multi-F0 estimation methods. An example of pianoroll-like representations obtained using the PreFEst is shown in Figure 5. In this figure, all elements of $\{w^{(t)}(F) \mid 0 \le t \le T, Fl \le F \le Fh\}$,

---

[3] http://hil.t.u-tokyo.ac.jp/index-e.html

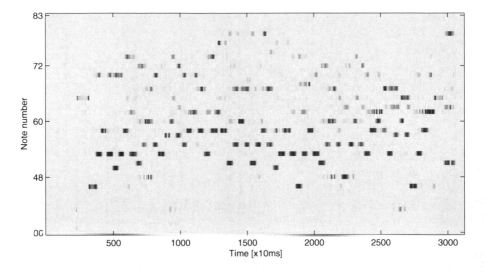

Fig. 5 Pianoroll-like representation of the same piece of trio music as that used in Figure 1, obtained with the PreFEst. Unlike Figure 1, each note is clearly represented because the overtones have been successfully suppressed.

where T is the end time of the signal, are represented as the intensity of color at every point in the time-frequency plane. Sparse coding and non-negative matrix factorization (NMF) [9] can also be used for obtaining a pianoroll-like representation (see [10] for detailed information).

The common principle of these techniques is to decompose a power spectrum x_t at every frame into a weighted sum of *basis functions*[4] b_n ($n = 1, \cdots, N$), that is,

$$x_t \approx \sum_{n=1}^{N} g_{n,t} b_n.$$

For T frames, this can be written in a matrix form as

$$X = BG,$$

where $X = [x_1, \cdots, x_T]$ is the *observed matrix* (typically a spectrogram), $B = [b_1, \cdots, b_N]$ is the matrix of the basis functions, and $G = [g_{n,t}]$ is the *gain matrix*. When the basis functions $b_1, \cdots, b_N$ represent a typical harmonic spectrum for every semitone, the gain $g_{n,t}$ for the basis function b_n represents the relative amplitude of the corresponding semitone at frame t. By imaging the gain matrix G, one can obtain a pianoroll-like representation.

This decomposition is, however, an ill-posed problem because B and G cannot be uniquely determined from the given X. To solve this ill-posed problem, a certain constraint for B or G is required. In the first versions of both specmurt analysis [8]

---

[4] They are actually vectors even though the name is *function*.

and PreFEst [6], B was pre-designed; it was not estimated from the given signal. Because this simplification affects the approximation accuracy, both methods were extended so that the quasi optimal B was estimated from the signal [5, 11]. On the other hand, sparse coding introduces the constraint that the gain matrix G is sparse, that is, as many components as possible are zero. NMF introduces the constraint that the all components of both B and G must be non-negative (i.e., equal to or greater than zero), as the name implies.

Instrument-specific Harmonic Atoms

Instrument-specific Harmonic Atoms is a sparse-based representation with explicit instrument labels, proposed by Leveau et al [12]. The key idea behind this is to prepare and train sinusoidal models for each target instrument. Let $x(t)$ be a given musical audio signal, then it is reprenseted as a weighted sum of *atoms* $h_\lambda(t)$ plus a residual $r(t)$ as follows:

$$x(t) = \sum_\lambda \alpha_\lambda h_\lambda(t) + r(t),$$

where λ is an atom index. The atom $h_\lambda(t)$ is taken from a dictionary $\mathscr{D} = \{h_\lambda(t)\}_\lambda$ consisting of sinusoidal representations of the sounds of each target instrument with every possible F0. The atom is, specifically, reprenseted as a weighted sum of sinu-soidal partials, where the weights (the relative amplitudes of the partials) are trained for each target instrument. After this training, the optimal description of a given signal is estimated using the Viterbi algorithm.

3.2 Representations of Harmonic Aspect

The highest-level representation of the harmonic aspect would be a sequence of chord symbols such as "Cm" and "G7". This representation is very useful espe-cially for popular music because it represents a harmonic progression with quite a limited amount of symbols. Obtaining a chord-symbol representation from an audio signal is a typical problem of pattern recognition and has obtained successful results (e.g., [13, 14, 15, 16, 17, 18]). Estimating chord symbols is easier than transcribing every note but estimation errors are still inevitable. Mid-level representations of the harmonic aspect would therefore be desired.

Chroma Vector

A *chroma vector*, also known as a *pitch class profile*, is a 12-dimensional vector where each element represents the cumulative magnitude of each pitch class. This has been widely used in almost all studies into chord recognition (e.g., [13, 14, 15, 16, 17, 18]) and other many tasks including detection of chorus sections [19]. In [19], the chroma vector is introduced as follows:

> The chroma vector is a perceptually motivated feature vector using the concept of *chroma* in Shepard's helix representation of musical pitch perception [20]. According

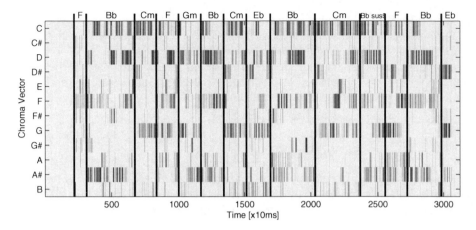

Fig. 6 Example of visualization of a sequence of chroma vectors. The figure shows that the elements corresponding to the component pitch classes of each chord tend to show high values.

to Shapard [20], the perception of pitch with respect to a musical context can be graphically represented by using a continually cyclic helix that has two dimensions, *chroma* and *height*. Chroma refers to the position of a musical pitch within an octave that corresponds to a cycle of the helix; i.e., it refers to the position on the circumference of the helix seen from directly above. On the other hand, height refers to the vertical position of the helix seen from the side (the position of an octave).

An example of the results of calculating chroma vectors is shown in Figure 6.

Psychophysical Representation of Harmony Perception

Although chroma vectors represent the acoustic characteristics of chords, they do not represent their perceptual properties, that is, the subjective impression that humans perceive from the chords. Fujisawa et al. [21] therefore proposed, according to their psychophysical study [22], a three-dimensional representation where each dimension corresponds to the dissonance, tension, and major/minor modality. For simplicity, we limit target chords to two-tone chords for calculating the dissonance and to three-tone chords for calculating the others in the following explanation.

The dissonance D between two complex tones A and B is calculated based on the interval x_{ij} between every pair of A's i-th partial and B's j-th partial as follows:

$$D = \sum_i \sum_j v_{ij}\gamma[\exp(-\alpha x_{ij}) - \exp(-\beta x_{ij})],$$

where v_{ij} is a coefficient based on the normalized intensities of A's i-th partial and B's j-th partial, and α, β, and γ are experimentally determined constants.

On the other hand, the tension T and major/minor modality M in three complex tones A, B, and C are calculated based on two kinds of intervals x_{ij} and y_{jk}; the

Table 1 Low-level timbral features commonly used in MIR studies

Spectral Centroid	Center of gravity of the magnitude spectrum of the STFT $C_t = \sum_{n=1}^{N} nM_t[n] / \sum_{n=1}^{N} M_t[n]$
Spectral Rolloff	Frequency R_t below which 85%, for example, of the magnitude distribution is concentrated R_t s.t. $\sum_{n=1}^{R_t} M_t[n] = 0.85 \sum_{n=1}^{N} M_t[n]$
Spectral Flux	Squared difference between the normalized magnitude of successive spectral distributions $F_t = \sum_{n=1}^{N} (N_t[n] - N_{t-1}[n])^2$
Zero Crossing Rate	Number of times that the time-domain signal changes its sign.
Mel-frequency Cepstral Coefficients (MFCCs)	Coefficients that collectively make up an mel-frequency cepstrum, which represents a coarse spectral envelope in the mel frequency scale

$M_t[n]$ and $N_t[n]$ denote the magnitude and normalized magnitude of the Fourier transform at frame t and frequency bin n. N is the number of the frequency bins.

definition of x_{ij} is the same as the above-mentioned one while y_{jk} is defined as the interval between B's j-th partial and C's k-th partial. That is, T and M are calculated as follows:

$$T = \sum_i \sum_j \sum_k v_{ijk} \exp\left[-\left(\frac{z_{ijk}}{\delta}\right)^2 \right],$$

$$M = \sum_i \sum_j \sum_k \left\{ -v_{ijk} \frac{2z_{ijk}}{\varepsilon} \exp\left[-\left(\frac{-z_{ijk}^4}{4}\right) \right] \right\},$$

while $z_{ijk} = y_{jk} - x_{ij}$, v_{ijk} is a coefficient determined by the normalized intensities of A's i-th, B's j-th, and C's k-th partials, and δ and ε are experimentally determined parameters.

They also developed a music mood visualizer that intuitively represents harmonic characteristics in the color determined by this three-dimensional representation [21].

3.3 Representations of Timbral Aspect

The most commonly used timbral representation in current MIR-related studies are low-level features, such as spectral and cepstral features, listed in Table 1 (e.g., [23, 24, 25, 26]). These features represent the characteristics of spectral envelopes, which consider one aspect of the timbre of musical instruments[5]. These features are extracted from the power spectrum at every frame and then are often summarized as their temporal means and variances (therefore the information about how these features evolve over time is lost). The spectra may contain the sounds of more-than-one sources but the features are extracted without separating the sources. Because a mixture of sounds is treated as if it has a single timbre, these features are sometimes

---

[5] The zero crossing rate is calculated in the time domain but it describes the amount of high-frequency energy in the signal (i.e., brightness) [27].

called *polyphonic timbre* [26]. They can capture the content of music to some extent with a low computational cost because it does not need sound source separation, but has a clear limit to capture higher-level content [25].

Strictly speaking, *timbre* is a very difficult concept because it does not clearly put on a physical scale. Timbre is considered multidimensional and has not been fully defined. In fact, the American Standards Association has defined timbre as follows:

> Timbre is the attribute of auditory sensation in terms of which a listener can judge that two sounds having the same loudness and pitch are dissimilar.

This definition is indirect and has serious limitations. In order for this definition to apply, for example, two sounds need to be able to be presented at the same pitch.

There are two acceptable standpoints for the definition of timbre. One is to consider timbre to be acoustical characteristics corresponding to all aspects of the impression that humans receive from sounds. In this case, timbre would be described verbally. The other is to consider timbre to be acoustical characteristics linked to differences between the sounds of different instruments. In this case, the names of the instruments can be used as labels for the timbres [28, 29].

From the latter standpoint, the names of musical instruments are the highest-level representation of the timbral aspect. Recognition of musical instruments from audio signals has therefore been studied in the last 20 years. Although most of the existing studies on musical instrument recognition dealt only with solo musical sounds (e.g., [30, 31, 32, 33, 34, 35, 36, 37]), the number of studies dealing with polyphonic music has been increasing in recent years [38, 39, 40, 41, 42, 43, 44, 45]. The common problem in these studies is to require preceding note (or F0) estimation process. For example, OPTIMA [38], Ipanema [39], Kinoshita *et al.*'s method [40], and Kitahara *et al.*'s method [45] identify the instrument for each note (*notewise processing*) and hence have to estimate the onset time and fundamental frequency (F0) of each note in advance. Eggink and Brown's methods [41, 42] identify instruments for each frame. Although they do not require onset detection, they still require the estimation of F0s of notes played at each frame. Because onset detection and F0 estimation are difficult in polyphonic music in general, the performance of instrument recognition in these studies are greatly suffered from their errors. In the experiments of most studies mentioned above, therefore, correct data on onset times and F0s were manually given.

Recently, new representations of the timbre or instrumentation (orchestration) of musical audio signals have been proposed. These representations can be obtained without requiring the deterministic estimation of the onset times or F0 of each note. They would therefore be obtained with accuracy for complex real-world musical audio signals.

Instrogram

An *Instrogram* [46, 47, 48] is a spectrogram-like graphical representation that enables users to find when which instruments are used in a musical piece. An Instrogram consists of several images each of which corresponds to each of the target

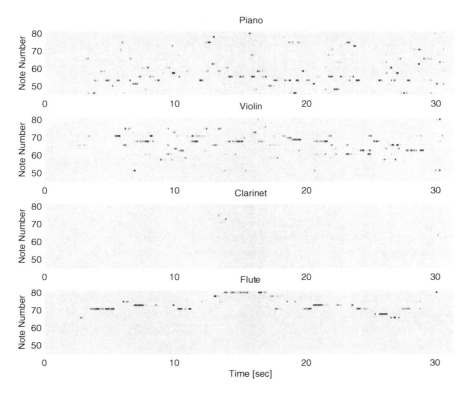

Fig. 7 Example of Instrograms. The rough content of the performance of each instrument can be seen from the Instrogram. An Instrogram can be considered a decomposition of a pianoroll-like representation, shown in Figure 5, into the representation for each instrument.

instruments. Each image has horizontal and vertical axes representing time and frequency, and the intensity of the color of each point (t, f) shows the probability $p(\omega_i; t, f)$ that the target instrument ω_i is used at time t and at F0 of f. An example is presented in Figure 7. This example shows the results of analyzing an audio signal of "Auld Lang Syne" played on the piano, violin, and flute. The target instruments for analysis were the piano, violin, clarinet, and flute. If the Instrogram is too detailed for some purposes, it can be simplified by dividing the entire frequency region into a number of subregions and merging the results within each subregion. A simplified version of Figure 7 is given in Figure 8. The original or simplified Instrogram shows that the melodies in the high (approx. note numbers 70–80), middle (60–75), and low (45–60) pitch regions are played on flute, violin, and piano, respectively.

TimbreGram

TimbreGrams, which have been proposed by Tzanetakis [49], map audio files to sequences of vertical color stripes where each stripe corresponds to a short slice of sound. The similarity of different files is shown as overall color similarity; similar

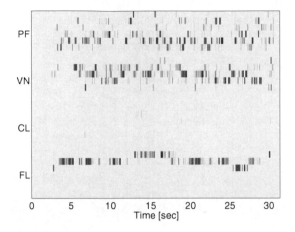

Fig. 8 Simplified (summarized) Instrogram for Figure 7. The melodic aspect was removed (accordingly the timbral or instrumental aspect was relatively enhanced) by the simplification.

musical pieces are represented in similar colors. The similarity is calculated based on low-level features. The concept of TimbreGrams is close to that of Instrograms but it is important to note that TimbreGrams are *relative* representation unlike Instrograms. The TimbreGram for each musical piece represents only the similarity to other musical pieces within a music collection. This fact means that the Timbre-Gram for a certain piece in a certain music collection may be different from the TimbreGram for the same piece in a different music collection. This is an important difference from the Instrogram representation.

3.4 Representations of Rhythmic Aspect

The research of rhythm recognition/classification has a long history. The simplest way for representing rhythm is to list all onset times or inter-onset intervals (IOIs). To obtain this type of representation from audio signals, onset detection has been conducted by various researchers [50]. From a musical point of view, the list of actual onset times is not always essential. The list of the onset times intended by the composer could rather be sometimes desired. This is represented as a list of note values, which is obtained by quantizing the actual onset times based on metrical structure. Beat tracking is necessary to do this and has been achieved with high accuracy for real-world CD recordings (e.g., [51]). The list of actual onset times and that of intended note values are equivalent to (and accordingly have the same levels of abstraction as) a MIDI sequence (or a piano roll) and a musical score, respectively.

The typical method of onset detection mainly consists of three steps: pre-processing, reduction, and peak-picking [50]. The pre-processing step aims to transform the original audio signal in order to accentuate the onsets of the notes in the signal. The reduction step calculates a *detection function*, also sometimes called a *novelty function*, which is a time series of values whose peaks are intended to coincide with the times of note onsets. In the final peak-picking step, the peaks of the detection function are extracted as the onset times. The design of the detection function is one of the most important issues in onset detection, and several functions have been proposed such as a high frequency content (HFC) function, a spectral difference function, and a phase deviation function (see [50] for the details).

Detection functions satisfy our two requirements for mid-level representations. They can therefore be used for various rhythm-related tasks as well as onset detection. In fact, Davies [52] used detection functions for beat tracking without extracting onset times from them.

Describing the onset times of all the notes or onset detection functions for the whole signal is an important approach, but the overall characteristics of rhythm are also an important description. In general, musical pieces of different genres have different characteristics of rhythm: rock music, for example, tends to have stronger beats than classical music. To describe or classify such characteristics, various features have been attempted. A beat histogram, used in [23], is one of the descriptors of the overall rhythmic characteristics. The beat histogram maps tempi (beats per minute, BPM) to beat strengths. When a beat histogram has two strong peaks at the BPM of 80 and 160, it implies that the main tempo is 160 and that the half-note-level beats are as strong as the quarter-note-level beats. When a beat histogram do not have particularly strong peaks in any BPM, it implies that the piece has no strong pulsed sounds at beat times as shown in classical music. As a similar representation, the histogram of IOIs (IOIH) is also used [53].

The beat histogram and IOIH are useful as concise representations but they discard important temporal information: two rhythmic patterns "x...x...x...x.x." and "x...x.x...x...x." are completely different from a musical point of view but they cannot be distinguished from their IOIHs. Dixon et al. [54] pointed this out and proposed a representation of rhythm patterns that does not discard the temporal information. Dixon et al.'s representation is basically an averaged within-measure amplitude envelope. The bar lines of a given signal are first detected and then the amplitude envelope within each measure is calculated. After outliers are removed, the average of the amplitude envelopes is finally calculated.

Paulus and Klapuri [55] dealt with the measurement of the similarity of rhythmic patterns. Based on the idea that rhythmic patterns are characterized by the sounds of percussive instruments, rhythmic patterns are represented after suppressing non-percussive (pitched) instrument sounds. The given musical signal is first approximated with a sinusoidal model. In the residual of the approximation, the drum sound is enough enhanced. From this drum-sound-enhanced signal, timbral features, such as MFCCs, are extracted, and finally, the similarity is calculated with the dynamic time warping (DTW).

Tsunoo et al. [56] proposed a new visual representation, called a *Rhythm Map*, that shows what kind of rhythm pattern is played in each measure from the beginning to the end of a musical signal. The Rhythm Map has the horizontal axis representing time and the vertical axis representing rhythm pattern indices. The index of the rhythm pattern being played in each measure is colored black. From the Rhythm

Map, one can see that, for example, Rhythm Pattern 1 is repeatedly played and once in four measures, Pattern 2 is played as a fill-in pattern, Pattern 3 is played in a bridge section, and this piece ends with Pattern 4. Similarly to Paulus and Klapuri's method [55], the Rhythm Map is also obtained by analyzing the spectrogram after suppressing the non-percussive sounds.

See also [57] for more details of rhythmic representation.

4   Instrogram: Probabilistic Representation of Instrumentation

An Instrogram is a mid-level representation of instrumentation developed by the author and collegues as introduced in the previous section. Here, we try a case study by describing the Instrogram representation, especially its formulation, calculation algorithm, and applications, in detail.

4.1   Formulation of Instrogram

Let $\Omega = \{\omega_1, \cdots, \omega_m\}$ be the set of target instruments. We then have to calculate the probability $p(\omega_i; t, f)$ that a sound of the instrument ω_i with an F0 of f exists at time t for every target instrument $\omega_i \in \Omega$. This probability is called the *instrument existence probability* (IEP). Here, we assume that multiple instruments are not being played at the same time and at the same F0, that is, $\forall \omega_i, \omega_j \in \Omega: i \neq j \Longrightarrow p(\omega_i \cap \omega_j; t, f) = 0$. Let ω_0 denotes the silence event, which means that no instruments are being played, and let $\Omega^+ = \Omega \cup \{\omega_0\}$. The IEP then satisfies $\sum_{\omega_i \in \Omega^+} p(\omega_i; t, f) = 1$. When the symbol "X" denotes the union event of every target instrument sounding, which stands for the existence of *some* instrument (i.e., $X = \omega_1 \cup \cdots \cup \omega_m$), the IEP for each $\omega_i \in \Omega$ can be calculated as multiplication of two probabilities:

$$p(\omega_i; t, f) = p(X; t, f)\, p(\omega_i | X; t, f),$$

because $\omega_i \cap X = \omega_i \cap (\omega_1 \cup \cdots \cup \omega_i \cup \cdots \cup \omega_m) = \omega_i$. Above, $p(X; t, f)$, called the *nonspecific instrument existence probability* (NIEP), is the probability that the sound of some instrument with an F0 of f exists at time t, while $p(\omega_i | X; t, f)$, called the *conditional instrument existence probability* (CIEP), is the conditional probability that, if the sound of some instrument with an F0 of f exists at time t, the instrument is ω_i. The probability $p(\omega_0; t, f)$ is given by $p(\omega_0; t, f) = 1 - \sum_{\omega_i \in \Omega} p(\omega_i; t, f)$.

4.2   Algorithm for Calculating Instrogram

Figure 9 shows an overview of the algorithm for calculating an Instrogram. Given an audio signal, the spectrogram is first calculated. The short-time Fourier transform (STFT) shifted by 10 ms (441 points at 44.1 kHz sampling) with an 8,192-point Hamming window is used in the current implementation. We next calculate the NIEPs and CIEPs. The NIEPs are calculated by analyzing the power spectrum at each frame (*timewise processing*) using the PreFEst [5], described in Section 3.1.

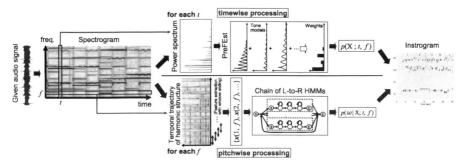

Fig. 9 Overview of the algorithm for calculating Instrograms

The CIEPs are, on the other hand, calculated by analyzing the temporal trajectory of the harmonic structure for every F0 (*pitchwise processing*). The trajectory is analyzed with a framework similar to speech recognition, based on left-to-right hidden Markov models (HMMs) [58]. This HMM-based temporal modeling of harmonic structures is important because temporal variations in spectra characterize timbres well. This is the main difference from framewise recognition methodologies [42, 59]. Finally, the NIEPs and CIEPs are multiplied.

The advantage of this technique lies in the fact that $p(\omega_i; t, f)$ can be estimated robustly because the two constituent probabilities are calculated independently and are then integrated by multiplication. In most previous studies, the onset time and F0 of each note were first estimated, and then the instrument for the note was identified by analyzing spectral components extracted based on the results of the note estimation. The upper limit of the instrument identification performance was therefore bound by the precedent note estimation, which is generally difficult and not robust for polyphonic music. Unlike such a notewise symbolic approach, our non-symbolic and non-sequential approach is more robust for polyphonic music.

Nonspecific Instrument Existence Probability

The NIEP is calculated by using the PreFEst, mentioned in Section 3.1. The PreFest calculates the relative dominance of the harmonic structure with an F0 of F at time t as the weight $w^{(t)}(f)$ for the corresponding tone model, we calculate the NIEP $p(X; t, f)$ by considering it equal to the weight $w^{(t)}(f)$.

Conditional Instrument Existence Probability

The CIEP is calculated based on left-to-right HMMs prepared for every semitone. In a typical speech recognition framework, the recognizer calculates which phoneme is likely spoken at every frame using HMMs. This framework cannot be straightly applied to instrument recognition in polyphonic music because the possibility of more-than-one sources existing is not considered. Once the harmonic structure for every semitone is separated, this framework can be applied, because we assume that

Table 2 Overview of 28 features extracted for calculating Instrograms

Spectral features		
1		Spectral centroid
2 – 10		Relative cumulative power (up to 9th partials)
11		Odd/even power ratio
12 – 20		Number of components having a duration that is $p\%$ longer than the longest duration $(p = 10, 20, \cdots, 90)$
Temporal features		
21		Gradient of straight line approximating power envelope
22 – 24		Temporal mean of differentials of power envelope from t to $t + iT/3$ $(i = 1, \cdots, 3)$
Modulation features		
25, 26		Amplitude and frequency of AM
27, 28		Amplitude and frequency of FM

multiple instruments are not being played at the same time and at the same F0. The details of the algorithm are as follows:

- **[Step 1] Harmonic structure extraction**

  The temporal trajectory of the harmonic structure with every semitone is extracted. This is represented as

  $$\mathcal{H}(t, f) = \{(F_i(t, f), A_i(t, f)) \mid i = 1, \cdots, h\},$$

  where $F_i(t, f)$ and $A_i(t, f)$ are the frequency of amplitude of i-th partial of the sound with F0 of f at time t.

- **[Step 2] Feature extraction**

  For every time t (every 10 ms in the implementation), we first excerpt a T-length bit of the harmonic-structure trajectory $\{\mathcal{H}_t(\tau, f) \mid t \leq \tau < t + T\}$ from the whole trajectory $\{\mathcal{H}(t, f)\}$ and then extract a feature vector $\boldsymbol{x}(t, f)$ consisting of 28 features listed in Table 2 from $\{\mathcal{H}_t(\tau, f)\}$. Then, the dimensionality is reduced to 12 dimensions using the principal component analysis with the proportion value of 95%. T is 500 ms in the current implementation.

- **[Step 3] Probability calculation**

  We train left-to-right HMMs, each consisting of 15 states, for target instruments $\omega_1, \cdots, \omega_m$, and then basically consider the time series of feature vectors, $\{\boldsymbol{x}(t, f)\}$, to be generated from a Markov chain of these HMMs. Then, the CIEP $p(\omega_i | X; t, f)$ is calculated as

  $$\begin{aligned} p(\omega_i | X; t, f) &= p(M_i | \boldsymbol{x}(t, f)) \\ &= \frac{p(\boldsymbol{x}(t, f) | M_i) p(M_i)}{\sum\limits_{i=1}^{m} p(\boldsymbol{x}(t, f) | M_i) p(M_i)}, \end{aligned}$$

where M_i is the HMM corresponding to the instrument ω_i, $p(\boldsymbol{x}(t,f)|M_i)$ is trained from data prepared in advance, and $p(M_i)$ is the *a priori* probability.

In the above formulation, $p(\omega_i|X;t,f)$ for some instruments may become greater than zero even if no instruments are played. Theoretically, this does not matter because $p(X;t,f)$ becomes zero in such cases. In practice, however, $p(X;t,f)$ may not be zero, especially when a certain instrument is played at an F0 of an integer multiple or factor of f. To avoid this, we prepare an HMM, M_0, trained with feature vectors extracted from silent signals (note that some instruments may be played at non-target F0s) and consider $\{\boldsymbol{x}(t,f)\}$ to be generated from a Markov chain of the $m+1$ HMMs $(M_0, M_1, \cdots, M_m)$. The CIEP is therefore calculated as

$$p(\omega_i|X;t,f) = \frac{p(\boldsymbol{x}(t,f)|M_i)p(M_i)}{\displaystyle\sum_{i=0}^{m} p(\boldsymbol{x}(t,f)|M_i)p(M_i)},$$

where we use $p(M_i) = 1/(m+1)$.

4.3 Examples of Instrogram Representation

Examples of Instrograms obtained from real performances of classical music are shown in Figure 10. For the lack of space, we show the simplified versions only. The piece for Figure 10 (a) is string music. The color for the violin only should therefore be deep. The obtained Instrogram shows a significantly deeper color for the violin than the other instruments although the color for the piano is also slightly deep in part. On the other hand, the piece for Figure 10 (b) is an ensemble of the piano and the strings, where the introduction is played on the piano only and the strings begin to play later. Also in the obtained Instrogram, the color for the piano only is deep at the beginning and that for the violin become deeper later, though the color for flute is also sometimes deep. Other examples are available at http://ist.ksc.kwansei.ac.jp/~kitahara/instrogram/.

4.4 Applications to Content-Based MIR

Instrumentation is an important factor in content-based MIR because it is deeply connected to listeners' impression. When the same musical piece is played on different instruments, listeners sometimes have different impressions on the piece. This shows a deep connection between instrumentation and listeners' impression. Here, we introduce some MIR applications based on instrumentation represented as Instrograms.

1. **Query-by-Example** (Figure 11)
   The first application of the Instrogram representation is a so-called Query-by-Example retrieval system. Query-by-Example is the retrieval approach that aims to search for musical pieces that are similar to the piece specified by the user.

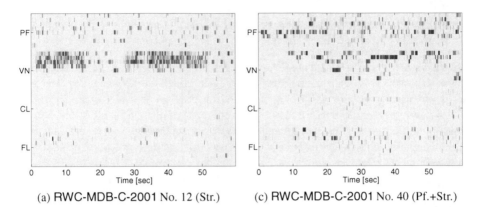

(a) RWC-MDB-C-2001 No. 12 (Str.)   (c) RWC-MDB-C-2001 No. 40 (Pf.+Str.)

Fig. 10 Results of calculating Instrograms from real-performance audio signals (simplified versions only)

This approach is useful because it does not require users to have special musical knowledge. As mentioned above, we consider that the similarity of the instrumentations between two musical pieces is deeply connected to the similarity of the impressions that listeners receive. This is why we adopt the similarity of Instrograms as a measure for musical similarity.

After searching for a musical piece, the user can listen to it while seeing the instrument existence probabilities evolving with time through two kinds of visualization of the Instrogram, one of which is a standard visualization like Figure 8 and the other of which is bar graphs moving up and down being synchronized with the instrument existence probabilities.

2. **Playlist generation with music thumbnailing function** (Figure 12)

The second application is an extension of the above Query-by-Example system. In the left-bottom panel of the window, the musical pieces that the user has are listed with Instrogram-based thumbnails. This thumbnail consists of the images of target musical instruments, in which the color depth of each image represents the possibility that the sound of the corresponding instrument exists. On the other hand, the right-bottom panel shows a playlist. The user can pick musical pieces from the left-bottom panel and drag and drop them on any items of the playlist. For example, the user can assign a piece with a brass section to the first item and a piano solo to the last item. The blank items of the playlist can be interpolated so that the instrumentation gradually changes [60].

5   Discussion

5.1   Common Properties among Mid-level Representation

Here we discuss the properties of the above-mentioned mid-level representations from several different perspectives.

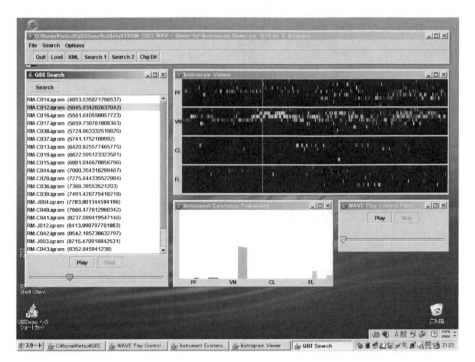

Fig. 11 First application of Instrogram-based MIR. Users (1) learn what kind of instruments are sounding by seeing two kinds of visualization of Instrograms, synchronized with the playback, and (2) search for musical pieces that have similar instrumentation to their favorite piece. The left window shows the result of similarity-based retrieval, where the values in the parentheses are the dissimilarity to the query piece.

- **Reproducibility**
  Although Ellis claimed that mid-level representations for CASA should be usable for regenerating sounds [4], almost no representations that have been reviewed here can regenerate them. This is because MIR does not require the reproduction of sound[6]. In addition, it has been pointed out that, from the viewpoint of CASA, the reproduction of each source from a mixture of sounds is not essential [62].
- **Correspondence to a Single Aspect**
  Correspondence to a single aspect is a very important property to achieve user-adaptive MIR systems. In general, different MIR users may pay attention to different aspects when listening to music. Some users may listen attentively to the melody or harmony while others to the timbre or rhythm. The criteria used for retrieval, typically music similarity measures, therefore, should adapt to such user preferences. It is, however, difficult if such different aspects are represented in a jumbled form. Because each of the above-mentioned representations basically

---

[6] On the other hand, representations that can reproduce sounds have recently been proposed such as [61]. This representation is close to what Ellis says [4] because it can reproduce the sound where interested part is enhanced.

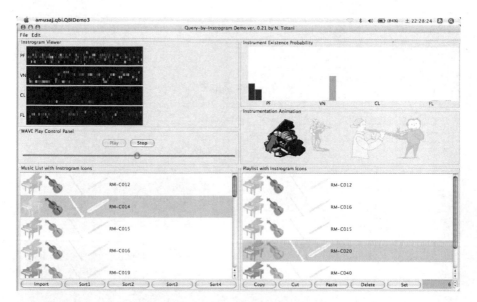

Fig. 12 Second application of Instrogram-based MIR. Users learn the instrumentations of the musical pieces that they have from the Instrogram-based thumbnails without need to listen to them. The system also has a function of automatic playlist generation where the system interpolates musical pieces in a playlist so that the instrumentation gradually changes.

corresponds to a single aspect, it would be usable for user-adaptive MIR by appropriately combining them.

- **Symbolizability**
  Although we mentioned the importance of non-symbolic representations, symbolic representations, such as musical scores and MIDI, are still important. Most of the mid-level representations reviewed here are applicable not only to MIR directly but also to obtain symbolic representations.

- **Temporality**
  Almost all representations reviewed here, except the beat histogram, are functions of time. We believe that it is very important to represent music without removing the time axis because music is an art form organized in time. In studies using spectral and cepstral features, in contrast, the time axis is often removed by calculating the temporal means and variances of these features. Although it is practically useful and effective in most cases, we believe that this process must lose an important nature of music.

5.2 Potential Applications

By using these mid-level representations, the characteristics of each aspect (e.g., the melody, harmony, timbre, and rhythm) can be separately calculated. This method can be used in various applications related to MIR.

The first one is user adaptation, as mentioned above. Suppose the situation when a user is trying to search for similar musical pieces to a certain piece. The retrieval system has to measure the similarity between the query piece and each piece contained in the collection. If the musical pieces are represented in a combination of different representations corresponding to different aspects, the similarity can be calculated by a weighted sum of the similarities measured in each representation. The user's preference (for example, he/she tends to listen attentively to harmony) can be considered by giving an appropriate weight to each similarity.

The second one is music visualization. Music visualization is an important subject because music is difficult to overview at once. Indeed, some of the mid-level representations reviewed here have already been applied to graphical visualization of music [21, 48, 49]. Achieving effective graphical visualizations, however, is not a trivial problem even if the mid-level representations are determined. Where what elements in a mid-level representation are displayed in what color is an important issue, and some attempts have been made [63, 64].

Music thumbnailing [65], another potential application, is also an important subject that is different from but related to music visualization. Music thumbnailing helps users choose musical pieces from a large-scale music list. We achieved music thumbnails that represent instrumentation by using Instrograms [60].

5.3 Support in MPEG-7

MPEG-7 is a well-known standard for distributing metadata for multimedia content including musical content. MPEG-7 provides various descriptors for representing audio content listed in Table 3, but it does not directly support such mid-level music representations as the ones mentioned above. The audio descriptors adopted in MPEG-7 can be classified into Low-level Audio Descriptors and High-level Description Tools. The Low-level Audio Descriptors include waveforms themselves, spectral envelopes, fundamental frequency, log attack times, and temporal and spectral centroids. The Musical Instrument Timbre Tool provides a way for describing perceptual features of sounds as a combination of low-level audio descriptors. The Melody

Table 3 Framework for describing audio content in MPEG-7

Low-level Audio Descriptors	Basic Descriptors
	Basic Spectral Descriptors
	Signal Parameter Descriptors
	Timbral Temporal Descriptors
	Timbral Spectral Descriptors
	Spectral Basis Descriptors
High-level Description Tools	Musical Instrument Timbre Tool
	Melody Description Tools
	General Sound Recognition and Indexing Description Tools
	Spoken Content Description Tools
	Audio Signature Description Scheme

Description Tools provide some symbolic representations of melodies. The General Sound Recognition and Indexing Description Tools provide a framework for describing sound recognizers. The Spoken Content Description Tools provide a framework useful for automatic speech recognition. The Audio Signature Description Scheme can be used in audio fingerprinting. However, mid-level representations are not directly supported in the current MPEG-7 standard. Hence establishing a standard supporting them is an important task for future research (see [66] for details).

6 Conclusion

In this chapter, we described various mid-level representations of music. In the early 2000s, it was common to use low-level features such as spectral and cepstral features extracted directly from polyphonic audio signals. This approach brings successful results to some extent, but it has been clarified that the performance of MIR has been mostly saturated; a significant further improvement will be difficult as long as low-level features are used [26]. Various researchers have therefore pointed out the importance of higher-level, musically meaningful representations and have been engaged in discovering such new music representations. The review here is a brief survey of the latest results of these attempts.

These representations have a potential to be applied to various tasks other than MIR. For example, they could be effective for developing a computational theory of music perception. Although existing music theories are one of the greatest inventions related to music, they have problems from the viewpoint of studying how humans perceive music. One of the major problems is that they are constructed based on musical scores. Obviously what we perceive from music is different from the score, and therefore new representations for describing music theories are required. Attempts to develop mid-level representations will bring us a new solution to this problem.

Acknowledgements. The research of Instrogram was conducted in collaboration with Professor Hiroshi G. Okuno (Kyoto University) and Dr. Masataka Goto (National Institute of Advanced Industrial Science and Technology). The second application of Instrogram, shown in Figure 12 was developed in collaboration with Mr. Naoyuki Totani. The result of melody estimation shown in Figure 4 was provided by Dr. Masataka Goto. Professor Haruhiro Katayose, Dr. Masataka Goto, and Dr. Shinichi Furuya gave the author valuable comments. The author appreciates their cooperation.

References

1. Good, M.: MusicXML: An internet-friendly format for sheet music. In: The XML 2001 Conf. Proc. (2001)
2. Bellini, P., Nesi, P.: WEDELMUSIC format: An XML music notation format for emerging applications. In: Proc. Int'l Conf. WEB Delivering of Music, pp. 79–86 (2001)

3. Klapuri, A., Davy, M. (eds.): Signal Processing Methods for Music Transcription. Springer, Heidelberg (2006)
4. Ellis, D., Rosenthal, D.F.: Mid-level representations for computational auditory scene analysis. In: Rosenthal, D.F., Okuno, H.G. (eds.) Computational auditory scene analysis, ch. 17, pp. 257–272. Lawrence Erlbaum, Mahwah (1998)
5. Goto, M.: A real-time music-scene-description system: Predominant-F0 estimation for detecting melody and bass lines in real-world audio signals. Speech Comm. 43(4), 311–329 (2004)
6. Goto, M.: A robust predominant-f0 estimation method for real-time detection of melody and bass lines in CD recordings. In: Proc. ICASSP, vol. II, pp. 757–760 (2000)
7. Marolt, M.: A mid-level melody-based representation for calculating audio similarity. In: Proc. ISMIR (2006)
8. Sagayama, S., Takahashi, K., Kameoka, H., Nishimoto, T.: Specmurt anasylis: A piano-roll-visualization of polyphonic music by deconvolution of log-frequency spectrum. In: Proc. SAPA (2004)
9. Abdallah, S.A., Plumbley, M.D.: Polyphonic music transcription by non-negative sparse coding of power spectra. In: Proc. ISMIR, pp. 318–325 (2004)
10. Virtanen, T.: Unsupervised learning methods for source separation in monaural music signals. In: Klapuri, A., Davy, M. (eds.) Signal Processing Methods for Music Transcription. Springer, Heidelberg (2006)
11. Saito, S., Kameoka, H., Nishimoto, T., Sagamaya, S.: Specmurt analysis of multi-pitch music signals with adaptive estimation of common harmonic structure. In: Proc. ISMIR, pp. 84–91 (2005)
12. Leveau, P., Vincent, E., Richard, G., Daudet, L.: Instrument-specific harmonic atoms for mid-level music representation. IEEE Trans., Audio, Speech, Lang., Process. 16(1), 116–128 (2008)
13. Fujishima, T.: Realtime chord recognition of musical sound: a system using common Lisp music. In: Proc. ICMC, pp. 464–467 (1999)
14. Sheh, A., Ellis, D.P.W.: Chord segmentation and recognition using EM-trained hidden Markov models. In: Proc. ISMIR (2003)
15. Yoshioka, T., Kitahara, T., Komatani, K., Ogata, T., Okuno, H.G.: Automatic chord transcription with concurrent recognition of chord symbols and boundaries. In: Proc. ISMIR, pp. 100–105 (2004)
16. Bello, J.P., Pickens, J.: A robust mid-level representation for harmonic content in music signals. In: Proc. ISMIR (2005)
17. Cabral, G., Pachet, F., Briot, J.-P.: Automatic X traditional descriptor extraction: The case of chord recognition. In: Proc. ISMIR (2005)
18. Lee, K., Slaney, M.: A unified system for chord transcription and key extraction using hidden Markov models. In: Proc. ISMIR (2007)
19. Goto, M.: Music scene description. In: Klapuri, A., Davy, M. (eds.) Signal Processing Method for Music Transcription, ch. 11, pp. 327–359. Springer, Heidelberg (2006)
20. Shepard, R.N.: Circularity in judgments of relative pitch. J. Acoust. Soc. Am. 36(12), 2346–2353 (1964)
21. Fujisawa, T.X., Tani, M., Nagata, N., Katayose, H.: Music mood visualization based on quantitative model of chord perception. IPSJ Journal 50(3) (2009) (in Japanese)
22. Cook, N.D., Fujisawa, T.X.: The psychophysics of harmony perception: Harmony is a three-tone phenomenon. Empirical Musicology Review 1(2), 106–126 (2006)
23. Tzanetakis, G., Cook, P.: Musical genre classification of audio signals. IEEE Trans. Speech Audio Process. 10(5), 293–302 (2002)

24. Lu, L., Liu, D., Zhang, H.-J.: Automatic mood detection and tracking of music audio signals. IEEE Trans. Audio, Speech, Lang. Process. 14(1) (2006)
25. Pampalk, E.: Computational Models of Music Similarity and their Application in Music Information Retrieval. PhD thesis, Technischen Universitat Wien (2006)
26. Aucouturier, J.-J., Pachet, F.: Improving timbre similarity: How high's the sky? Journal of Negative Results in Speech and Audio Sciences (2004)
27. Herrera-Boyer, P., Klapuri, A., Davy, M.: Automatic classification of pitched instrument sounds. In: Klapuri, A., Davy, M. (eds.) Signal Processing Methods for Music Transcription. Springer, Heidelberg (2006)
28. Bregman, A.S.: Auditory Scene Analysis: The Perceptual Organization of Sound. MIT Press, Cambridge (1990)
29. Namba, S.: Definition of timbre. J. Acoust. Soc. Jpn. 49(11), 823–831 (1993) (in Japanese)
30. Martin, K.D.: Sound-Source Recognition: A Theory and Computational Model. PhD thesis, MIT (1999)
31. Brown, J.C.: Computer identification of musical instruments using pattern recognition with cepstral coefficients as features. J. Acoust. Soc. Am. 103(3), 1933–1941 (1999)
32. Eronen, A., Klapuri, A.: Musical instrument recognition using cepstral coefficients and temporal features. In: Proc. ICASSP, pp. 735–756 (2000)
33. Fujinaga, I., MacMillan, K.: Realtime recognition of orchestral instruments. In: Proc. ICMC, pp. 141–143 (2000)
34. Marques, J., Moreno, P.J.: A study of musical instrument classification using Gaussian mixture models and support vector machines. CRL Technical Report Series CRL/4, Compaq Cambridge Research Laboratory (1999)
35. Kitahara, T., Goto, M., Okuno, H.G.: Musical instrument identification based on F0-dependent multivariate normal distribution. In: Proc. ICASSP, vol. V, pp. 421–424 (2003)
36. Livshin, A.A., Peeters, G., Rodet, X.: Studies and improvements in automatic classification of musical sound samples. In: Proc. ICMC, pp. 171–174 (2003)
37. Essid, S., Richard, G., David, B.: Musical instrument recognition by pairwise classification strategies. IEEE Trans. Audio, Speech, Lang. Process. 14(4), 1401–1412 (2006)
38. Kashino, K., Nakadai, K., Kinoshita, T., Tanaka, H.: Application of the Bayesian probability network to music scene analysis. In: Rosenthal, D.F., Okuno, H.G. (eds.) Computational Auditory Scene Analysis, pp. 115–137. Lawrence Erlbaum Associates, Mahwah (1998)
39. Kashino, K., Murase, H.: A sound source identification system for ensemble music based on template adaptation and music stream extraction. Speech Comm. 27, 337–349 (1999)
40. Kinoshita, T., Sakai, S., Tanaka, H.: Musical sound source identification based on frequency component adaptation. In: Proc. IJCAI CASA Workshop, pp. 18–24 (1999)
41. Eggink, J., Brown, G.J.: A missing feature approach to instrument identification in polyphonic music. In: Proc. ICASSP, vol. V, pp. 553–556 (2003)
42. Eggink, J., Brown, G.J.: Application of missing feature theory to the recognition of musical instruments in polyphonic audio. In: Proc. ISMIR (2003)
43. Vincent, E., Rodet, X.: Instrument identification in solo and ensemble music using independent subspace analysis. In: Proc. ISMIR, pp. 576–581 (2004)
44. Essid, S., Richard, G., David, B.: Instrument recognition in polyphonic music based on automatic taxonomies. IEEE Trans. Audio, Speech, Lang. Process. 14(1), 68–80 (2006)
45. Kitahara, T., Goto, M., Komatani, K., Ogata, T., Okuno, H.G.: Instrument identification in polyphonic music: Feature weighting to minimize influence of sound overlaps. EURAIP J. Adv. Signal Processing 2007(51979), 1–15 (2007)

46. Kitahara, T., Goto, M., Komatani, K., Ogata, T., Okuno, H.G.: Instrogram: A new musical instrument recognition technique without using onset detection nor F0 estimation. In: Proc. ICASSP, vol. V, pp. 229–232 (2006)

47. Kitahara, T., Goto, M., Komatani, K., Ogata, T., Okuno, H.G.: Instrogram: Probabilistic representation of instrument existence for polyphonic music. IPSJ Journal 48(1), 214–226 (2007); (also published in IPSJ Digital Courier, vol.3, pp.1–13)

48. Kitahara, T.: Computational Musical Instrument Recognition and Its Application to Content-based Music Information Retrieval. PhD thesis, Kyoto University (2006)

49. Tzanetakis, G.: Manipulation, Analysis and Retrieval Systems for Audio Signals. PhD thesis, Princeton University (2002)

50. Bello, J.P., Daudet, L., Abdallah, S., Duxbury, C., Davies, M., Sandler, M.B.: A tutorial on onset detection in music signal. IEEE Trans. Audio, Speech, Lang. Process. 13(5), 1035–1047 (2005)

51. Goto, M.: An audio-based real-time beat tracking system for music with or without drums. J. New Music Res. 30(2), 159–171 (2001)

52. Davies, M.E.P., Plumbley, M.D.: Comparing mid-level representations for audio based beat tracking. In: Proc. DMRN Summer Conf., pp. 36–41 (2005)

53. Dixon, S., Pampalk, E., Widmer, G.: Classification of dance music by periodicity patterns. In: Proc. ISMIR (2003)

54. Dixon, S., Gouyon, F., Widmer, G.: Towards characteristics of music via rhythmic patterns. In: Proc. ISMIR, pp. 509–516 (2004)

55. Paulus, J., Klapuri, A.: Measuring the similarity of rhythmic patterns. In: Proc. ISMIR (2002)

56. Tsunoo, E., Ono, N., Sagayama, S.: Rhythm map: Extraction of unit rhythmic patterns and analysis of rhythmic structure from music acoustic signals. In: Proc. ICASSP, pp. 185–188 (2009)

57. Gouyon, F., Dixon, S.: A review of automatic rhythm description systems. Computer Music Journal 29(1), 34–54 (2005)

58. Rabiner, L.R.: A tutorial on hidden Markov models and selected applications in speech recognition. Proc. IEEE 77(2), 257–286 (1989)

59. Eggink, J., Brown, G.J.: Extracting melody lines from complex audio. In: Proc. ISMIR, pp. 84–91 (2004)

60. Totani, N., Kitahara, T., Katayose, H.: Music player with music thumbnailing and playlist generation functions based on instrumentation. In: Proc. Interaction (2008) (in Japanese)

61. Itoyama, K., Goto, M., Komatani, K., Ogata, T., Okuno, H.G.: Integration and adaptation of harmonic and inharmonic models for separating polyphonic musical signals. In: Proc. ICASSP, vol. I (2007)

62. Goto, M.: Analysis of musical audio signals. In: Wang, D., Brown, G.J. (eds.) Computational Auditory Scene Analysis, ch. 8, pp. 251–295. Wiley Interscience, Hoboken (2006)

63. Gomez, E., Bonada, J.: Tonality visualization of polyphonic audio. In: Proc. ICMC (2005)

64. Mardirossian, A., Chew, E.: Visualizing music: Tonal progressions and distributions. In: Proc. ISMIR (2007)

65. Yoshii, K., Goto, M.: Music thumbniler: Visualizing musical pieces in thumbnail images based on acoustic features. In: Proc. ISMIR, pp. 212–216 (2008)

66. Kimi, H.-G., Moreau, N., Sikora, T.: MPEG-7 Audio and Beyond. Wiley, Chichester (2005)

The Music Information Retrieval Evaluation eXchange: Some Observations and Insights

J. Stephen Downie, Andreas F. Ehmann, Mert Bay, and M. Cameron Jones

Abstract. Advances in the science and technology of Music Information Retrieval (MIR) systems and algorithms are dependent on the development of rigorous measures of accuracy and performance such that meaningful comparisons among current and novel approaches can be made. This is the motivating principle driving the efforts of the International Music Information Retrieval Systems Evaluation Laboratory (IMIRSEL) and the annual Music Information Retrieval Evaluation eXchange (MIREX). Since it started in 2005, MIREX has fostered great advancements not only in many specific areas of MIR, but also in our general understanding of how MIR systems and algorithms are to be evaluated. This chapter outlines some of the major highlights of the past four years of MIREX evaluations, including its organizing principles, the selection of evaluation metrics, and the evolution of evaluation tasks. The chapter concludes with a brief introduction of how MIREX plans to expand into the future using a suite of Web 2.0 technologies to automated MIREX evaluations.

Keywords: MIREX, Music Information Retrieval, Evaluation.

1 Introduction

Since 2005, a special set of sessions has convened at the annual International Conference on Music Information Retrieval (ISMIR). At these special sessions, which include a poster exhibition and a plenary meeting, Music Information Retrieval (MIR) researchers from around the world come together to compare, contrast and discuss the latest results data from the Music Information Retrieval Evaluation eXchange (MIREX). MIREX is to the MIR community what the Text Retrieval Conference (TREC) is to the text information retrieval community: A set of

J. Stephen Downie · Andreas F. Ehmann · Mert Bay · M. Cameron Jones
International Music Information Retrieval Systems Evaluation Laboratory
Graduate School of Library and Information Science,
University of Illinois at Urbana-Champaign,
501 East Daniel Street, Champaign, Illinois, USA 61820
jdownie@illinois.edu

Z.W. Raś and A.A. Wieczorkowska (Eds.): Adv. in Music Inform. Retrieval, SCI 274, pp. 93–115.
springerlink.com   © Springer-Verlag Berlin Heidelberg 2010

community-defined formal evaluations through which a wide variety of state-of-the-art systems, algorithms and techniques are evaluated under controlled conditions. MIREX is managed by the International Music Information Retrieval Systems Evaluation Laboratory (IMIRSEL) at the University of Illinois at Urbana-Champaign (UIUC).

This chapter builds upon and extends Downie [5]. While some overlap is unavoidable, the reader is strongly encouraged to consult the earlier work for it covers important issues that will be missing detailed discussion in the current chapter. Unlike Downie [5], the present chapter will focus on the evolution of MIREX throughout the years. It will highlight some specific issues, problems and challenges that have emerged during the running of MIREX.

Section 2 reviews the history, special characteristics and general operations of MIREX. Section 3 examines the important primary metrics used to evaluate the algorithms. Section 4 takes an in-depth look at two closely related tasks, Audio Music Similarity (AMS) and Symbolic Melodic Similarity (SMS), as a kind of case study in the evolution of MIREX. Section 5 summarizes the chapter and briefly introduces the NEMA (Networked Environment for Music Analysis) project which is designed to support and strengthen MIREX and the ongoing formal evaluation of MIR systems.

2 History and Infrastructure

While MIREX officially began in 2005, it took a considerable amount of time and collective effort to make it a reality. In 1999 Downie led the *Exploratory Workshop on Music Information Retrieval* as part of the 1999 ACM Special Interest Group Information Retrieval (SIGIR) Conference at Berkeley, CA. One of the primary goals of this workshop was the exploration of "(…) consensus opinion on the establishment of research priorities, inter-disciplinary collaborations, evaluation standards, test collections (…) and TREC-like trials" [2]. In 2001, the attendees of ISMIR 2001 at Bloomington, IN passed a resolution calling for the establishment of formal evaluation opportunities for MIR researchers. This resolution helped garner modest feasibility study grants from the Andrew W. Mellon Foundation and the National Science Foundation. With the funds provided by the grants, a sequence of workshops was convened and a collection of evaluation white papers was compiled [3]. The recommendations based upon these workshops and white papers were subsequently published as [4] and worked into several grant applications. In late 2003, both the Andrew W. Mellon Foundation and the NSF awarded the substantial grants that were to make MIREX possible. ISMIR 2004 was held in Barcelona, ES and was hosted by The Music Technology Group (MTG) of the University Pompeu Fabra. At this meeting the MTG convened an evaluation session called the Audio Description Contest (ADC) [1]. While more limited in scope than MIREX, it is from the ADC that MIREX learned many valuable lessons. MIREX made its debut as part of ISMIR 2005, held at Queen Mary College, University of London, in September, 2005.

TREC, ADC and MIREX share a common intellectual foundation. All three are predicated upon the standardization of:

1. Test collections of considerable size;
2. Tasks and/or queries to be performed against the collections; and,
3. Evaluation procedures to compare performances among systems.

It is important to note that MIREX differs from TREC in one significant way. Unlike TREC, where the evaluation datasets are sent out to the participant labs, MIREX operates under an "algorithm-to-data" model. This means that algorithms are sent to IMIRSEL to be run on IMIRSEL equipment by IMIRSEL personnel and volunteers. While the algorithm-to-data model puts a considerable burden on IMIRSEL resources in terms of workload, data management and equipment, it is currently the only feasible solution to working within the boundaries of the highly restrictive and litigious legal environment surrounding music intellectual property law.

Beyond computational infrastructure, IMIRSEL also hosts the basic communications infrastructure for MIREX which includes the MIREX wikis[1] and the MIREX mailing lists. The MIREX wikis serve two purposes. First, during the spring and summer each year, they are used by the community to define the task sets, evaluation metrics and general rules for the year's upcoming MIREX. Second, the wikis are used to publish and archive the raw and summarized results data for each task and associated algorithms just prior to ISMIR convening each autumn. These results data are used by participants to help them put together their mandatory poster presentations and for further use in follow up publications. The MIREX "EvalFest"[2] mailing list is the general purpose mailing list that is used to solicit task ideas and collections and to foster broad discussions about evaluation issues. On a case-by-case basis, IMIRSEL also creates task-specific mailings lists through which finely detailed discussions and debates about metrics, collections and input/output formats, etc. are undertaken.

Table 1 MIREX Descriptive Statistics 2005-2008

	2005	2006	2007	2008
Number of Task (and Subtask) "Sets"	10	13	12	18
Number of Individuals	82	50	73	84
Number of Countries	19	14	15	19
Number of Runs	86	92	122	169

MIREX has enjoyed considerable growth over its four year history. As Table 1 indicates, the number of task sets (including subtasks) has grown 80% from 10 (2005) to 18 (2008). The number of individual algorithms evaluated has similarly grown 95% from 86 (2005) to 169 (2008). In total, MIREX has evaluated 469 algorithm runs.

The range of MIREX tasks broadly reflects the varied interests of the MIR research community. Many tasks, such as Audio Artist Identification (AAI), Symbolic

---

[1] See http://music-ir.org/mirexwiki
[2] Subscription information at https://mail.isrl.illinois.edu/mailman/listinfo/evalfest

Genre Classification (SGC), Audio Genre Classification (AGC), Audio Mood Classification (AMC), and Audio Tag Classification (ATC), represent classic machine learning train-test classification evaluations. These classification tasks accounted for 28% (129) of MIREX's 469 evaluation runs. Other tasks, such as Audio Beat Tracking (ABT), Audio Chord Detection (ACD), Audio Melody Extraction (AME), Audio Onset Detection (AOD) and Multiple F_0 Estimation (MFE), etc. are "low-level" tasks that

Table 2 The MIREX Tasks and Number of Runs per Task 2005-2008

KEY	TASK NAME	2005	2006	2007	2008
AAI	Audio Artist Identification	7		7	11
ABT	Audio Beat Tracking		5		
ACD	Audio Chord Detection				15*
ACC	Audio Classical Composer ID			7	11
ACS	Audio Cover Song Identification		8	8	8
ADD	Audio Drum Detection	8			
AGC	Audio Genre Classification	15		7	26*
AKF	Audio Key Finding	7			
AME	Audio Melody Extraction	10	10*		21**
AMC	Audio Mood Classification			9	13
AMS	Audio Music Similarity		6	12	
AOD	Audio Onset Detection	9	13	17	
ATC	Audio Tag Classification				11
ATE	Audio Tempo Extraction	13	7		
MFE	Multiple F_0 Estimation (Frame Level)			16	15
MFN	Multiple F_0 Note Detection			11	13
QBSH	Query-by-Singing/Humming		23*	20*	16*
QBT	Query-by-Tapping				5
SF	Score Following		2		4
SGC	Symbolic Genre Classification	5			
SKF	Symbolic Key Finding	5			
SMS	Symbolic Melodic Similarity	7	18 **	8	
* task comprised two subtasks ** task comprised three subtasks					

represent tools and techniques upon which many MIR systems depend. For example, many systems use melody extractors as a first step toward building searchable indexes. The development and evaluation of low-level MIR subsystems is important to the MIR community as this category of evaluation comprised 201 (43%) of the MIREX evaluation runs. Audio Cover Song Identification (ACS), Audio Music Similarity (AMS), Query-by-Singing/Humming (QBSH), Query-by-Tapping (QBT) and Symbolic Melodic Similarity (SMS) are those tasks related to what most people would consider to be MIR, that is, the idea of searching for music given some type of music query. QBSH has been the single most evaluated task with over 59 individual runs evaluated over 2006-2008 (or 13% of runs evaluated). As a category, these searching tasks represented 139 or 30% of all MIREX runs.

3 Overview of MIREX Primary Evaluation Metrics

Table 3 Top MIREX Scores and Primary Evaluation Metrics 2005-2008 (normalized 0-1)

KEY	PRIMARY METRIC	2005	2006	2007	2008
AAI	Average Accuracy	0.72		0.48	0.48
ABT	Average Beat P-Score		0.41		
ACD1	Overlap Score				0.66
ACD2	Overlap Score				0.72
ACC	Average Accuracy			0.54	0.53
ACS	Average Precision		0.23	0.52	0.75
ADD	Average F-Measure	0.67			
AGC1	Average Hierarchical Accuracy	0.83		0.68	0.66
AGC2	Average Accuracy				0.65
AKD	Average Hierarchical Accuracy	0.90			
AMC	Average Accuracy			0.62	0.64
AME1	Average Accuracy	0.71	0.73		0.70
AME2	Average Accuracy		0.83		0.85
AME3	Average Accuracy				0.76
AMS	Average Fine Score		0.43	0.57	
AOD	Average F-Measure	0.80	0.79	0.81	
ATE	Average F-Measure				0.28
ATE	Average Tempo P-Score	0.69	0.81		
MFE	Average Accuracy			0.62	0.67
MFN	Average F-Measure			0.61	0.61
QBSH1	Average Precision (Mean Reciprocal Rank)		0.93	0.93	0.93
QBSH2	Average Precision		0.93	0.94	0.94
QBT	Average Precision (Mean Reciprocal Rank)				0.52
SF	Average Precision		0.83		0.67
SGC	Average Hierarchical Accuracy	0.77			
SKD	Average Hierarchical Accuracy	0.91			
SMS1	Average Dynamic Recall [Binary Score][3]	0.66	0.72 [0.73]		
SMS2	Average Dynamic Recall [Binary Score]		0.82 [0.44]		
SMS3	Average Dynamic Recall [Binary Score]		0.78 [0.83]		
SMS4	Average Dynamic Recall [Fine Score]			0.72 [0.56]	

---

[3] The SMS tasks used two different metrics as "primary" scores. Each year the Average Dynamic Recall (ADR) score was reported along with either the Binary Score or the Fine Score (presented in square brackets). More information about these metrics found in Sections 3.6 and 4.2.

   The selection or creation of appropriate evaluation metrics is crucial to the proper scientific evaluation of MIR system performance. The selection of evaluation metrics also has a strong emotional component as participants strive to show off the success of their algorithms and systems in the best possible terms. Thus, it is not surprising that the selection/creation of MIREX evaluation metrics undergoes a great deal of sometimes heated discussion while the tasks are being developed by the community. Table 3 summarizes the primary evaluation metrics used in each task along with the top-ranked score using that metric for each year. These are the primary, or "official," metrics used to rank order the MIREX results each on the master MIREX results poster published each year at ISMIR. It is important to note, however, that most tasks are actually evaluated using a wide variety of metrics. For each task, the results using the other metrics are summarized and posted on each respective results wiki page. Notwithstanding all the debate over which metric should become the "official" metric, we are discovering a general trend among the tasks that the rank order of system performances within a task appears remarkably stable regardless of the metric chosen [14]. As one can see in Table 3, MIREX uses many different primary metrics; some are used over a range of related tasks, others are tailored specifically to one. We will now discuss the primary evaluation metrics used by MIREX in evaluating MIR performance along with some of the justifications for using the metrics.

3.1 Average Accuracy and Hierarchical Accuracy

In classification tasks such as Audio Artist Identification (AAI), performance can be measured using classification accuracy. Given N_{total} pieces to be classified, the average accuracy of a classifier, Acc, can be defined as

$$Acc = \frac{N_{correct}}{N_{total}} \tag{1}$$

where $N_{correct}$ is the number of correctly classified instances. Average accuracy is also applicable to such tasks as Audio Melody Extraction (AME) where accuracy measures the number of analysis frames where fundamental frequencies (F_0s) are correctly estimated versus the total number of frames in a piece.

   For Multi-F_0 Estimation (MFE) accuracy is calculated as

$$Acc = \frac{TP}{TP + FP + FN} \tag{2}$$

where TP is the count of true positives, FP is the count of true negatives and FN is the count of false negatives. In the MFE task, where the number of active F_0s in the ground-truth changes per frame, TP is the number of correctly detected F_0s per frame summed over all frames. FP is the number of detected F_0s that are not in the ground-truth list for that frame summed over all frames and FN is the number of F_0s in the ground-truth list minus the number of detected F_0s for that frame summed over all frames.

In some cases, misclassifications can occur that are not as "erroneous" or "offensive" as others. For instance, it is generally consider "better" to misclassify a Baroque work as Classical than Heavy Metal. Similarly, having a system misclassify a hard-driving Blues song as a Rock & Roll song is quite understandable as it is the same kind of "error" that many humans might make. In the cases of the Symbolic and Audio Key Detection (SKD and AKD) tasks, misclassifying a key with its perfect fifth is also considered a somewhat acceptable mistake (i.e., many humans make the same error). For these reasons, some tasks are also evaluated using hierarchical accuracies, which discount certain acceptable errors. For example, in the two key finding tasks, correct keys were awarded 1.0 point, perfect fifth errors were given 0.5 points, relative major/minor errors 0.3 points, and parallel major/minor errors 0.2 points. Therefore, the hierarchical accuracy, Acc_H can be expressed as

$$Acc_H = \frac{N_{correct} + 0.5E_{fifth} + 0.3E_{relative} + 0.2E_{parallel}}{N_{total}} \tag{3}$$

where E_{fifth} represents the number of perfect-fifth errors, $E_{relative}$ the number of relative major/minor errors, and $E_{parallel}$ the number of parallel major minor errors. Similarly, for the Audio Genre Classification (AGC) and Symbolic Genre Classification (SGC) tasks, a genre hierarchy was employed such that errors between similar genres were only discounted a half point (e.g., Jazz and Blues, Classical and Romantic, etc.).

3.2 Precision, Recall, and F-Measure

Consider a system that when given a query, returns a list of documents/items that it "believes" are the proper responses to the query. If an item is returned, and it is relevant, it can be considered a true positive, *TP*. If a document is returned and it is not relevant it is a false positive, *FP*. If a document is not returned, but is relevant it is a false negative, *FN*. Finally if a document is not returned and is not relevant, it is a true negative, *TN*. Using this system of *TP*, *FP*, *TN* and *FN* documents we can now define the two "classic" information retrieval evaluation metrics whose use predates the use of computers: *precision* and *recall*.

We can define the precision, *P*, as

$$P = \frac{TP}{TP + FP} \tag{4}$$

Put simply, precision is the ratio of relevant returned documents to the total number of returned documents. Recall, *R*, on the other hand is the ratio of relevant returned documents to the total number of relevant documents available in a system, and is expressed as

$$R = \frac{TP}{TP + FN} \tag{5}$$

These two measures can be combined into a single measure called the *F*-Measure, *F*, which is the harmonic mean of precision and recall:

$$F = \frac{2PR}{P+R} \tag{6}$$

As an example of how *precision*, *recall*, and *F*-Measure are appropriate to a MIR task, consider the Audio Onset Detection (AOD) task. AOD concerns itself with finding the start times of all sonic events in a piece of audio. In the AOD task, each system outputs its predicted onset times for a piece of audio, and these predictions are compared to a ground-truth of manually annotated onsets. Assume, for example, the ground-truth annotation of an audio snippet contains 100 onsets of audio events. Furthermore, let us assume an algorithm returns 80 onsets, 60 of which are correct. In this case, the precision would be 60/80 (0.75) and the recall 60/100 (0.60).

To better understand why we are interested in the *F*-Measure, let us now consider some extreme situations. Assume an algorithm predicts that there are one million onsets and, because these predicted onsets are so densely spaced, it has subsequently managed to predict the locations of the 100 true onsets in the piece. In this case, because all of the onsets that were in the ground-truth were correctly recalled, we have a case of perfect *recall*, i.e., 100/100 (1.0). However, only 100 of the one million returned onsets were correct, causing *precision* to drop to 100/1,000,000 (0.0001). Conversely, let us assume that an algorithm only returns one single onset, which is correct. In this case the *precision* is perfect 1/1 (1.0), but the *recall* is quite small, 1/100 (0.01). In general, *recall* and *precision* are traded at the expense of the other. The two measures are combined in the *F*-Measure. If either *recall* or *precision* are very low, the *F*-measure will be as well. Therefore, *F*-Measure rewards those systems that have the best balance of simultaneously high *recall* and *precision* scores.

3.3 Mean Reciprocal Rank

In query-based tasks such as Query-by-Singing/Humming (QBSH), system performance can be measured with mean reciprocal rank. Consider a system that returns a ranked list of results given a query. For example, QBSH systems are designed to return a ranked list of songs in response to a user singing a melody into a microphone. If the desired response to a query such a "Twinkle Twinkle Little Star" is third in the returned list (i.e., rank 3), the reciprocal rank is 1/3. If we take the mean of the reciprocal ranks over all queries, we arrive at the mean reciprocal rank (MRR) which can be formally expressed as

$$MRR = \frac{1}{N} \sum_{n=1}^{N} \frac{1}{rank_n} \tag{7}$$

where N is the number of queries, and $rank_n$ is the rank of the correct response of query n in the returned response list. This metric rewards systems for placing the desired items near the top of the ranked lists and quickly penalizes those systems that returned the desired items lower in the list.

3.4 Audio Tempo and Audio Beat P-Score

In some MIREX tasks, more heuristic measures, derived from the outcomes of real-world user experiments, are used for evaluation. For example, in the Audio Tempo Extraction (ATE) task, two dominant tempi were required to be returned by each algorithm. This two tempi approach was adopted based on the work of McKinney and Moelants [13] which showed that perceived tempi vary among real listeners in a relatively discrete and integer-based manner (i.e., some persons hear a given song at 60 beats per minute (bpm) while others hear the same song at 120 bpm). The ATE ground-truth contained the two true dominant tempi, as well as a salience of the first tempo. Denoting this salience as α, we compute the P_{tempo} score as

$$P_{tempo} = \alpha T_1 + (1 - \alpha)T_2 \qquad (8)$$

where T_1 is 1 if the first tempo is identified within ± 8% of the ground-truth tempo value and 0 otherwise. Likewise, T_2 takes the value 1 if the second tempo was identified and 0 otherwise.

In the Audio Beat Tracking (ABT) task, the work of Moelants and McKinney also was used as a basis for the choice of metric. For the ABT task each evaluated system provided beat times extracted from a piece of audio, akin to the tapping of a foot along with the music. These extracted beat times are compared to 40 ground-truth beat tracks for each musical piece collected from humans tapping to the beat of the piece. The beat times for the algorithms and each ground-truth are converted to an impulse train (at a 100 Hz sampling rate), and a cross correlation of the algorithm output impulse train, $y[n]$, and the ground-truth, $a_s[n]$, is measured. The cross correlation is calculated across a small window of possible delays between $-W$ and $+W$ (where W is 1/5[th] of the beat length). The correlations are then averaged across the S ($S=40$) ground-truths. Therefore, the beat P-Score, P_{beat}, can be expressed as

$$P_{beat} = \frac{1}{S}\sum_{s=1}^{S}\frac{1}{N_p}\sum_{m=-W}^{+W}\sum_{n=1}^{N}y[n]a_s[n-m] \qquad (9)$$

where N_p is a normalization factor for the number of beats and is calculated as

$$N_p = \max(\sum y[n], \sum a_s[n]) \qquad (10)$$

3.5 Chord Detection Overlap Score

The chord detection overlap score was specifically designed for the MIREX Audio Chord Detection (ACD) task. For the ACD task, each system had to return the chord names along with their associated onset and offset times within a piece of music. A good performance criterion in such a situation is to measure the amount of overlap in time between the detected chords and the ground-truth. A system's returned chords can be represented as

$$C_{dt}(c,t) = \begin{cases} 1, & \text{if chord } c \text{ is active at time } t \\ 0, & \text{else} \end{cases} \tag{11}$$

Then the overlap score can be calculated as

$$\text{Overlap Score } = \frac{\int_t C_{dt}(c,t) \cdot C_{gt}(c,t)dt}{\int_t C_{gt}(c,t)dt} \tag{12}$$

where $C_{dt}(c,t)$ and $C_{gt}(c,t)$ are the detected and the ground-truth chords.

In the 2008 MIREX ACD task, the systems were restricted to return one of 25 different chord types rooted on the 12 pitches of the chromatic scale (i.e., 12 major chords, 12 minor chords, and no chord). Therefore, c is a discrete variable. Moreover, systems were only allowed to return one active chord at any given time.

3.6 Average Dynamic Recall

For such similarity tasks as Audio Music Similarity (AMS) and Symbolic Melodic Similarity (SMS) where systems return relevant items as a list ranked according to "similarity" to a given query, the underlying relevance measure is subjective because it is based on the biases, tastes, levels of expertise and foibles of the human assessors providing the assessments. For example, in the MIREX 2005 Symbolic Melodic Similarity (SMS05) task, systems returned a ranked list of songs whose melodies were "similar" to the query song. The ground-truth for SMS05 task was generated by humans manually scoring every query against every returned result in a set of pre-MIREX experiments conducted by researchers at Utrecht University [16]. It was Rainer Typke, then a graduate student at Utrecht University, who took the lead on proposing and organizing the SMS05 task. Using the *Répertoire International des Sources Musicales (RISM). Serie A/II, manuscrits musicaux après 1600* collection [15] of incipits, Typke and his colleagues at Utrecht created a ground-truth set of similarity judgments. For each of 11 queries, the Utrecht team had 35 music experts rank order the pre-filtered individual results (about 50 per query) based on similarity to the original query. The median ranks assigned to the retrieved incipits were subjected to the Wilcoxon rank sum test. This statistical testing procedure allowed the Utrecht team to create groups of results that contained items of comparable similarity while at the same time being able to order the groups themselves with regard to similarity to the query. It was these 11 lists of group-ordered incipits that formed the SMS05 ground-truth set.

Because the ground-truth generation involved highly subjective human judgments, it is quite reasonable to treat the ground-truth rank list not in an absolute item-by-item sense but rather in a more relativistic sense of ranked groups of equivalently similar items. For example, the first and the second items in a list might have slightly different scores but for all intents and purposes they are equivalently similar to a given query. To reflect this state of affairs Typke and his Utrecht colleagues developed the Average Dynamic Recall (ADR) metric [17].

ADR measures the average recall over the first n documents with a dynamic set of relevant documents where n is the number of documents in the ground-truth list. The set of relevant documents grows with the position in the list but not just by one item. For example if the ground-truth has two groups of equally relevant items such as $<(1, 2), (3, 4, 5)>$, then at position number 2 there are 2 relevant documents whereas, at position number 3 there are a total of 5 relevant documents. However, at each position in the results list the recall is calculated by dividing the number of relevant items with the position number, not the total number of relevant documents. Thus, the above definition can be written formally as

$$r_i = \#\{R_w \mid w \le i \wedge \exists j, k : j \le c \wedge R_w = G_k^j\}/i \tag{13}$$

$$ADR = \frac{1}{n}\sum_{i=1}^{n} r_i \tag{14}$$

where $< R_1, R_2, ... >$ is the returned results list. The ground-truth list has g groups such as $< (G_1^1, G_2^1, ..., G_{m1}^1), (G_1^2, G_2^2, ..., G_{m2}^2), ...(G_1^g, G_2^g, ..., G_{mg}^g) >$. The ranking does not matter within each group. For example we do not know if $rank(G_i^j)$ is less than $rank(G_k^j)$. However, we do know that $rank(G_i^j)$ is less than $rank(G_i^k)$ given that $j < k$. Also c in the above equation is the group number that contains the ith item in its group. The key point to remember about ADR is that it represents an attempt to mitigate the distorting effects of relying solely on absolute (mostly minute) differences in human-generated similarity scores by allowing for grouping of functionally similar items.

4 Evolution of Similarity Evaluation Tasks

The set of similarity tasks, Audio Music Similarity (AMS) and Symbolic Melodic Similarity (SMS), comprises those tasks that most closely resemble a classic information retrieval scenario. That is, for a given piece of music submitted as a query, the systems under evaluation are expected to return a ranked list of music pieces that are deemed to be similar to the query. In many ways, it is this scenario that most people think of when they think of MIR systems in real-world deployments. In this section we explore this issues raised in running the set of similarity tasks run by MIREX between 2005 and 2007. We will also examine how responding to these issues caused the structure of the similarity tasks to evolve over time.

4.1 2005 Symbolic Melodic Similarity

In 2005, there was no running of an AMS task as the community could not decide
how it would set up the ground-truth for such a task. However, as mentioned in
Section 3.6, it was Rainer Typke who was the intellectual leader of SMS05 because he had the ground-truth set in hand that he and his Utrecht colleagues had
already created (SMS1 in Table 3). SMS05 had 7 algorithms evaluated. The best
algorithm had an ADR score of 66% while the worst had a 52% ADR score.
Overall, participants were pleased with the 2005 running of SMS. However, several important issues arose that would influence future similarity task runs at
MIREX. First, it was obvious that the 11 queries contained in the Utrecht ground-
truth set was not a big enough set upon which to make broad generalizations about
system behaviors. Second, there was some concern that Utrecht's pre-filtering step
might have removed potentially relevant items from the ground-truth. Third, generating new ground-truth data for MIREX 2006 using the Utrecht method would
not be possible given the time and manpower constraints of MIREX. Fourth, the
RISM collection does not encompass a wide enough range of music to represent
all the musical tastes, genres and styles of interest to MIR developers and users.
Fifth, no one in the MIREX community could come up with a feasible way to rep-
licate the Utrecht method to generate ground-truth for a meaningful collection of
audio-based music files.

4.2 2006 Symbolic Melodic Similarity and Audio Music Similarity

Given the issues raised after the running of SMS05, the IMIRSEL team worked
with the MIREX community to define a general framework for MIREX 2006 that
could be used to construct both the SMS06 and Audio Music Similarity (AMS06)
tasks. The principal difference that would set the 2006 tasks apart from the 2005
task would be the adoption of a more TREC-like evaluation scenario. That is, rather than creating and using a pre-compiled ground-truth set, the evaluation of retrieved results would be conducted *post-hoc* using human judges (or "graders") to
score the similarity between queries and their respective returned items.

Acceptance of the *post-hoc* set up was not controversial. However, as with
many things community-based, three issues became hotly debated. The first
point of contention was the choice of evaluation metric. The second was the
number of graders that would be used to evaluate the results. The third issue was
basic feasibility.

Many community members, including the authors of this chapter, favoured a
simple binary measure of similarity/relevance. That is to say, a returned piece (also
known as a "candidate") was, or was not, similar/relevant to the query piece. The
binary approach would make the use of classic precision (see Section 3.2) easy to
calculate. Others fought vigorously for some type of graduated "broad" judgment
(e.g., Not Similar (NS), Somewhat Similar (SS), Very Similar (VS)) to better reflect the nuances in perceptions of similarity. Deciding upon methods for weighting
the relative importance of NS, SS and VS opened up another debate thread. A
continuous fine-scaled approach was also suggested to more "accurately" capture

subtle differences in similarity. This fine-scale could be represented using some kind of slider that a grader could position between 0.0 (NS) and 10.0 (VS). The sum or averages of the slider-generated values could then be used to evaluate the success of each system for each query/candidate pair.

The number of graders per query/candidate pair was another contentious issue. One group, including this set of authors, wanted to mimic the traditional TREC approach of one grader per query. Others argued strongly that music relevance and text relevance were not equivalent and that music similarity most likely required a number of judges to overcome personal biases with regard to expertise and taste.

Issues of feasibility became intertwined with the number-of-graders debate. As stated before, MIREX has an algorithm-to-data model that means the bulk of the evaluation work has to performed and managed by the IMIRSEL team. Under ethics guidelines as advised by the University of Illinois' Institutional Review Board, volunteers cannot be expected to perform more than 3-4 hours of work without compensation (and added administrative rights and protections). This range of 180-240 minutes set an upper bound on the scope of the similarity tasks. Thus, the IMIRSEL team had to juggle a set of factors that would influence the scope (and hence the workload) of the similarity evaluations. These factors included:

1. Number of *A*lgorithms submitted
2. Number of *Q*ueries
3. Number of *C*andidates returned per query
4. Number of *M*inutes spent evaluating each query/candidate pair
5. Number of graders per query/candidate *P*air
6. Number of *G*raders available

These combined to form the basic feasibility equation outlined below.

$$(A * Q * C * M * P) / G \leq 240 \text{minutes (per Grader)} \tag{15}$$

As one can see, there are many trade-offs involved. When developing the guidelines for the 2006 similarity tasks under these conditions it was also problematic that one does not know in advance with any certainty such things as how many algorithms will actually be submitted (A), how many people will volunteer to be graders (G) and/or how long it will actually take to evaluate each query/candidate pair (M).

So, given all the debate over the issues of metrics, graders and feasibility, what decisions were actually made? On the metric issue, Table 4 shows the breakdown of the set of new *post-hoc* similarity metrics tabulated for both AMS06 and SMS06. Fine-scaled scoring (FINE in Table 4) and graduate BROAD scores (NS, SS, and VS) were both included. Binary scoring (Greater0 and Greater1) was accomplished by treating Somewhat Similar (SS) scores as either Not Similar (NS) or Very Similar (VS). A trio of different broad score weighting schemes was created (PSum, WCsum and SDsum) to emphasize the relative importance of systems returning Very Similar (VS) results (see Table 4). As Table 5 shows, both SMS06 and AMS06 tasks ended up having 3 graders per query/candidate pair. Most importantly, the combination of factors outlined in the feasibility equation yielded an average number of query/candidate pairs per grader of 205 for SMS06 and 225 for AMS07. Under the relatively realistic assumption of 1 minute per evaluation, this brought the time commitments of the graders within our upper bound of 240 minutes.

Table 4 Basic Metrics used in the 2006 Similarity Tasks

METRIC	COMMENT	RANGE or VALUE
FINE	Sum of fine-grained human similarity decisions	0-10
PSum	Sum of human broad similarity decisions	NS=0, SS=1, VS=2
WCsum	'World Cup' scoring (rewards Very Similar)	NS=0, SS=1, VS=3
SDsum	'Stephen Downie' scoring (strongly rewards Very Similar)	NS=0, SS=1, VS=4
Greater0	Binary relevance judgment	NS=0, SS=1, VS=1
Greater1	Binary relevance judgment using only Very Similar	NS=0, SS=0, VS=1

Table 5 Summary Statistics for the AMS and SMS Tasks 2006-2007

	SMS06	AMS06	SMS07	AMS07
Number of algorithms	7	6	6	12
Number of queries	17	60	30	100
Total number of candidates returned	1360	1800	2400	6000
Total number of unique query/candidate pairs graded	905	1629	799	4832
Number of graders available	20	24	6	20
Number of evaluations per query/candidate pair	3	3	1	1
Number of queries per grader	15	7~8	1	5
Number of candidates returned per query	10	5	10	5
Average number of query/candidate pairs per grader	225	205	133	242
Number of grading events logged	23491	46254	3948	21912

For SMS06, each system was given a query and asked to return the 10 most melodically similar songs from a given collection. The collections were *RISM* (monophonic; 10,000 pieces; SMS1 in Table 3), *Karaoke* (polyphonic; 1,000 pieces; SMS2), *Mixed* (polyphonic; 15,741 pieces; SMS3). There were 6 *RISM* queries, 5 *Karaoke* queries and 6 *Mixed* queries for a total of 17 queries. Then, for each query, the returned results from all participants were grouped and anonymized into collections of query/candidate pairs. These pairs were evaluated by human graders, with each query/candidate pair being evaluated by the 3 different graders. Each grader was asked to provide a categorical BROAD score with 3 categories: NS, SS, VS as explained previously, and one FINE score (in the range from 0 to 10). Along with the basic FINE and BROAD scores, Utrecht's Average Dynamic Recall (ADR) was also calculated to provide some continuity with SMS05.

For AMS06, each system was given 5000 songs chosen from IMIRSEL's *USPOP*, *USCRAP* and *CoverSong* collections. The *USPOP* collection was donated to IMIRSEL by Dan Ellis's Lab Rosa at Columbia University and represents hit pop songs from 2002. The *USCRAP* collection was an eclectic mix of tracks bought by IMIRSEL from a music wholesaler specializing in remaindered CDs (thus the music quality was quite varied). The 330 member *CoverSong* collection is described in [6] and contains a set of 30 titles each of which is represented by 11 variants. The *CoverSong* collection was included in the AMS06 data set to test an ancillary hypothesis that explored whether standard spectral-based similarity techniques were suitable to detect the musically similar but acoustically disimilar cover songs.[4] Each system returned a 5000x5000 distance matrix that recorded the similarities between each song in the collection. After all the matrices were submitted, IMIRSEL randomly selected 60 songs to act as "queries." The first 5 most highly ranked songs out of the 5000 were extracted for each query from each system's matrix (after filtering out the query itself, returned results from the same artist and members of the *CoverSong* collection). Then, for each query, the returned results from all participants were grouped and anonymized into collections of query/candidate pairs. Like SMS, each query/candidate pair was evaluated by 3 different graders using the same set of BROAD and FINE scoring metrics.

It is interesting to note here that the AMS community did not adopt the ADR metric as its primary evaluation metric. The differences in primary metrics between SMS06 and AMS06 reflects the fact that the symbolic and audio research communities are quite independent of each other and hold different worldviews on the notion of what the similarity task is meant to achieve. As the respective task names suggest, the SMS community is interested in notions of "melodic" similarity regardless of such externalities as timbre or orchestration while the AMS community is interested in notions of "musical" similarity which does include timbre and orchestration as contributory factors.

In order to collect grader scores and manage the whole grading process for both SMS06 and AMS06, the IMIRSEL team developed the Evalutron 6000 (E6K) [7, 9, 10]. The E6K (Figure 1) was coded by IMIRSEL team member Anatoliy Gruzd using the "CMS Made Simple" open-source content management system[5] which both reduced the development time and simplified system management. As a web-based application, E6K adheres to a Client-Server model: the client consists of HTML, CSS and JavaScript; and, the server – PHP and MySQL. E6K's web-based approach has the benefit of allowing graders to use the system from anywhere they have a browser and an Internet connection. This was particularly important for MIREX given the international scope of its participants. E6K employs the popular Web 2.0 programming technique referred to as AJAX (Asynchronous JavaScript and XML)[6] to save similarity/relevance judgments and other interaction events in real time, allowing graders to leave the

---

[4] The answer to this hypothesis was a resounding no [5].
[5] See http://www.cmsmadesimple.org
[6] See http://en.wikipedia.org/wiki/Ajax_(programming).

system and come back as they wish. The ability of graders to come and go as they saw fit was a formal requirement mandated by UIUC's research ethics board. As a side benefit, however, adoption the AJAX approach helped to prevent data loss during unexpected service interruptions or system failures.

The E6K gives graders a choice of three audio players: Flash, Windows Media Player, and Quicktime to ensure cross-platform usability. All players draw from a common set of query/candidate MP3 files. The E6K tracks and records all user-interactions with the system. For MIREX 2006, this consisted of 69,745 logged events across both SMS06 and AMS06.

To better understand the effects that the various evaluation design decisions had on running both the AMS06 and SMS06 tasks, the IMIRSEL team performed a range of analyses on the results data [10]. One such analysis examined the inter-grader reliability (or inter-subjectivity) of the judgments made by trio of graders assigned to each query/candidate pair. The IMIRSEL team chose Fleiss's Kappa to evaluate the inter-grader reliability as it was a metric that it had worked with before. It is a measure of inter-grader reliability for nominal data and is based upon Cohen's two-grader reliability Kappa but is intended for use with an arbitrary number of graders [8]. Fleiss's Kappa is defined as

$$\kappa = \frac{\bar{P} - \bar{P}_e}{1 - \bar{P}_e} \tag{16}$$

where:

$$\bar{P} = \frac{1}{Nn(n-1)} \sum_{i=1}^{N} \sum_{j=1}^{k} n_{ij}(n_{ij} - 1) \tag{17}$$

$$\bar{P}_e = \sum_{j=1}^{k} \left(\frac{1}{Nn} \sum_{i=1}^{N} n_{ij} \right)^2 \tag{18}$$

In the context of the AMS06 and SMS06 tasks, N is the total number of query/candidate pairs that need grading; n is the number judgments per query/candidate pair; k is the number of BROAD response categories (here equal to 3); and, n_{ij} is the number of graders who assigned the i-th query/candidate pair to the j-th BROAD category. Fleiss's Kappa scores range from 0.0 (no agreement) to 1.0 (complete agreement).

Two sets of BROAD category configurations were evaluated. First, we computed the Kappa score using all three BROAD categories (VS, SS, NS). Second, we computed the Kappa score after collapsing the VS and SS categories into a single "similar" (S) category to create a classic binary classification scheme of Similar (S) and Not Similar (NS). Table 6 below presents the Kappa scores for AMS06 and SMS06 under the 3-level and 2-level schemes.

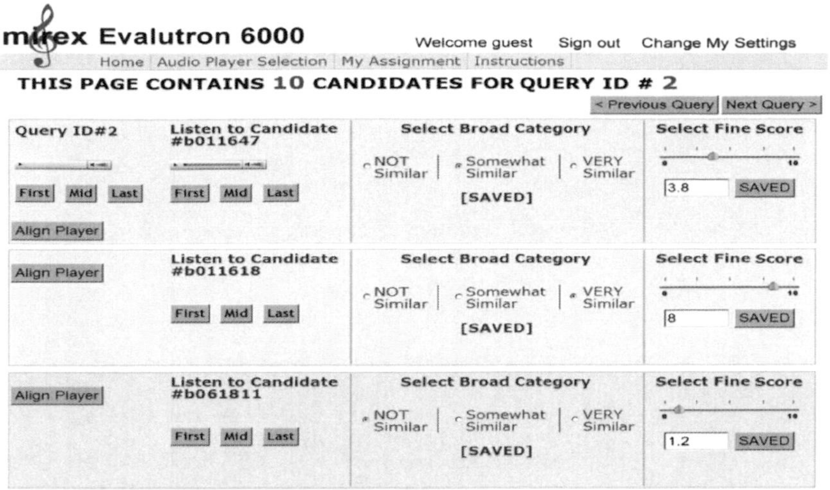

Fig. 1 Screenshot of the Evalutron 6000 (E6K) interface

Table 6 Fleiss's Kappa Scores for SMS06 and AMS06

	3-level (VS, SS, NS)	2-level (S, NS)
SMS06	0.3664	0.3201
AMS06	0.2141	0.2989

According to Landis and Koch [11] who studied the consistency of physician diagnoses of patient illnesses and then derived a scale for interpreting the strength of agreement indicated by Fleiss's Kappa, the scores reported in Table 6 show a "fair" level of agreement. While not ideal, the fair level of agreement is remarkable given the diversity in levels of music skill, tastes and cultural backgrounds of the MIREX 2006 graders. Also noteworthy is the difference in score agreements between the SMS06 and AMS06 tasks with the SMS06 graders being in stronger agreement with each other than the AMS06 graders. For example, when compared under the 3-level scheme, 7.1% of the query/candidate pairs had no agreement among the graders while only 2.2% percent of the SMS06 query/candidate pairs had no agreement. Partial disagreements (where two graders agreed and the third did not) occurred in 51.9% of the SMS06 cases and 62.8% of the AMS06 cases. 30.1% of the query candidate/pairs for AMS06 had perfect agreement among the three graders while the SMS06 grader reached perfect agreement on 45.9% of their query candidate pairs. Under the 2-level scheme, things even out between the two tasks as the AMS06 graders reached perfect agreement on 48.3% of the query/candidate pairs and the SMS06 graders reached perfect agreement on 49.8% of the query/candidate pairs. We believe this overall disparity in agreement levels between the two tasks is attributable to the different notions of similarity held by the two task communities.

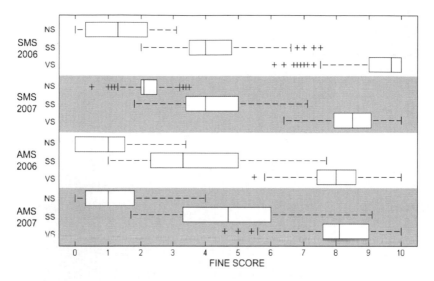

Fig. 2 Distribution of FINE scores within BROAD categories for the SMS06, SMS07, AMS06 and AMS07 tasks. The Box-and-Whisker plots show the median FINE score bounded by the first and third quartiles in the box, with whiskers extending to one-and-half times the inter-quartile range (i.e., the distance between the first and third quartiles) and outliers denoted with +'s.

To measure the consistency between the BROAD category and FINE scores, we calculated the distribution of FINE scores within each BROAD category. Figure 2 shows box-and-whisker plots for both SMS and AMS tasks from 2006 and 2007. The boxes have vertical lines at the 1st, 2nd, and 3rd quartiles. The whiskers bound the minimum and maximum values which fall within 1.5 times the interquartile range (IQR), outliers are denoted by + symbols. The box-and-whisker plots illustrate the relative differences between tasks in terms of the assignment of FINE and BROAD scores to musical works. Not only do they reveal the distribution of FINE score responses for a given BROAD score for a given task, they also speak to the variation among graders of what constitutes two pieces being not similar (NS), somewhat similar (SS), or very similar (VS).

In all four of the AMS06, SMS06, AMS07 and SMS07 tasks, the NS and VS categories have the most compact distributions of FINE scores. Similarly, the SS category has the largest IQR in both sets of SMS and AMS tasks. SS scores overlap with the NS values from both task sets. In AMS, however, note how SS greatly overlaps **both** the NS and VS values. The data presented in Figures 2 lead us to two important observations. First, and again, there appears to be a fundamental difference in the interpretations of "similarity" between the SMS and AMS task communities. Second, the SS category, regardless of task, appears problematic. The overly broad term "somewhat similar" (SS) is open to many interpretations and meanings, allowing graders to judge two pieces "similar" at any number of ranks. This is not only what would be expected given our natural intuitions about labels like "somewhat similar", but is also evidenced in the data which clearly illustrate a wide distribution of FINE scores corresponding to the SS category for all tasks.

4.3 2007 Symbolic Melodic Similarity and Audio Music Similarity

The single biggest difference between the running of the 2006 similarity task set and its 2007 running was the adoption of a single grader per query/candidate pair model for both SMS07 and AMS07. This single grader model was adopted for two reasons. First, it greatly reduced the administrative overhead of finding and managing a large number of graders needed. This lessening of load allowed the AMS07 community to significantly increase the number of queries that could be evaluated from 60 in AMS06 to 100 in AMS07. Similarly, the SMS07 query set increased to 30 from its 2006 17 queries. Second, the two communities were convinced that general state of agreement among graders shown in the analyses of the 2006 data indicated that there was less of a need to control for inter-grader variance by having multiple graders than they had originally thought.

Both the AMS07 and SMS07 tasks kept the FINE and the 3-level BROAD categories scoring systems available on the E6K system. Notwithstanding the problematic nature of the SS category, the 3-level system was kept in part to have consistency between years, in part to see if any differences could be noted across years, and in part because it really cost very little to collapse the VS and SS categories into a single S category to create a binary relevance score. The SMS07 community also kept ADR as its primary evaluation metric while AMS07 community continued to ignore it in favour of average FINE score.

SMS07 differed from SMS06 in several significant ways. First, the underlying dataset chosen (SMS4 in Table 3) was changed to 5274 pieces drawn from the Essen Associative Code and Folksong Database.[7] Second, for each query, four classes of error-mutations were created to test the fault-tolerance of the systems. Thus the query set comprised the following 5 query classes:

1. No errors
2. One note deleted
3. One note inserted
4. One interval enlarged
5. One interval compressed

For each query (and its 4 mutations), the returned candidates from all systems were then grouped together (query set) for evaluation by the human graders. The graders were provided, via the E6K system, with the perfect version to represent the query. It was against this perfect version that the graders evaluated the candidates. Graders did not know whether the candidates came from a perfect or mutated query. Each query/candidate pair was evaluated by 1 individual grader. Furthermore, each query was the sole responsibility of only 1 grader to ensure uniformity of results within that query.

---

[7] More information about the Essen Collection available at http://www.esac-data.org

While the basic premise of AMS07 was kept, there were several subtle changes from AMS07. First, the size of the returned matrices increased to 7000 X 7000 as the dataset also included additional new music drawn from IMIRSEL's *Classical* and *Sundry* collections. Second, 30 second clips were used for AMS07 rather than the full songs used in AMS06 to speed up processing and to ensure that the graders would listen to the exact same music as the systems. This change came about after analyzing the 2006 AMS E6K time-on-task data where it became apparent that many graders were making their judgments without listening to the entire pieces of music provided. Second, 100 songs were randomly selected from the returned matrices with the constraint that the 100 songs equally represent each of the 10 genres found in the database (i.e., 10 queries per genre). The 5 most highly ranked songs were then returned per query (after filtering for the query song and songs from the same artist). As with SMS07, a query became the sole responsibility of 1 grader to ensure uniformity of scoring within each query.

The actual running of the 2007 similarity tasks was much less stressful on the IMIRSEL team primarily because the need to manage multiple graders per query/candidate pair was eliminated. The SMS07 ADR highest score of 0.72 along with the highest FINE score of 0.56 indicates that the best performing SMS systems are quite tolerant of query input errors. These scores compare favourably with the SMS1 scores (found in Table 3) from 2005 (0.66 ADR) and 2006 (0.72 ADR). The AMS07 highest average FINE score increased to 0.57 from 0.43 for AMS06. We believe this increase in score is jointly attributable to improvements in the algorithms (primarily) and the shortening of the query/candidate pair lengths to 30 seconds to ensure the synchronization of "listening" between the systems and the graders (secondarily).

There was no running of either an SMS or AMS task during MIREX 2008. The first reason for this was a general consensus that developers needed more time to make non-trivial improvements to their systems. The second and perhaps more daunting reason is the data shortage issue. For SMS in particular, there is an acute shortage of available trustworthy symbolic music from which to build meaningful, large-scale test collections. Even with AMS, where more data is available, it still takes considerable effort and expense to acquire and then prepare the audio files (i.e., cutting to length, normalizing data formats, etc.) for use in a proper test collection. Of course, it is possible to re-use the datasets from previous years. However, constant reuse of data will most likely lead to the "overfitting" of algorithms to the data leading to meaningless apparent improvements in results.

If the two communities can decide on how to deal with the data issue, the IMIRSEL team looks forward to running future iterations of both the AMS and SMS similarity tasks. Based on the smoothness with which both SMS07 and AMS07 ran, we recommend keeping the basic 2007 format for each task with one caveat. As Figure 2 shows, the SS BROAD category continues to be problematic with its wide variance in both the AMS and SMS tasks. We need to encourage these communities to think hard about either how to reduce this variability or to consider eliminating the SS BROAD category altogether, particularly with regard to the AMS task.

5 Summary and Future Directions

In this chapter we have examined the history and infrastructure of MIREX. MIREX is a community-led endeavour that reflects the wide range of research streams being undertaken by MIR researchers from all around the world. The growth of MIREX over the period from 2005 to 2008 has been remarkable with the number of algorithms evaluated increasing steadily each year. We have discussed the numerous primary evaluation metrics used to "officially" report the performance results of each tasks. We have highlighted that the choice and/or creation of particular primary metric is influenced by a mixture of scientific imperatives, previous empirical research, traditions of practice, participant desires to succeed and the pragmatics of actually getting the evaluations completed within constraints of acceptable time and effort expenditures. By looking at the two similarity tasks, AMS and SMS, we have illustrated how MIREX tasks have evolved through time as the MIREX research community builds upon its successes and addresses the problems it encounters.

We have also suggested that one major problem facing the MIREX community is the lack of useable data upon which to build realistic test collections. In an effort to solve this problem, along with the potential problem of having IMIRSEL overcome by the increase in the number of algorithm submissions, a new international research collaboration called the Networked Environment for Music Analysis (NEMA) has been formed. NEMA comprises research labs from the Universities of Waikato, Illinois at Urbana-Champaign, Southampton, London (both Goldsmiths and Queen Mary Colleges), and McGill. One important goal of NEMA is the construction of a web-service framework that would make MIREX evaluation tasks, test collections and automated evaluation scripts available to the

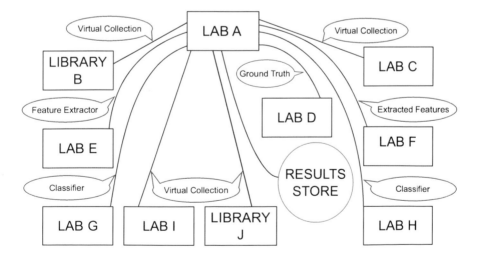

Fig. 3 An illustration of how NEMA will be used to gather remote resources for evaluation

community on a year-round basis. It would also expand the availability of data by developing the On-demand Metadata Extraction Network (OMEN) [12]. OMEN is designed to acquire metadata and features from remote music repositories while at the same time respecting copyright laws. NEMA hopes to incorporate OMEN within a system that would allow researchers to locate and use bits and pieces of algorithms from other researchers. Thus, as Figure 3 illustrates, a team at Lab A could build a hybrid experimental system quickly from the classifiers and feature extractors from other participating labs. It could run a MIREX-based evaluation from the comfort of its lab any time it chose using its new system. It could then report its findings and, if successful, redeposit its new hybrid algorithm into the NEMA repository for others to build upon.

Acknowledgments. The IMIRSEL team wishes to thank the Andrew W. Mellon Foundation for its support. This work has also been support by the National Science Foundation Grant IIS 0328471.

References

1. Cano, P., Gomez, E., Gouyon, F., Herrera, P., Koppenberger, M., Ong, B., Serra, X., Streich, S., Wack, N.: ISMIR 2004 audio description contest. MTG Technical Report, MTG-TR-2006-02 (Music Technology Group, Barcelona, Spain) (2004)
2. Downie, J.S. (ed.): The exploratory workshop on music information retrieval. Graduate School of Library and Information Science, Champaign (1999)
3. Downie, J.S. (ed.): The MIR/MDL evaluation project white paper collection, edition #3: Establishing music information retrieval (MIR) and music digital library (MDL) evaluation frameworks: Preliminary foundations and infrastructures. Graduate School of Library and Information Science, Champaign (2003)
4. Downie, J.S.: The scientific evaluation of music information retrieval systems: Foundations and future. Computer Music Journal 28(3), 12–23 (2004)
5. Downie, J.S.: The Music Information Retrieval Evaluation Exchange (2005-2007): A window into music information retrieval research. Acoustical Science and Technology 29(4), 247–255 (2008)
6. Downie, J.S., Bay, M., Ehmann, A.F., Jones, M.C.: Audio cover song identification: MIREX 2006-2007 results and analyses. In: Proceedings of the 9th International Conference on Music Information Retrieval (ISMIR 2008), pp. 468–473 (2008)
7. Downie, J.S., Lee, J.H., Gruzd, A.A., Jones, M.C.: Toward an understanding of similarity judgments for music digital library evaluation. In: Proceedings of the 7th ACM/IEEE Joint Conference on Digital Libraries (JCDL 2007), pp. 307–308 (2007)
8. Fleiss, J.L.: Measuring nominal scale agreement among many raters. Psychological Bulletin 76(5), 378–382 (1971)
9. Gruzd, A.A., Downie, J.S., Jones, M.C., Lee, J.H.: Evalutron 6000: Collecting music relevance judgments. In: Proceedings of the 7th ACM/IEEE Joint Conference on Digital Libraries (JCDL 2007), p. 507 (2007)
10. Jones, M.C., Downie, J.S., Ehmann, A.F.: Understanding human judgments of similarity in music information retrieval. In: Proceedings of the Eighth International Conference on Music Information Retrieval (ISMIR 2007), pp. 539–542 (2007)

11. Landis, J.R., Koch, G.G.: The measurement of observer agreement for categorical data. Biometrics 33, 159–174 (1977)
12. McEnnis, D., McKay, C., Fujinaga, I.: Overview of OMEN. In: Proceedings of the Seventh International Conference on Music Information Retrieval (ISMIR 2006), pp. 7–12 (2006)
13. McKinney, M.F., Moelants, D.: Tempo perception and musical content: What makes a piece fast, slow or temporally ambiguous? In: Proceedings of the International Conference on Music Perception and Cognition, Evanston, IL, USA (2004)
14. Pampalk, E.: Audio-based music similarity and retrieval: Combining a spectral similarity model with information extracted from fluctuation patters. In: MIREX 2006, Submission Abstracts (2006), http://www.music-ir.org/evaluation/MIREX/2006_abstracts/AS_pampalk.pdf
15. Répertoire International des Sources Musicales (RISM). Serie A/II, manuscrits musicaux après 1600. München. K. G. Saur Verlag, Germany (2002)
16. Typke, R., den Hoed, M., de Nooijer, J., Wiering, F., Veltkamp, R.C.: A ground truth for half a million musical incipits. Journal of Digital Information Management 3(1), 34–39 (2005)
17. Typke, R., Veltkamp, R., Wiering, F.: A measure for evaluating retrieval techniques based on partially ordered ground truth lists. In: IEEE International Conference on Multimedia and Expo (ICME 2006), pp. 1793–1796 (2006)

Part II
Harmony

Chord Analysis Using Ensemble Constraints

David Gerhard and Xinglin Zhang

Abstract. Many applications in music information retrieval require the analysis of the harmonic structure of a music piece. In Western music, the harmonic structure can be often be well illustrated by the chord structure and sequence. This chapter presents a technique of disambiguation for chord recognition based on *a priori* knowledge of probabilities of voicings of the chord in a specific musical medium. The main motivating example is guitar chord recognition, where the physical layout and structure of the instrument, along with human physical and temporal constraints, make certain chord voicings and chord sequences more likely than others, and make some impossible. Pitch classes are extracted, and chords are then recognized using pattern recognition techniques. The chord information is then analyzed using an array of voicing vectors indicating likelihood for chord voicings based on constraints of the instrument. Chord sequence analysis is used to reinforce accuracy of individual chord estimations. The specific notes of the chord are then inferred by combining the chord information and the best estimated voicing of the chord.

1 Introduction

Traditional Western music is performed by instruments, including the human voice, and instruments are constrained. All the instruments have their characteristic ranges, timbres, playing styles and techniques. Each instrument (including voice parts) has a standard *range* of notes that are playable. For example, the modern piano has a total of 88 keys, ranging from A0 to C8. Instruments also have corresponding standard playing techniques, which can often be derived from the physical structure

David Gerhard
Department of Computer Science, University of Regina, Regina, SK Canada
e-mail: david.gerhard@uregina.ca

Xinglin Zhang
Department of Computer Science, University of Regina, Regina, SK Canada
e-mail: xinglinzh@gmail.com

Z.W. Raś and A.A. Wieczorkowska (Eds.): Adv. in Music Inform. Retrieval, SCI 274, pp. 119–142.
springerlink.com   © Springer-Verlag Berlin Heidelberg 2010

and layout of the instruments as well as the physical abilities of human player, and are sometimes related to the style of music being played. For example, the guitar is commonly played in one of two styles: chording, where a group of strings is sounded simultaneously, and picking, where individual notes are played to create a melody.

1.1 Prescriptive and Descriptive Constraints

A distinction should be made between *prescriptive* and *descriptive* constraints. Prescriptive means that there is a rule prescribing a musical constraint. The rule can be about composition, technique or any other aspect of the musical characteristics of a piece, but what makes a rule prescriptive is that it can be broken. It is a common practice, put in place by earlier composers and it suggests ways of making music which "sounds good". An example is the prescription to avoid the motion of parallel fifths. Composers have found that motion of parallel fifths can be distracting, can lead to a reduced sense of the perception of the key or chord pattern of the piece, and can make musical critics unhappy. The prescriptive constraint comes from real or imagined reasons for not using a particular construct, but there is no physical reason a composer cannot use motion of parallel fifths. A renegade composer needing just that distraction or wanting to upset critics in this way, is not physically prevented from using parallel fifths. It is a convention, not a requirement. The term "prescribe" means "to write before" and prescriptive constraints are rules which are written before the composer begins to write the song.

A descriptive constraint, on the other hand, is a physical condition of the instrument, ensemble or player which is being asked to produce sounds. And some constraints are so rigid that they cannot be broken. For example, no matter how convincing the request is, a clarinet can never produce more than one note. No matter what enticements or threats are brought to bear, a trumpet cannot play a C two octaves below middle C. And no human can play a chord containing notes that are five octaves apart using a single hand on a piano. Descriptive constraints detail characteristics of the music which cannot be changed usually. There are other descriptive constraints that can be broken by using some certain techniques. For example, the musical *range* of an instrument is an example of a descriptive constraint. This is a standard measure of the notes that an instrument can play, and are so rigid that they are often programmed into musical composition software[1]. However, it is possible, for example, to use some overtone techniques to produce higher pitches that are theoretically "out of range" for a guitar. Another example is on an alto saxophone, where it is possible to achieve a note lower than concert D♭3 by partially blocking the bell against the player's leg, both causing a resonance node and slightly elongating the effective pipe length, producing a lower pitch. The term "describe" means "to write down" and descriptive constraints are used to describe the instrument itself and the "standard" playing techniques ascribed to that instrument.

---

[1] For example, the *Sibelius* composing software (http://www.sibelius.com/) colors a note red when it is out of range for the staff instrument.

In the context of music information retrieval, prescriptive and descriptive constraints can both be used as additional information to enhance analysis and disambiguate results. Developers must take care when using these constraints, because while prescriptive constraints yield an increased or decreased probability of occurrence from the norm, depending on the rule, rigid descriptive constraints theoretically lead to musical events or characteristics with zero probability. In practice, however, it should be noted that musicians and composers are, by nature, creative, and often seek ways to overcome constraints both prescriptive and descriptive, as mentioned above. The piano, however, has no such techniques and the playable range (A0 to C8) is a true descriptive constraint.

1.2 Application of Constraints in Music Information Retrieval Research

Although these constraints limit the extent of the sounds which can be produced, they can be used by researchers studying music to narrow down the possible answers to the questions they are asking. Researchers have made use of prescriptive and descriptive constraints in music information retrieval in the past, but the distinction between the types of constraints is rarely made, and estimated probability distribution measures are often used in place of identified constraints. Automatic generation of musical instrument fingerings is a research area in which algorithms are used to calculate an optimal fingerings (hand positions and movements) for a sequence of given notes, either monophonic or polyphonic.

Tuohy and Potter [20] present a genetic algorithm for the automatic generation of playable guitar tablature through the use of a fitness function that assesses the playability of a given set of fretboard positions. Though not explicitly using the term "constraints", their fitness function takes into account some physical characteristics of the guitar. "Hand movement" and "Hand manipulation" are considered by the fitness function, which favors easier situations and penalizes complicated situations. In [9], a system for generation of piano fingerings is proposed. The system constrains fingerings by requiring that the same finger be used as long as a note persists and no finger substitution is allowed[2]; each finger may only depress one note. What's more, they use "vertical cost", "horizontal cost" and "user specification of cost function" to measure the playability of the underlying fingering. The three costs take into consideration different kinds of constraints.

Radicioni et al [18] explicitly use the term "constraints-based approach" in their system for modeling guitar performers' gestures and annotating a musical score. They believe that the physical gestures used to operate musical instruments are responsible for the characteristics of the sound being produced in a performance. They make use of the highly constrained nature of performers' gestures to build a model of music performance, coupled with a strategy aimed at maximizing the gestural comfort of performers. They draw on physical and bio-mechanical constraints for their model, from the implications of the fact that notes have certain positions on

---

[2] Finger substitutions happen in reality for long notes.

the guitar fretboard and human's hands have a certain range of span, making certain fingerings more frequent than others.

When the music score is known *a priori*, an interesting problem is to discover the fingerings for automatic performance environments [2], learning aid systems [13, 22] and so on. These approaches make use of constraints to achieve this goal. In the situation that we do not know the music scores, *i.e.* that we only have the recorded audio signal, we are interested in discovering both the score and the fingering information. If a system already exists that can discover fingerings based on a score, one can concentrate on the transcription of music signals into scores which can then be passed to the fingering system. Since the audio signals contain sounds which are produced by instruments according to a score (written or not) using specific fingerings which provide us with useful information for the possible combination of notes, we can use the fingering information to assist our transcription. Moreover, fingerings have a constrained nature, thus a constraint-based approach can be applied to chord detection.

This chapter is based around a detailed example of constraint-based chord analysis and tracking, using descriptive constraints based on chordal strumming technique for a standard-tuned guitar with six strings (E4, B3, G3, D3, A2, E2), considering the physical layout of the instrument along with human physical and temporal constrains. The following subsection introduces current research on chord detection in general. Several standard chord-detection techniques will be employed in our constraint-based approach.

1.3   Chord Detection

The harmonic structure of a musical work depicts the content of the music in a high level, illustrating how the music is organized. Many applications in music information retrieval such as semantic analysis of music, finding similar music and segmentation into characteristic parts, require the harmonic structure as a mid-level representation of the musical piece under analysis. The *chord*, which is defined as several notes played simultaneously, is used to represent the harmonic structure in the form of a sequence, or temporal arrangement, of chords depicting the overall structure of the music. Thus the recognition of chords plays an important role in music information retrieval. Many researchers have expended great effort on the chord detection task, and many MIREX[3] tasks begin with chordal analysis. Audio chord detection was a separate task in the 2008 MIREX competition.

The most popular method used for chord detection is a pattern classification approach, which first extracts low-level features describing harmonic content, *e.g.*, pitch class profile (PCP) vectors (introduced in Sect. 2), or, in other words chroma

---

[3] The Music Information Retrieval Evaluation eXchange (http://www.music-ir. org/mirex/2008) is a competition between researchers to compare the accuracy of algorithms written to solve common music information retrieval problems. It takes places as a part of the International Conference on Music Information Retrieval (ISMIR).

vectors, then uses a classifier such as a hidden Markov model (HMM) to perform the recognition. PCP and HMM techniques are described in more detail in Sect. 2.

Ryynänen and Klapuri [19] follow the general method, but instead of one PCP vector, they use two-pitch class profiles, one for low-register notes D1–C♯3 and one for high-register notes D3–C♯5. They argue that the low-register profile captures the bass notes contributing to the chord root whereas the high-register profile has more clear peaks for the major/minor third and the fifth. Instead of estimating chord profiles for all chords, they estimate the profiles only for major and minor triads to prevent the problem of insufficient training data for particular chords. A chord HMM with 24 states is defined, twelve states each for both major and minor triads[4]. The observation likelihoods for each chord are calculated by comparing the low and high-register profiles with the estimated trained profiles. Then Viterbi decoding through the chord HMM produces chord labeling for each analysis frame.

Ellis and Poliner [4] models the distribution of the chroma vector using a single Gaussian Model. They make use of beat tracking, and extract the chroma vector one per beat. Because chords usually change at the beginning of a beat, features extracted using beat boundaries are believed to be more confident. They also use an ergodic[5] HMM with states corresponding to chord labels. It is worth noting that chord changes on beat boundaries is a prescriptive constraint, rather than a descriptive constraint, however, it is a very common composition practice in Western music.

Instead of using an ergodic chord HMM, where the hidden states represent chords and all possible transitions between states are allowed, Khadkevich and Omologo [10] create a separate model for each chord. Their approach also differs from others in that they use 512 Gaussian mixtures (a weighted sum of guassian distributions) representing the chroma vector probabilities rather than the standard 12. In order to prevent difficulties from lack of training data, similar to [19], they also only train 2 models: a major profile and a minor profile. The parameters of the HMMs are obtained using expectation maximization, which iteratively estimates the maximum likelihood estimates of parameters in a probabilistic model.

Weil and Durrieu [23] add a preprocessing step which attenuates the main melody of the musical piece. They believe that the main melody often contains intentionally anharmonic notes. This is an example of breaking a prescriptive constraint, that of ensuring melody notes fall within the notes of the underlying chord structure. While these anharmonic notes are crucial for the perceived richness of the global timbre, they also blur the accompanying harmonies and make chord detection difficult, which is why they attempt to remove these notes with pre-processing. To make the chroma vectors robust, they also estimate the offset of the tuning frequency relative to A4=440Hz[6]. In this system, they use a system of tonal centroid vectors [8],

---

[4] According to the Audio Chord Detection task in MIREX20008, the chord vocabulary for the task is restricted to 12 major triads (Cmaj, C♯maj,...,Bmaj) and 12 minor triads (Cm,C♯m,...,Bm).

[5] not sensitive to initial conditions.

[6] A4 or Concert A is the 440 Hz tone that serves as the standard for musical pitch. A4 is the musical note A above middle C (C4).

where a 6-D vector is obtained from the 12-D PCP vector as features, and an ergodic HMM is trained.

While the abovementioned systems have to be trained to get the HMM parameters, Papadopoulos and Peeters [14, 15] derive the HMM parameters manually, taking into account the presence of high harmonics of pitch notes and some music knowledge. In this approach, no training is needed. They represent each chord profile as a vector which contains the theoretical amplitude values of the notes and their harmonics comprising a specific chord. The observation possibilities are then obtained by computing the correlation between the observation vectors and a set of chord profiles. An ergodic HMM is also used.

Pauwels *et al* [16] use a novel feature extractor which first uses multiple pitch tracking techniques to couple the higher harmonics to their fundamental frequency and then compute the chroma vector from these harmonics. Different from [19], they require that the bass-notes with fundamental frequencies lower than 100Hz are not allowed to contribute to the chroma vector. They argue that although bass-notes could make a significant addition to the chord, which agrees with [19], bass notes are often duplicates of notes in the higher register, or they do not contribute to the chord (*e.g.* a walking bass).

There are many other works that also have great contribution to this problem. Uchiyama *et al* [21] use pre-processing step which eliminates percussive sounds from audio, because percussive sounds are non-harmonic and they interfere with chord detection. Lee [11, 12] builds key-dependent HMMs for chord transcription and key extraction, using an HMM for each of the 24 keys, thus detecting the key and chord sequence concurrently. Bello and Pickens [1] present a system for detecting harmonic content in music signals, using chroma vectors and HMMs.

Although some researchers are not explicitly using the term constraint, it is clear that constraints are a common theme in chord detection research. Playability constraints, physical layout constraints for both the instruments and human abilities, and stylistic constraints have appeared in the literature. The "standard" model for chord detection, that of feature extraction using a chroma-type technique combined with some form of pattern classification system, makes several assumptions that many researchers use because they are considered standard. In the following sections, we will describe a new approach in detail, including many of the low-level details and why certain decisions are appropriate in this domain. Before that, there are a few common terms in the chord detection research field that should be explained.

2   Term Explanation

Several concepts in the area of chord recognition are well known to researchers immersed in the work but may not be as familiar to occasional readers. We present here a detailed description of some of these concepts. Readers familiar with chord recognition and pattern analysis techniques may be inclined to bypass this section.

2.1   Pitch Class Profile (PCP) Vector

A played note gives us many perceptual properties: one is pitch, which corresponds to the frequency of the note: the higher the frequency, the higher the pitch. The second property, related to pitch, is the note name regardless of octave, and corresponds to one of the twelve pitch classes. This property can be termed the "color" of the note. Although chords are made up of notes perceived as absolute pitches, the sensation of a chord is more often that of a coherent sound, with the color property of the notes being dominant over the pitch. Thus, "color" can be used as a term to represent a chord based on its root, corresponding to one of the twelve pitch classes, *i.e.* C, C♯/D♭, D, D♯/E♭, E, F, F♯/G♭, G, G♯/A♭, A, A♯/B♭, B. If we know the power distribution of a chord in each of the twelve pitch classes, which can be represented as a twelve-dimensional vector, the color of a chord can be quantitatively represented. This vector is called the chroma vector or Pitch Class Profile (PCP) vector, which maps the notes in several octaves into 12 bins of pitch classes. The PCP vector technique was first proposed by Fujishima [5] in 1999 for the representation of audio and it is widely used today to represent the features of a chord for analysis and classification.

2.2   Artificial Neural Networks

An artificial neural network (ANN), often just called neural network for short, is a mathematical model simulating biological networks of neurons. It is composed of a group of interconnected processing elements. Each such "neuron" is in fact a function, taking some input and producing an output based on the input, usually as a type of summative threshold function. A nerual network can be used for pattern classification [3] through a learning process. The most common type of ANN used for pattern classification is a three-layer feedforward network. Feedforward means connections between units do not form a directed cycle. The information moves in only one direction, forward from the input layer to the output layer through the hidden layer. Although a complete description of ANNs is beyond the scope of this work, the interested reader will find a complete description in many pattern recognition or information theory textbooks, including [3].

2.3   Hidden Markov Models and the Viterbi Algorithm

Hidden Markov models are the pattern recognition technique most commonly used for detecting patterns in temporal sequences. The sequence of chords in a musical piece is an example of a temporal sequence. HMMs work by building a model of an underlying system the states of which are unobserved (hidden) but which produces a series of observations based on the internal hidden states. HMMs are used for classification by lining up an observed sequence with the possible observed sequences generated by the model. Often, a number of HMMs are compared and the one most likely to have produced the observed sequence is judged to best model the internal structure of the system being classified.

The Viterbi algorithm is a standard process for finding the sequence of hidden states with the highest likelihood in a particular HMM. Given a set of hidden states and transition probabilities between them, the Viterbi algorithm proceeds by finding the most likely state at each time increment, and maintaining a list of the most likely historical sequences of states that would lead to the current state. Again, a complete description of both HMMs and the Viterbi algorithm is beyond the scope of this work, and interested readers are encouraged to consult [17] for details.

3   Chord Analysis in Guitar Music

The remainder of the chapter will present, in detail, the analysis and classification of guitar chords using constraints based on the physical construction of the instrument. The chords in the experiments are played in an acoustic guitar. We call this technique "voicing constraints" because it can identify different chord voicings based on constraints of the instrument and chord voicings which might be less easy to play or even impossible.

Our approach deviates from the approaches presented in the introduction section in several key areas. The use of voicing constraints (described below) is the primary and fundamental difference, but our low-level analysis is also somewhat different from current work. First, current techniques will often combine PCP with hidden Markov models. Our approach analyzes the PCP vector using Neural Networks since Neural Networks are a more stable method and are more capable of capturing the probability distribution of the PCP vectors than a single Gaussian or a Gaussian Mixture Model (GMM) and using a pseudo-Viterbi algorithm to model chord sequences in time. Second, current techniques normally use small window size. Our technique makes use of comparatively large window sizes (500ms). The description and justification of these methods is presented in the following sections. Although the constraints and system development are based on guitar music, similar constraints (with different values) may be determined for other ensemble music.

3.1   Large Window Segmentation

Choosing the size of the analysis window for feature extraction is always a challenge. Small windows are able to localize higher-frequency events in time, while larger windows can localize longer-time events in frequency. Depending on the application, a larger or smaller window may be appropriate, but it should be noted that no window size is appropriate for all applications. A type of "uncertainty principle" exists between time-localization and frequency-localization: in order to know the exact instant when an event takes place, down to the sample, one cannot know anything about the frequency content of the event, since the event consists of a single sample. Likewise, in order to fully analyze the frequency content of an event, the event must be of infinite length[7]. In chord recognition, some constraints can be identified which will allow us to use a larger window than most researchers do, and therefore produce a more detailed frequency analysis.

Guitar music varies widely, but common popular guitar music maintains a tempo of 80–120 beats per minute. Because chord changes typically happen on the beat or on beat fractions (a prescriptive constraint as mentioned earlier), and because of physical limitations of the way guitar chords are played, we can see that statistically, the time between chord onsets is typically 600–750 ms. Segmenting guitar chords is not a difficult problem, since the onset energy is large compared to the release energy of the previous chord. The chord pitch class profile pattern does not change from onset to onset (although local relative changes will occur), even taking into account effects such as slides and hammer-ons, since those would produce small but measurable onsets themselves. Because of this, the entire signal from one onset to the next can be taken as a single frame when calculating the pitch class. This results in a more accurate pitch class analysis than for small window sizes. Further, using a larger window like this has the effect of blurring the analysis of percussive (fast, high-frequency) events, which makes them easier to ignore.

Experimentation has shown that onset detection, while a useful addition to the algorithm, is not entirely necessary. Universal 500ms frames provide sufficient accuracy when applied to guitar chords for a number of reasons. First, if a chord change happens near a frame boundary, the chord will be correctly detected because the majority of the frame is a single pitch class profile, as shown in Fig. 1. If the chord change happens in the middle of the frame, the chord will be incorrectly identified because contributions from the previous chord will contaminate the reading. However, if sufficient overlap between frames is employed (*e.g.* 75%), then only one in four chord readings will be inaccurate, and the chord sequence rectifier (see Sect. 3.4) will take care of the erroneous measure.

The advantage of the large window size is the accuracy of the pitch class profile analysis, and, combined with the chord sequence rectifier, this outweighs the possible drawbacks of incorrect analysis when a chord boundary is in the middle of a frame. The disadvantage of such a large window is that it makes real-time processing impossible. At best, the system will be able to provide a result half a second after a note is played. Offline processing speed will not be affected, however, and will be comparable to other frame sizes. In our experience, real-time guitar chord detection is not a problem for which there are many real-world applications.

3.2 PCP with Neural Networks

We have employed an Artificial Neural Network to analyze and characterize the pitch class profile vector and detect the corresponding chord. A network was first constructed to recognize seven common chords for music in the keys of C and G, for which the target chord classes are [C, Dm, Em, F, G, Am, D]). It is common practice

---

[7] Theoretically, Fourier analysis requires an infinite-length signal, however, in practice, we add 0 outside.

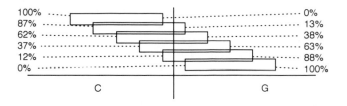

Fig. 1 Large frames at 75% overlap across a chord boundary. Frames that cross the boundary will either be dominated by one chord and successfully recognized, or contain similar contributions from both chords and possibly be mis-recognized.

for chord detection systems under development to begin with this or a similar subset of all possible chords. These chords were chosen as common chords for "easy" guitar songs. The network architecture was set up in the following manner:

- 1 input layer, 2 hidden layers, and 1 output layer;
- 12 input cells corresponding to 12 elements the of PCP vector;
- 10 cells for each of the hidden layer;
- 7 output cells corresponding to the 7 chosen chords.

With the encouraging results from this initial problem (described in Sect. 4), the vocabulary of the system was expanded to recognize chords in the seven roots (C, D, E, F, G, A, B). The system was trained to recognize Major (I, III, V), Minor (I, iii, V) and Seventh (I, III, V, vii) chords for each key, totaling 21 chords. Recognition rates were lower than with the seven-chord system, as may be expected, but still very good. The complete set of major, minor and seventh chords for all 12 chord roots would include 36 chords. With the multitude of complex and colorful chords available, it is unclear whether it is possible to have a "complete" chord recognition system which uses specific chords as recognition targets, however a limit of 4-note chords would provide a reasonably complete and functional system[8].

3.3 Supervised Training with Real-World Data

Unlike Gagnon, Larouche and Lefebvre [6], who use synthetic chords to train the network, we use real recordings of chords played on a guitar. In the research discussed in the introduction section which adopts HMMs as the classifier, the training material is a labelled chord sequence, including many chords in one training sample, typically a song with the chord labels. However, our system is trained using separate chords, *i.e.* one training sample contains only one chord. A guitar is connected to the audio input of the computer, and chords are recorded using Audacity[9], an open source audio editor. These recorded chords are then labeled and fed into the system as training data.

---

[8] Only Major, Minor and Seventh Chords are considered in this system.

[9] http://audacity.sourceforge.net/

3.4   Sequence Rectifier

Chord recognition rarely operates on a single isolated chord. This is primarily a prescriptive constraint, since there is no physical reason why any chord can't follow any other. The only descriptive aspect of this constraint is the time it takes to move from one chord to another. Chord transitions that are faster that the human hand can move would be disallowed. Some chord transitions are easier to make, for example, while maintaining a common finger position, and some transitions are unlikely to happen, for example briefly switching to a relative minor chord and then to the subdominant chord. These constraints, combined with a measure of the certainty of the initial chord detection result, can increase the recognition accuracy of individual chords.

After the chord detection system is trained, it can be used to classify the chord for each frame. Because the frame length is usually smaller than the time interval during which a chord is played, we are likely to have several instances of the same chord recognized in a sequence. For example, if a C chord is played for 1 second, and assuming a 500ms frame overlapped at 75% as justified above, the correct output should contain 6 instances of the C chord. If a chord boundary falls in the middle of one of the frames, or if noise or other difficulties are present, the network might erroneously produce an Am chord (for example) instead of a C chord. Based on the confidence level of the chord recognition as well as changes in analyzed feature vectors from one frame to the next, we construct a *sequence rectifier* which will select the second-most-likely chord if it fits better into the sequence. In this way, the rectifier improves the overall accuracy of the system.

For each frame, the neural network gives a rank-ordered list of the possible chord candidates, each with a confidence value in the range of [0 1]. The sequence rectifier algorithm is:

1. Estimate the chord transition possibilities for each key pair (Major and relative minor) through large musical database.
2. The Neural Network provides a matrix S, which has N rows and T columns. Each column gives the chord candidates with ranking values for each frame. N is the size of the chord dictionary. T is the number of frames.
3. Based on the first row of matrix S, calculate the most probable key for the entire piece of music. For the 24 possible keys, the key corresponding to the maximum number of the chords in the first row of S wins. This is an example of *key finding*, another common MIREX task.
4. Using the estimated key, construct the transition matrix A from step 1.
5. Calculate the best sequence from S and A using the Viterbi Algorithm.

3.5   Voicing Constraints

Many chord recognition systems assume a generic chord structure with any note combination as a potential match, or assume a chord "chromaticity", assuming all chords of a specific root and "color" (see Sect. 2.1) are the same chord. For example,

Table 1 Typical note ranges for SATB choir

Voice	Range
Soprano	C4-C6
Alto	E3-E5
Tenor	C3-C5
Bass	C2-C4

a system like this would identify [C4-E4-G4] as identical to [E4-G4-C5], the first inversion[10] of the C Major triad. Although these chords have the same chromaticity (all contain C, E, and G components), they will sound different to the ear. A system which not only identifies [C-E-G] as a C Major triad, but also can identify a unique C Major triad depending on whether the first note was middle C (C4) or C above middle C (C5), would be preferable in most circumstances.

For a chord, allowing *any* combination of notes regardless of the *voicings*[11] provides too many similar categories which are difficult to disambiguate, and allowing a single category for all versions of a chord does not provide complete information since equivalent chords in different octaves are not disambiguated. What is necessary, then, is a compromise which takes into account statistical, musical, and physical likelihood constraints for chord patterns.

The goal of our system is to constrain the available chords to the common voicings available to a specific instrument or set of instruments. The experiments that follow concentrate on guitar chords, but the technique would be equally applicable to any instrument or ensemble where there are specific constraints on each note-producing component. As an example, consider a SATB choir, with typical note ranges as shown in Table 1.

In this example, then, some chord voicings are more likely than others, depending on the key of the piece, the chord progression, and the melodic movement. Further, compositional practice (a prescriptive constraint) means that depending on the musical context, certain voicings may be more common, for example, it is common compositional practice to have the Bass singing the root (I), Tenor singing the fifth (V), Alto singing the third (III or iii) and Soprano doubling the root (I) when the chord being sung is the root of the key.

This *a priori* knowledge can be combined with statistical likelihood based on measurement to create a Bayesian analysis resulting in greater classification accuracy using fewer classification categories. A similar analysis can be performed on any well-constrained ensemble, for example a string quartet, and on any single instrument with multiple variable sound sources, for example a piano. At first, the

---

[10] An *inversion* of a chord is an arrangement of notes where the triad begins with the root (root position), the third (first inversion), or the fifth(second inversion).

[11] A chord *voicing* is a specific way of arranging the notes which make up the chord. An inversion is a special case of a voicing.

piano does not seem to benefit from this method, since any combination of notes is possible, and likelihoods are initially equal. However, if one considers musical expectation and human physiology (hand-span, for example), then similar voicing constraints may be constructed.

One can argue that knowledge of the ensemble may not be reasonable *a priori* information—will we really know if the music is being played by a wind ensemble or a choir? The assumption of a specific ensemble is a limiting factor, but is not unreasonable: metadata may tell us the instrumentation, and timbre analysis methods can be applied to detect whether or not the music is being played by an ensemble known to the system, and if not, PCP combined with Neural Networks can provide a reasonable chord approximation without voicing or specific note information.

For a chord played by a standard 6-string guitar, we are interested in two features: what chord is it and what voicing of that chord is being used. The PCP vector describes the chromaticity of a chord, hence it does not give any information on specific pitches present in the chord. Given knowledge of the relationships between the guitar strings, however, the voicings can be inferred based the *voicing vectors* (VV) in a certain category. VVs are produced by studying and analyzing the physical, musical and statistical constraints (both prescriptive and descriptive) on an ensemble. Here, the process was performed manually for the guitar chord recognition system but could be automated based on large annotated musical databases.

Thus the problem can be divided into two steps: determine the category of the chord, and then determine the voicing. The chord category is determined using the PCP vector combined with Artificial Neural Networks, as described previously. Chord voicings are determined by matching harmonic partials in the original waveform (extracted using Fourier or Constant-Q transforms, for example) to a set of voicing templates.

When the chord is strummed, it is possible that not all the strings are sounded. For example, we may strum the first 4 strings or the middle 3 strings in a chord. Because it is impossible to identify which strings may be missing in a particular chord, we must take into account that the VVs against which we are matching may be missing one or more feature values. This kind of problem can be described as pattern recognition with incomplete feature vectors. Standard methods are available for this type of problem.

4 Guitar Chord Recognition System

The general chord recognition ideas presented above have been implemented here for guitar chords. Figure 2 provides a flowchart for the system. The feature extractor provides two feature vectors: a PCP vector which is fed to the input layer of the neural net, and a voicing vector which is fed to the voicing detector. The rectifier corrects the errors, which are marked in purple.

Table 2 gives an example of the set of chord voicing arrays and the way they are used for analysis. The fundamental frequency (f_0) of the root note is presented along with the f_0 for higher strings as multiples of the root f_0.

Table 2 Chord pattern array, including three forms of five of the natural-root chords in their first voicings (root always the lowest). S1–S6 are the relative f_0 of the notes from the lowest to highest string, and H1–H6 are the first harmonic partial of those notes. For example, in the E chord, String 2 plays the $\approx$123.6Hz B ($1.5 \times f_0$), and the first harmonic is twice that($3 \times f_0$). See text for further explanation of boxes and symbols.

Chord / f_0, in Hz	S1 / H1	S2 / H2	S3 / H3	S4 / H4	S5 / H5	S6 / H6
E	1	1.5	2	2.52	3	4
82.4	2	3	4	φ	φ	φ
Em	1	1.5	2	2.38	3	4
82.4	2	3	4	φ	φ	φ
E7	1	1.5	1.78	2.52	3	4
82.4	[2]	3	3.56	φ	φ	φ
F	1	1.5	2	2.52	3	4
87.31	2	3	4	φ	φ	φ
Fm	1	1.5	2	2.38	3	4
87.31	2	3	4	φ	φ	φ
F7	1	1.5	1.78	2.52	3	4
87.31	[2]	3	[3.56]	φ	φ	φ
G	1	1.26	1.50	2	2.52	4
98	2	2.52	[3]	4	φ	φ
Gm	1	1.19	1.50	2	3	4
98	2	[2.38]	3	4	φ	φ
G7	1	1.26	1.50	2	2.52	3.56
98	2	2.52	φ	φ	φ	φ
A	—	1	1.5	2	2.52	3
110		2	3	φ	φ	φ
Am	—	1	1.5	2	2.38	3
110		2	3	φ	φ	φ
A7	—	1	1.5	1.78	2.52	3
110		[2]	3	φ	φ	φ
B	—	1	1.5	2	2.52	3
123.5		2	3	φ	φ	φ
Bm	—	1	1.5	2	2.38	3
123.5		2	3	φ	φ	φ
B7	—	1	1.26	1.78	2	3
123.5		2	[2.52]	φ	φ	φ
C	—	1	1.26	1.50	2	2.52
130.8		2	2.52	φ	φ	φ
Cm	—	1	1.19	1.5	2	—
130.8		2	φ	φ	φ	φ
C7	—	1	1.26	1.78	2	2.52
130.8		2	2.52	φ	φ	φ
D	—	—	1	1.5	2	2.52
146.8			2	φ	φ	φ
Dm	—	—	1	1.5	2	2.38
146.8			2	φ	φ	φ
D7	—	—	1	1.5	1.78	2.52
146.8			[2]	φ	φ	φ

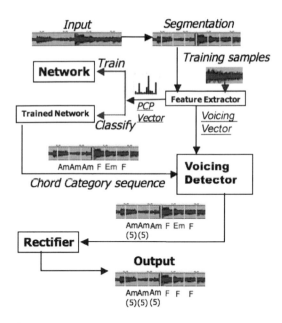

Fig. 2 Flowchart of the chord recognition system. Note: the yellow lines showing segmentation do not show overlap, but overlap is done as shown in Fig. 1.

The Guitar has a note range from E2 (82.41Hz, open low string) to C6 (1046.5Hz, 20th fret on the highest string). Guitar chords that are above the 10th fret of which the highest note is D5 are rare, thus we can restrict the chord position to be lower than the 10th fret, that is, the highest note would be 10th fret on the top string, *i.e.* D5, with a frequency of 587.3Hz. Thus if we only consider the frequency components lower than 600Hz, the effect of the high harmonic partials would be eliminated. And because the guitar has only six strings, if the six strings are all strummed, we only have 6 fundamental frequencies. The relationship between fundamental frequencies of the notes in the chord can be used to identify the voicing.

Each chord entry in Table 2 provides both the frequencies of all sounded notes on each string (or an indication that the string is not played) as well as an indication of the first harmonic of each note. "Standard" chords such as Major, Minor and Seventh, contain notes for which f_0 is equal to the frequency of harmonic partials of lower notes, providing consonance and a sense of harmonic relationship. Indeed, this is why these chords are pleasant to hear. This can be seen as a liability, since complete harmonic series are obscured by overlap from harmonically related notes. Current systems attempt to overcome this by reinforcing, re-interpolating or duplicating harmonics, but our system takes advantage of this by observing that a specific pattern of harmonic partials equates directly to a specific voicing of a chord. Table 2 shows this by detailing the pattern of string frequencies and first harmonic partials for the root voicings of these chords. Harmonic partials above 600Hz are ignored, since there is no possibility to overlap the fundamental frequency of higher notes (as described above). These are indicated by the symbol "ϕ" In this way, we construct a pattern of components that are expected to be present in a specific chord as played on the guitar, and similarly for other voicings.

4.1 Harmonic Coefficients and Exceptions

If we make an assumption that the lowest sounded note on the guitar is the root of the chord (which is prescriptive but not unreasonable), it can be seen from Table 2 that there are three main categories of chords on the guitar, based on the frequency of the second note in the chord. The patterns for the three categories are:

- (1.5), where the second note is (V): F, Fm, F7, E, Em, E7, A, Am, A7, B, Bm, D, Dm, D7
- (1.26), where the second note is (III): B7, C7, G, G7
- (1.19), where the second note is (iii): Cm, Gm.

Thus, from the first coefficient (the ratio of the first harmonic peak to the second) we can identify which group a certain chord belongs to. After identifying the group, we can use other coefficients to distinguish the particular chord.

In some situations (*e.g.*, F and E; A and B), the coefficients are identical for all notes in the chord, thus the chords themselves cannot be distinguished in this manner. Here, the chord result will be disambiguated based on the result of the Neural Network and the f_0 analysis of the root note. Usually, all first harmonic partials line up with f_0 of higher notes in the chord. When the first harmonic falls between f_0 of higher notes in the chord, they are indicated by $\boxed{\text{boxed}}$ coefficients.

Underlined coefficients correspond to values which may be used in the unique identification of chords. In these cases, there are common notes within a generic chord pattern, for example the root (1) and the fifth (1.5). String frequencies corresponding to the Minor Third (1.19, 2.38) and Minor Seventh (2.78) are the single unique identifiers between chord categories in many cases.

4.2 Feature Extractor

The feature extractor in Fig. 2 includes two parts: a PCP extractor and a voicing extractor.

4.2.1 PCP Extractor

As introduced in Sect. 2, a PCP vector has 12 dimensions, corresponding to the 12 pitch classes. The value for each dimension represents the energy in the corresponding pitch class, and this value is usually normalized with the largest value equal to 1. PCP begins from a frequency representation of the audio of a chord, for example, in our implementation, we use the fast Fourier transform (FFT). To calculate the PCP vector, first we use FFT to get the frequency components of the frame and then map the frequency components into the 12-bin pitch classes. After the frequency components have been calculated, we get the corresponding notes of each frequency component and find its corresponding pitch class. Then we add the power of each frequency component to the corresponding pitch class. In this way, PCP provides a profile of the frequency components in an audio signal, regardless of octave.

4.2.2 Voicing Extractor

The voicing extractor uses FFT to get the frequency components for each frame, and then a peak finding algorithm[12] is used to find the most evident peaks from the frequency components that are lower than 600Hz. After the peak frequencies are obtained, the lowest frequency among them is selected as the bass note and then the voicing vector is obtained by dividing all the frequency component by the bass frequency.

5 Accuracy and Obstacles

In this section we describe the accuracy of this implementation of a chord constraints system, taking into account the relative severity of common chord errors, and we compare this system to an off-the-shelf system.

5.1 *Common Chord Errors*

It is important to recognize that chord detection errors do not all have the same level of what can be called "severity". A major chord (*e.g.* C) may be recognized as the relative minor of the same chord (*e.g.* Am) since they are based around the same set of accidentals, and many of the harmonic partials are the same because they share two notes. In many musical situations, although the Am chord is incorrect, it will not produce dissonance if played with a C chord. This can be seen as analogous to the problem of octave errors in pitch detection, where the pitch is incorrect but would not produce dissonance if played with the correct note. Relative Minor chords are perceptually more similar to a Major chord than is the corresponding same-root Minor chord. Mistaking an F chord for an Fm chord, for example, is a significant problem. Although the chords again differ only by one note, the note in question differs in more harmonic partials. Further, it establishes the mode of the scale being used, and if played at the same time as the opposing mode, will produce dissonance.

5.2 *Comparison*

Chord Pickout[13] is a popular off-the-shelf chord recognition system. Although the algorithm used in the Chord Pickout system is not described in detail by the

---

[12] http://terpconnect.umd.edu/ toh/spectrum/PeakFindingandMeasurement.htm
[13] http://www.chordpickout.com/

developers, it is not unreasonable to make a comparison with our system since Chord Pickout is a commercial system with good reviews. We applied the same recordings to both systems and identified the accuracy of each system. We were more forgiving with the analysis for Chord Pickout in order to better detail the types of errors that were made. If Chord Pickout was able to identify the root of the chord, ignoring Major, Minor or Seventh, it is described as "correct root". If the chord and the chord type are both correct, it is described as "correct chord". Inconsistencies between the correct root and correct chord include Major-Minor, Major-Seventh, and Minor-Major confusions.

We also make a comparison with *CLAM Chorddata*[14], which implements a chord detector using the algorithm proposed by Christopher Harte [7]. For each frame, CLAM gives several chord candidates. If the correct chord is contained in the candidates, we regard this as a correct recognition.

For our system, all chord errors are treated as incorrect, regardless of severity. The complete results for 5 trials are presented in Table 3. The results of the inversion constraints are of the same order of magnitude as other modern techniques, which is encouraging but not particularly impressive. It is important to keep in mind that this new system is also able to determine specific voicings of a chord, as will be described below. The main result is that we are able to identify these voicings while maintaining overall chord detection accuracy.

Table 3 Comparison of our Inversion Constraints system to *CLAM Music Annotator* and *Chord Pickout*

		Inversion Constraints		CLAM		Chord Pickout			
Trial	Frames	Correct	Rate	Correct	Rate	Root	Rate	Chord	Rate
1	281	255	90.8%	240	85.4%	190	67.6%	42	14.9%
2	322	286	88.8%	301	93.4%	172	53.4%	72	22.4%
3	405	356	88.0%	368	90.8%	225	55.6%	56	13.8%
4	466	396	84.9%	410	87.9%	293	62.9%	50	10.7%
5	472	387	81.9%	403	85.3%	321	68.0%	101	21.4%

5.3 Independent Accuracy Trials

To evaluate the overall accuracy of our system, independent of a comparison with other systems, we presented a set of chord exemplars (50 of each type) to the system and evaluated its recognition accuracy. Two systems were trained for specific subsets of chord detection, and the results are presented in three tables. The first set of results, presented in Table 4 shows the recognition accuracy of a system trained to detect chords appropriate to the key of C and D, as discussed above. The system used seven chord classification targets, and produced 93.2% accuracy over all trials. Misclassifications in this case were normally toward adjacent chords in the scale.

---

[14] http://clam.iua.upf.edu/index.html

Table 4 Recognition results for common chords. Overall accuracy is 93.2%

Chord	C	Dm	Em	F	G	Am	D
Rate	50/50	48/50	43/50	50/50	48/50	42/50	45/50

The second system was trained to recognize Major, Minor and Seventh chords of all seven natural-root keys, resulting in 21 chord classification targets. Classification results are presented for Major versus Minor comparisons as shown in Table 5, which produce good results (86.8% accuracy). The most common errors were between a Major chord and its relative Minor, although errors between the Major chord and the same-root Minor were also detected. Table 6 provides a confusion matrix for chord recognition between Major and Seventh chords. The accuracy is reduced in these results, for two reasons: in some cases the first three notes (and correspondingly the first three harmonic partials detected) are the same between a chord and its corresponding Seventh. Also, in some cases the first harmonic of the root note does not line up with the fundamental frequency of a note an octave above, and thus contributes to the confusion of the algorithm.

Table 5 Recognition results for natural-root Major and Minor chords. Overall accuracy is 86.8%

Chord	C	Cm	D	Dm	E	Em
Rate	50/50	39/50	48/50	41/50	46/50	38/50

Chord	F	Fm	G	Gm	A	Am
Rate	47/50	42/50	50/50	36/50	49/50	35/50

Recognition accuracy is higher for the Major chords and lower for Seventh chords. Taking E7 for example, there are 9 samples where the E7 chord is recognized as E Major. Examining Table 2, the voicing vectors for E Major and E7 are $[1, 1.5, 2, 2.52, 3, 4]$ and $[1, 1.5, 1.78, 2.52, 3, 4]$. Since the first element in E7 produces a harmonic at twice the fundamental, the detected vector is $[1, 1.5, 1.78, 2, 2.52, 3, 4]$. If the third element 1.78 is too weak for the peak picking algorithm to detect, we may erroneously detect $[1, 1.5, 2, 2.52, 3, 4]$ instead, which is exactly the voicing vector for E major. Since E Major and E7 have the same root, we recognize E7 as an E Major in this situation.

For the B7 chord, in addition to recognizing it as the corresponding Major chord, 20% are recognized as C7, one semitone above B7. The voicing vector for B7 is $[1, 1.26, 1.78, 2, 3]$ in which the second element 1.26 produces a second harmonic peak at 2.52, leading to the array $[1, 1.26, 1.78, 2, 2.52, 3]$, the first five elements of which are the same as that of C7. Moreover, since there a small difference $(130.8Hz - 123.5Hz = 7.3Hz)$ between the root f_0 of B7 and C7, if the guitar is tuned slightly higher, B7 might be recognized wrongly as C7. And we notice

Table 6 Confusion Matrix for Major and Seventh chords of natural-root keys. Overall accuracy is 78.6%.

chord was recognized as:

Chord	Rate	C	C7	D	D7	E	E7	F	F7	G	G7	A	A7	B	B7
C	50/50	50													
C7	35/50	12	35												3
D	50/50			50											
D7	26/50			14	26			3	6	1					
E	45/50					45	2	3							
E7	26/50					9	26	3	12						
F	48/50					1	1	48							
F7	32/50					2	2	14	32						
G	50/50									50					
G7	35/50									15	35				
A	48/50											48	2		
A7	31/50	2										17	31		
B	45/50								2					45	3
B7	29/50			10										11	29

the same error happens for E7, where the difference between the frequencies of the roots of E7 and F7 is 4.91 Hz. This indicates that correct tuning as well as correct peak identification are significant problems for chord recognition, and will manifest more obviously in chords which differ by a semitone (B-C and E-F) than chords which differ by a tone.

Another difficult case is with D7, which contains only 4 notes (the two lowest notes are not sounded for D chords in the root position), and the first note produces a harmonic that does not correspond to a higher played string. From Table 2, we can see that the string f_0 multiplier pattern for D7 is [1, 1.5, 1.78, 2.52], and the first harmonic partial of the root note inserts a 2 into the sequence, producing [1, 1.5, 1.78, 2, 2.52] for the sequence of harmonics within the range [82.4 Hz–600 Hz] as previously justified. This is very similar to the sequence for F7, which is why the patterns are confused. It would be beneficial, in this case, to increase the weight ascribed to the fundamental frequency when the number of strings played is small. Unfortunately, detecting the number of sounded strings in a chord is a difficult task. Instead, f_0 disambiguation can be applied when a chord with fewer strings is one of the top candidates from the table, if that information is known. Further, the confusion matrix adds to the measure of certainty of the chord recognition, indicating when other methods should be employed to increase confidence.

5.4   Recognition of Voicings

After the chord labels are recognized, our system then gives the voicing information. For the guitar chords, different voicings correspond to different hand positions on the guitar neck. Figure 3 shows 3 voicings of the C Major chord and 3 of the F

Major chord, which correspond to positions that are usually used by guitar players. An empty circle means an open string while a filled circle means the string is pressed on the corresponding fret. Numbers in the left-top represent the fret. For example, "'5'" denotes the 5th fret. If there is no number, it means the first fret by default.

It is reasonable to assume that all the strings involved in the chord are played, but sometimes the bass note might be ignored. Thus we present 2 voicing vectors for a particular voicing, the first one including all the strings and the second one with the bass note missed. For voicing C1, the first voicing vector is [1 1.33 1.68 2 2.67 3.36] with a bass frequency 98 Hz. The second voicing vector is [1.33 1.68 2 2.67 3.36]. To keep a constant no. of elements in the voicing vector, we add a sixth element "4", which is caused by the third harmonic of 1.33 and the second harmonic of 2. Normalizing it, we get [1 1.26 1.5 2.0 2.52 3] with the bass frequency $98 \times 1.33 = 130.8$ Hz. Table 7 shows the voicing vectors for the C Major and F Major Voicings.

A total of 55 recorded samples played by 3 persons for each voicing were recorded and tested, the strumming style was used and they were not told whether to play the bass note or not. Table 8 and Table 9 show the confusion matrix for the disambiguation of the voicings for C Major and F Major respectively. The system

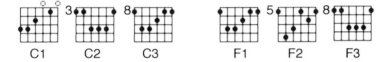

Fig. 3 Three different possible voicings of the C Major and F Major chords

Table 7 Voicing vectors For C Major voicings and F Major voicings, including root-played and root-missing

Voicing	Bass (Hz)	Voicing Vectors					
C1	98	1	1.33	1.68	2	2.67	3.36
	130.8	1	1.26	1.5	2	2.52	3
C2	98	1	1.33	2	2.67	3.36	4
	130.8	1	1.5	2	2.52	3	4
C3	130.8	1	1.5	2	2.52	3	4
	196	1	1.33	1.68	2	2.67	3
F1	87.3	1	1.5	2	2.52	3	4
	130.8	1	1.33	1.68	2	2.67	3
F2	110	1	1.59	2	2.38	3.17	4
	174.6	1	1.26	1.5	2	2.52	3
F3	130.8	1	1.33	2	2.68	3.36	4
	174.6	1	1.5	2	2.52	3	4

Table 8 Confusion Matrix for Disambiguating C Major Voicings. Multiple recognitions are possible.

recognized as:

Voicing	C1	C2	C3
C1	48	15	1
C2	45	55	24
C3	0	52	55

Table 9 Confusion Matrix for Disambiguating F Major Voicings. Multiple recognitions are possible.

recognized as:

Voicing	F1	F2	F3
F1	51	4	30
F2	0	55	0
F3	2	5	55

compares the detected voicing vector with the vectors in Table 7 and recommends the most likely voicings. The sum of each row in the confusion matrices is larger than the number of testing samples because the system can recommend more than one voicing in certain circumstances. Table 8 shows that the system is able to recommend the right voicings at a rate of $(48 + 55 + 55)/3 \times 55$=95.7%, though there are extra recommendations. There are many circumstances where C2 is recognized as C1 because of the similarity between the second vector of these voicings and the same bass notes. Also, many C3 chords are recognized as C2 because the first vector for C3 is the same as the second vector for C2. Table 9 shows that the system works better for disambiguating F Major voicings, with a correct recommendation rate of $(51 + 55 + 55)/(3 \times 55)$=97.6% and fewer overlapping incorrect recommendations.

6 Conclusions

When performing music information retrieval on recorded audio, applying constraints can greatly reduce the number of candidate recognition results which must be considered. Such constraints can be either prescriptive or descriptive. Descriptive constraints usually indicate physical constraints of the player or instrument which are quite unlikely to be broken. Prescriptive constraints usually indicate stylistic or cultural choices which can be broken but are statistically relevant.

Chord analysis in particular can benefit from applying physical constraints. Not every chord can be played by a specific polyphonic instrument or monophonic instrument ensemble. Depending on the key, style, instrumentation and other parameters, certain chord candidates can be discarded.

Current chord analysis techniques often disregard specific note information in favor of a chord color or, in other words pitch class profile technique. Pitch class profiles cannot disambiguate between inversions or voicing of a single chord, nor can they identify where in the musical range a chord may have been played. By enumerating possible and common chord voicings, it is possible to improve standard Pitch class profile techniques by identifying the chord voicing.

These techniques are demonstrated in a chord detection system we developed which makes use of voicing constraints to increase accuracy of chord and chord sequence identification. Although the system is developed for guitar chords specifically, similar analysis could be performed to apply these techniques to other constrained ensembles such as choirs or string, wind, or brass ensembles, where specific chords are more likely to appear in a particular voicing given the constraints of the group.

References

1. Bello, J.P., Pickens, J.: A robust mid-level representation for harmonic content in music signals. In: Proceedings of the 6th International Conference on Music Information Retrieval (ISMIR 2005), London, UK (September 2005)
2. Cabral, G., Zanforlin, I., Lima, R., Santana, H., Ramalho, G.: Playing along with d'Accord guitar. In: Proceedings of the 8th Brazilian Symposium on Computer Music (2001)
3. Duda, R.O., Hart, P.E., Stork, D.G.: Pattern Classification. Wiley-Interscience Publication, Hoboken (2000)
4. Ellis, D.P.W., Poliner, G.E.: Identifying 'cover songs' with chroma features and dynamic programming beat tracking. In: Proceedings of the IEEE International Conference on Acoustics, Speech and Signal Processing, ICASSP 2007, April 15-20, vol. 4, pp. IV–1429–IV–1432 (2007)
5. Fujishima, T.: Real-time chord recognition of musical sound: A system using Common Lisp Music. In: Proceedings of the International Computer Music Conference, Beijing, China, pp. 464–467 (1999)
6. Gagnon, T., Larouche, S., Lefebvre, R.: A neural network approach for preclassification in musical chords recognition. In: Conference Record of the Thirty-Seventh Asilomar Conference on Signals, Systems and Computers, November 9-12, vol. 2, pp. 2106–2109 (2003)
7. Harte, C.A., Sandler, M.: Automatic chord identification using a quantised chromagram. In: Proc. of the 118th Convention of the AES (2005)
8. Harte, C., Sandler, M., Gasser, M.: Detecting harmonic change in musical audio. In: AMCMM 2006: Proceedings of the 1st ACM workshop on Audio and music computing multimedia, pp. 21–26. ACM, New York (2006)
9. Kasimi, A., Nichols, E., Raphael, C.: A simple algorithm for automatic generation of polyphonic piano fingerings. In: Proceedings of the 8th International Conference on Music Information Retrieval, pp. 355–356 (2007)
10. Khadkevich, M., Omologo, M.: Mirex audio chord detection (abstract). In: Music Information Retrieval Exchange (2008), http://www.music-ir.org/mirex/2008/abs/khadkevich_omologo_final.pdf (accessed January 2008)

11. Lee, K.: A System for Acoustic Chord Transcription and Key Extraction from Audio Using Hidden Markov Models Trained on Synthesized Audio. PhD thesis, Stanford University (March 2008)
12. Lee, K., Slaney, M.: Acoustic chord transcription and key extraction from audio using key dependent HMMs trained on synthesized audio. IEEE Transactions on Audio, Speech, and Language Processing 16(2), 291–301 (2008)
13. Miura, M., Hirota, I., Hama, N., Yanagida, M.: Constructing a system for finger-position determination and tablature generation for playing melodies on guitars. Syst. Comput. Japan 35(6), 10–19 (2004)
14. Papadopoulos, H., Peeters, G.: Large-scale study of chord estimation algorithms based on chroma representation and HMM. In: Proceedings of the International Workshop on Content-Based Multimedia Indexing, CBMI 2007, June 25-27, pp. 53–60 (2007)
15. Papadopoulos, H., Peeters, G.: Simultaneous estimation of chord progression and downbeats from an audio file. In: Proceedings of the IEEE International Conference on Acoustics, Speech and Signal Processing, ICASSP 2008, March 31-April 4, pp. 121–124 (2008)
16. Pauwels, J., Varewyck, M., Martens, J.-P.: Audio chord extraction using a probabilistic model (abstract). In: The Music Information Retrieval Exchange (2008), http://www.music-ir.org/mirex/2008/abs/mirex2008-audio_chord_detection-ghent_university-johan_pauwels.pdf (accessed January 2008)
17. Rabiner, L.: A tutorial on HMM and selected applications in speech recognition. Proceedings of the IEEE 77(2), 257–286 (1989)
18. Radicioni, D.P., Lombardo, V.: A constraint-based approach for annotating music scores with gestural information. Constraints 12(4), 405–428 (2007)
19. Ryynänen, M.P., Klapuri, A.P.: Automatic transcription of melody, bass line, and chords in polyphonic music. Comput. Music J. 32(3), 72–86 (2008)
20. Tuohy, D.R., Potter, W.D.: A genetic algorithm for the automatic generation of playable guitar tablature. In: Proceedings of the International Computer Music Conference, Barcelona, Spain, September 2005, pp. 499–502 (2005)
21. Uchiyama, Y., Miyamoto, K., Sagayama, S.: Automatic chord detection using harmonic sound emphasized chroma from musical acoustic signal (abstract). In: The Music Information Retrieval Exchange (2008), http://www.music-ir.org/mirex/2008/abs/khadkevich_omologo_final.pdf (accessed January 2008)
22. Viana, A.B., Cavalcanti, J.H.F., Alsina, P.J.: Intelligent system for piano fingering learning aid. In: Proceedings of the Fifth International Conference on Control, Automation, Robotics & Vision, ICARCV 1998 (1998)
23. Weil, J., Durrieu, J.-L.: An HMM-based audio chord detection system: Attenuating the main melody (abstract). In: The Music Information Retrieval Exchange (2008), http://www.music-ir.org/mirex/2008/abs/Mirex08_AudioChordDetection_Weil_Durrieu.pdf (accessed January 2008)

BREVE: An HMPerceptron-Based Chord Recognition System

Daniele P. Radicioni and Roberto Esposito

Abstract. Tonal harmony analysis is a sophisticated task. It combines general knowledge with contextual cues, and it is concerned with faceted and evolving objects such as musical language, execution style and taste. We present BREVE, a system for performing a particular kind of harmony analysis, *chord recognition*: music is encoded as a sequence of sounding events and the system should assing the appropriate chord label to each event. The solution proposed to the problem relies on a conditional model, where domain knowledge is encoded in the form of Boolean features. BREVE exploits the recently proposed algorithm CarpeDiem to obtain significant computational gains in solving the optimization problem underlying the classification process. The implemented system has been validated on a corpus of chorales from J.S. Bach: we report and discuss the learnt weights, point out the committed errors, and elaborate on the correlation between errors and growth in the classification times in places where the music is less clearly asserted.

Keywords: Chord Recognition; Machine Learning; Music Analysis.

1 Introduction

The musical domain has always exerted a strong fascination on researchers from diverse fields. In the last few years a wealth of research effort has been invested to analyze music, under an academic and industrial pressure [31]. Techniques of intelligent music search and analysis are crucial to devise systems for various purposes, such as for music identification, for deciding on music similarity, for music

Daniele P. Radicioni
Dipartimento di Informatica, Università di Torino, Corso Svizzera 185, 10149, Torino
e-mail: radicion@di.unito.it

Roberto Esposito
Dipartimento di Informatica, Università di Torino, Corso Svizzera 185, 10149, Torino
e-mail: esposito@di.unito.it

Z.W. Raś and A.A. Wieczorkowska (Eds.): Adv. in Music Inform. Retrieval, SCI 274, pp. 143–164.
springerlink.com   © Springer-Verlag Berlin Heidelberg 2010

classification based on some set of descriptors, for algorithmic playlist generation, and for music summarization. Indeed, recent technological advances significantly enhanced the way automatic environments compose music [5], expressively perform it [12], accompany human musicians [23], and the way music is sold and bought through web stores [11].

Music *analysis* is a necessary step for composing, performing and –ultimately– understanding music, for both human beings and artificial environments (see, e.g., the works in [30] and [18]). Within the broader area of music analysis, we single out the task of *chord recognition*. This is a challenging problem for music students, who spend considerable amounts of time in learning tonal harmony, as well as for automatic systems. It is an interesting problem, and a necessary step towards performing a higher-level structural analysis that considers the main structural elements in music in their mutual interconnections. In Western tonal music at each time point of the musical flow (or *vertical*) one can determine which chord is sounding: chord recognition typically consists in indicating the fundamental note (or *root*) and the mode of the chord (Figure 1).

In our present approach the analysis task is cast to a supervised sequential learning (SSL) problem. From a methodological viewpoint, we transport to the musical domain the state-of-the-art machine learning *conditional models* paradigm, originally devised for the part-of-speech (POS) tagging problem [4]. A set of Boolean *features* has been designed in the attempt to encode the main cues used by human experts to analyze music.

One particular difficulty in dealing with sequential prediction is the computational complexity of inference algorithms. Intuitively, this problem can be cast to a path finding problem over a layered graph (described below) with vertices representing chord labels. In particular, given T layers and K label classes per layer, the graph representing the corresponding search space is a $T \times K$ graph. The problem of finding an optimum path over such graph is customarily solved by using the Viterbi algorithm, a dynamic programming algorithm having $\Theta(TK^2)$ time complexity [29]. Instead, to perform such decoding step our system relies on a recently developed decoding algorithm, CarpeDiem, that finds the *optimal* path in $O(TK\log(K))$ time in the best case, degrading to Viterbi complexity in the worst case [9].

We presently illustrate BREVE, a system for chord recognition that takes as input musical pieces encoded as MIDI files, extracts the corresponding sequence of music events, and computes the corresponding sequence of chord labels. Our approach puts together various insights from the fields of Machine Learning, Cognitive Science and Computer Music in an interdisciplinary fashion. The work is structured as follows. We start by formulating the problem of chord recognition (Section 2). Subsequently, we survey some related works on the problem of chord recognition (Section 3). In Section 4 we introduce the system BREVE: we illustrate how musical information is represented (Section 4.1), we introduce tonal harmony analysis as a sequential problem (Section 4.2), and motivate the adoption of a conditional model (Section 4.3). In particular we introduce the Boolean features framework (Section 4.4), and we illustrate the set of implemented features (Section 4.5). In Section 4.6 we briefly summarize the functioning of the CarpeDiem algorithm

Fig. 1 The chord recognition problem consists of indicating for each vertical which chord is currently sounding. Excerpt from Beethoven's Piano Sonata Opus 31 n.2, 1st movement.

and then report the results of the experimentation in Section 5. We both examine in detail the computed weights to analyze which sorts of knowledge they overall describe (Section 5.1) and elaborate on the errors committed (Section 5.2). Finally, we draw some conclusions on future directions of BREVE and on the technologies currently adopted.

2   Chord Recognition Problem

The task of chord recognition consists in indicating a chord label for each music event in a music piece. Equivalently, one could individuate segments as portions of the piece with same harmonic content, and then assign the appropriate label to each segment [20, 27]. Several types of notation can be adopted in analysing music, such as Figured Bass, Roman numeral notation, classical letter, Jazz notation, and different representations for musical chord symbols are possible [13]. Furthermore, chord recognition is the first step toward a higher level structural analysis, mainly concerned with individuating the main building blocks of a composition along with the structural relationships underlying whole pieces [26].

We define the problem of chord recognition as follows. A *chord* is a set of (three or more) notes sounding at the same time. *Chord recognition* consists in indicating the fundamental note (or *root*) and the *mode* of the chord, e.g., *CMaj* or *Fmin*, at each time point of the musical flow (Figure 1). Given a score, we individuate sets of simultaneous notes (*verticals*), and associate to each vertical a *label* ⟨*fundamental note, mode* ⟩. Additionally, we individuate the added notes that possibly enrich the basic harmony: we handle the cases of seventh, sixth and fourth. By considering 12 root notes × 3 possible modes (see below) × 3 possible added notes, we obtain 108 possible labels.

Tonal harmony theory encodes two key aspects in music: how to build chords (that is, which simultaneous sounds are admissible)[1] and how to build successions of chords (that is which chord sequences are admissible). In the following we refer to them as *vertical* and *horizontal* information, respectively. For example, three

---

[1] Admissible are sounds perceived as *consonant* ones within a given musical style.

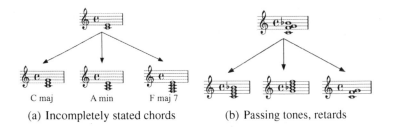

(a) Incompletely stated chords   (b) Passing tones, retards

Fig. 2 (a) Cases in which triads are incompletely stated, and (b) in which triads are stated together with further notes like passing tones and retards can make chord recognition a harder problem.

main kinds or *modes* of chords are defined: *major, minor* and *diminished* chords. This sort of information is proper to states, and we denote it as *vertical* information. Moreover, tonal harmony theory encodes rules to concatenate chords, thus describing which successions are acceptable: for example, after a *CMaj* chord one could expect *FMaj* or *GMaj*, rather than *C♯Maj*. Typically, this sort of information can be represented as proper to transitions between states, and we denote it as *horizontal* information.

If we assign a label y to an event x and we only use information about the notes sounding around x, then our analysis relies on vertical information. By converse, if to predict the label y we consider only the previous label regardless the notes currently sounding, we are using only horizontal information.

Chord recognition is a hard task that requires integrating both kinds of information. In fact, music harmony can be incompletely stated (i.e., we are given only 2 elements of a chord, as in Figure 2(a)), or it can be stated by *arpeggio* (i.e., one note at a time); moreover, *passing tones, retards*, etc., can further complicate chord recognition (Figure 2(b)). Even fully stated chords can be ambiguous: let us consider, e.g., a chord composed of the notes A-C-D-F. This set of notes can be labeled as a *Dmin7* or *FMaj6*, depending on the inversion, on the harmonic flow of surrounding events, and on voicing information. In addition, one has to handle ambiguous cases, where the composer aims at violating the listener's expectation, deliberately contravening "grammatical" rules [3].

3 Related Works

Much work has been carried out in the field of automatic tonal analysis: since the pioneering grammar-based work by Winograd [32], a number of approaches have been proposed that address the issue of tonal analysis. A survey of past works is provided by Barthelemy and Bonardi [1]; we focus on the closest approaches.

One of the preference-rules systems described by Temperley [28] is devoted to harmonic structure analysis. It relies on the Generative Theory of Tonal Music [17], providing that theory with a working implementation. In this approach, preference rules are used to evaluate possible analyses along a given dimension, such as harmony, also exploiting meter, grouping structure, and pitch spelling information. A major feature of Temperley's work concerns the application of high level domain knowledge, such as "Prefer roots that are close to the roots of nearby segments on the circle of fifths", which leads to an explanation of results.

The system by Pardo & Birmingham [21] is a simple template matching system. It performs tonal analysis by assigning a score to a set of 72 templates (that are obtained from the combination of 6 original templates transposed over 12 semitones of the chromatic scale). The resulting analysis is further improved by 3 tie-resolution rules, for labeling still ambiguous cases. This approach has been recently extended in the COCHONUT system by taking into account sequential information, in the form of chord sequence patterns [27].

Raphael & Stoddard [24] proposed a machine learning approach based on a Hidden Markov Model that computes Roman numeral analysis (that is, the higher level, functional analysis mentioned above). A main feature of their system is that the generative model can be trained using unlabeled data, thus determining its applicability also to unsupervised problems. In order to reduce the huge number of parameters to be estimated, they make a number of assumptions, such as that the current chord does not affect the key transitions. Also, the generative model assumes conditional independence of notes (the observable variables) given the current mode/chord.

Lee & Slaney [16] propose a system where 24 distinct HMMs are trained from acoustic signals. Two aspects are interesting in their work. First, by modelling 24 distinct HMMs they account for the differences in the chord transition probabilities that characterize different keys. Second, they train the HMMs using an input synthesized starting from a MIDI input. The MIDI files they use are downloaded from the Internet and annotated using Melisma music analyzer [28]. This approach has the advantage that large annotated corpora can be easily produced, however it can be argued that the final system learns the Melisma way of annotating music instead of the supposedly correct and unbiased way.

4   BREVE: An SSL System for Chord Recognition

In current Section we illustrate the design choices and the working of BREVE. We introduce the input representation, the Boolean-features framework, the motivations behind the sequential classification approach and the actual algorithms implemented.

4.1   Encoding

BREVE takes as input music pieces encoded as MIDI files. The fundamental representational structure processed is the music *event*; whole pieces are represented as *event lists*. An event is a set of *pitch classes* sounding at the same time; each new onset or offset determines a new event. Each event may be *accented* or *unaccented*. For each event we retain information about the bass.

Pitch classes are categories such that any two pitches one or more octaves apart are members of the same category. Their psychological reality received solid experimental evidence (e.g., see [6, 28]). Provided that we take as input MIDI files where pitch information is encoded as a number, pitch classes are computed by means of modulo-12 of MIDI pitches. E.g., if we consider the notes G3 and G4 corresponding to MIDI pitches 55 and 67, respectively, they both are mapped onto the same pitch class: $55 \equiv 67 \bmod 12$. Then, a vertical composed of the notes *C4-E4-G4* corresponding to the MIDI pitch numbers 60-64-67 is converted into an event composed of the pitch classes 0-4-7. Although loosing some information, pitch classes still permit to grasp the differences between chords. Also, they allow one to better characterize different modes. In fact, the characterizing aspect of a particular mode is the distance between the pitch classes that are present in the chord. For instance, distances intervening in $\langle 0,4,7 \rangle$, $\langle 0,3,7 \rangle$ and $\langle 0,3,6 \rangle$, correspond to major, minor and diminished modes respectively. In other terms, a chord mode is invariant under *rotations* of its constituting pitch classes (see Figure 3). If we consider major chords, whose pitches are 4 and then 3 semitones apart, we see that by rotating a major chord like *C Major* two 2 steps clockwise, we obtain the *D major* triad; by further rotating *D Major* 2 steps clockwise we obtain the *E Major* triad. The same holds if the chords are enriched with one or more added notes (Figure 3).

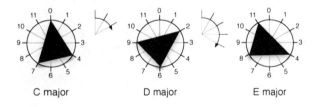

| C major | D major | E major |

Fig. 3 The three main *modes* of chords

In our representation, for each event we retain some information about the event *duration* as well: if a note i is held while a new note j is played, we consider an event containing i, and an event composed of both i and j, in that also the held note i affects the harmonic content of the vertical (see the *C4* and *F4* spanning over the second and third event in Figure 4). Most of the information required to perform tonal harmony analysis lies in the notes currently sounding and in their metrical salience: in particular, since harmonic content is mainly conveyed by accented events [17], we also annotate whether an event is accented or not. Meter estimation is based on the work of Temperley [28]. We annotate input events also with bass information, that provides valuable insights about inversions and more in general on harmonic flow. The final input representation and the information actually used through the analysis process is illustrated in Figure 4.

Our present representation disregards some musically relevant though secondary aspects, among which doubled notes, absolute pitches, pitch spelling, actual durations and voicing information are the most prominent.

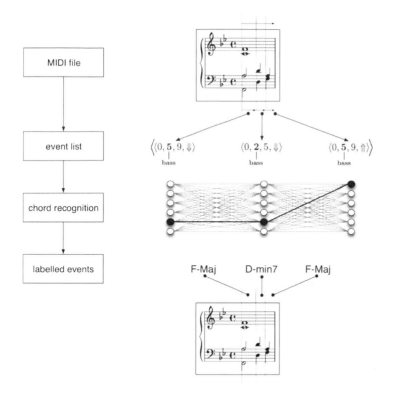

Fig. 4 The main steps performed by BREVE: it takes as input MIDI files, it extracts the event list with annotated bass and metrical accents ($\Downarrow$ indicates an accented event, while $\Uparrow$ indicates an unaccented event) and then it assigns a chord label to each event

4.2 Tonal Harmony Analysis as a Sequential Classification Problem

To analyze the harmonic structure of a piece is a sequential task, where contextual cues are widely acknowledged to play a fundamental role [2]. We have earlier argued that vertical and horizontal (that is, sequential) information grasp distinct though connected aspects of the musical flow. It follows that standard classification approaches such as decision trees, naive bayes, etc. are arguably less likely to produce good classification hypotheses, since they do not take into account the contextual information. This information is provided by surrounding observations, as well as nearby labeling.

In our view, the sequential aspects in the harmonic flow require to consider the problem of tonal harmony analysis as a sequential one. In particular, we adopt a Supervised Sequential Learning (SSL) approach. The task, known as the *supervised sequential learning* task, is not novel to the machine learning community. In recent years a wealth of research has been invested in developing algorithms to solve this

kind of problem, and a number of interesting algorithms have been proposed [7, 4].
The SSL task can be specified as follows [8]:

Given: A set L of training examples of the form (X_m, Y_m), where each $X_m = (x_{m,1}, \ldots, x_{m,T_m})$ is a sequence of T_m feature vectors and each $Y_m = (y_{m,1}, \ldots, y_{m,T_m})$ is a corresponding sequence of class labels, $y \in \{1, \ldots, K\}$.

Find: A classifier H that, given a new sequence X of feature vectors, predicts the corresponding sequence of class labels $Y = H(X)$ accurately.

In our case, each X_m corresponds to a particular piece of music; $x_{m,t}$ is the information associated to the event at time t; and $y_{m,t}$ corresponds to the chord label (i.e., the chord root and mode) associated to the event sounding at time t. The problem is, thus, to learn how to predict accurately the chord label *given* the information on the music event.

4.3   Conditional Models Approach

The SSL problem can be solved with several techniques, such as Sliding Windows, Hidden Markov Models, Maximum Entropy Markov Models [19], Conditional Random Fields [15], and Collin's adaptation of the Perceptron algorithm to sequencial problems [4] (henceforth, HMPerceptron). All these methods, with the exception of Sliding Windows, are generalizations and improvements of Markovian sequence models, culminating in Conditional Random Fields and HMPerceptrons. Conditional Random Fields (CRFs) are state-of-the-art conditional probabilistic models, which improve on Maximum Entropy models [15, 7] while maintaining most of the beneficial properties of conditional models.

The major benefit of using conditional models is that fewer parameters need to be estimated at learning time. In fact, while generative models need to estimate the "complete" distribution governing the random variables involved in the problem, conditional models only need to estimate the distribution of the output variables given the observed variables. This allows the learning algorithm to disregard many aspects of the problem that are not directly needed for the current classification task. On the other hand, in contrast with generative models which can be used to solve any conceivable inference problem, conditional models are specifically targeted to a given inference problem and cannot be extended beyond that.

CRFs, Maximum Entropy Markov Models and the HMPerceptron exploit the same way of interacting with the data. The "Boolean features" they use are functions of the current sequence. They are devised to return 1 if the label currently predicted for a variable of interest is coherent with the data, and to return 0 otherwise. Not only does this simplify the specification of the system, but also it allows domain knowledge to be easily plugged into the system. The algorithm, which is an extension to sequential problems of Rosenblatt's Perceptron algorithm [25], is reportedly at least on par with Maximum Entropy and CRFs models from the point of view of classification accuracy [4]. To the best of our knowledge, a direct comparison of HMPerceptron and CRFs has not been provided, even though they both were applied to the same Part-Of-Speech tagging problem, with analogous results [4, 15].

Therefore, on the basis of the above-mentioned literature, we have chosen the HMPerceptron as the main learning algorithm for the harmonic labeling prediction task. We briefly introduce the main facts about the working of the HMPerceptron, which has been investigated in [10].

The hypothesis acquired by the HMPerceptron has the form:

$$H(X) = \arg \max_{Y' = \{y'_1 \ldots y'_T\}} \sum_t \sum_s w_s \phi_s(X, y'_t, y'_{t-1})$$

where ϕ_s is a *Boolean feature*, i.e., it is a function of the sequence of events X and of the previous and current labels. The HMPerceptron has been defined within the Boolean features framework [19]. In this setting, the learnt classifier is built in terms of a linear combination of Boolean features. Each feature reports about a salient aspect of the sequence to be labelled in a given time instant. More formally, given a time point t, a Boolean feature is a $1/0$-valued function of the whole sequence of feature vectors X, and of a restricted neighborhood of y_t. The function is meant to return 1 if the characteristics of the sequence X around time step t support the classifications given at and around y_t.

4.4   The Boolean Features Framework and the HMPerceptron Algorithm

The ϕ_s functions are called *features* and allow the algorithm to take into consideration different aspects of the sequence being analyzed. To devise properly the set of features is a fundamental step of the system design, since this is the place where both the domain knowledge and the model simplifications come into play. The w_s weights are the parameters that need to be estimated by the learning algorithm. To these ends, the HMPerceptron applies a simple scheme: it iterates over the training set updating the weights so that features correlated to correct outputs receive larger values and those correlated with incorrect ones receive smaller ones.

This is actually the same kind of strategy adopted by Rosenblatt's perceptron algorithm [25], the main difference between the two algorithms is in the way the hypothesis is evaluated. In the classification problem faced by the perceptron algorithm, in fact, it is sufficient to enumerate all the possible labels and to pick the best labelling. In the case of the SSL problem, however, this cannot be efficiently done: the labelling of a single "example" is a sequence of T labels, the number of such labelling is thus exponential in T. To enumerate all the possible label sequences and to pick the best is clearly unfeasible. In order to tame the complexity of the hypotheses evaluation, a first order Markov assumption can be made. In a conditioned model, this amounts to assume that the probability of observing any given label y_t depends only on the previous label y_{t-1}, given the observations: $\Pr(y_t|y_1, \ldots y_{t-1}, X) = \Pr(y_t|y_{t-1}, X)$.

The first order Markov assumption has two important consequences. The first one is that features cannot exploit any information about labels besides that currently assigned and the previous one. Vice versa, they can collect information from the whole sequence of observations. Secondly, the *Viterbi decoding* [22] can be used to decide about the best possible sequence of labels with a computational complexity in the order of $\Theta(TK^2)$. Viterbi can be interpreted as an optimum path finding algorithm, suitable for particular kinds of graphs. In our setting, the graph is constructed starting from $K \times T$ nodes: one for each label/time position pair (Figure 5). All vertices that correspond to a given time instant are fully connected to the nodes that correspond to the following time event (the absence of all other edges in the graph is the fingerprint of the first order Markov assumption). A left-to-right path in this graph corresponds to computing T labels, i.e., a labelling of T events. In this setting, the classifier acquired by the HMPerceptron can be though of as an assignment of weights to the vertices and the edges in the graph such that the best scoring path corresponds to the most likely sequence of labels.

The HMPerceptron algorithm estimates the weights associated to the features so to maximize H accuracy over the training set. The algorithm is very similar to Rosenblatt's Perceptron [25]: W is initialized to the zero vector; then, for each example (X_m, Y_m) in the training set, $H(X)$ is evaluated using the current W. Two situations may occur: the sequence of labels predicted by H is identical to Y_m or this is not the case, and a number of errors are committed. In the first case nothing is done for that example, and the algorithm simply jumps to the following one. In the second case the weight vector is updated using a rule very similar to the Perceptron update rule:

$$w_s = w_s + \sum_{t=1}^{T} \phi_s(X, y'_t, y'_{t-1})(I_{y'_t = y_t} - I_{y'_t \neq y_t}).$$

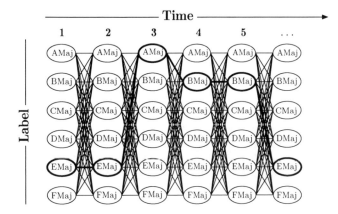

Fig. 5 Graphical representation of the labelling problem for $K = 6$

In the formula ϕ_s represents a Boolean feature, y_t represents the t-th correct label, y'_t denotes currently predicted t-th label, and I_P represents the function that returns 1 in case P is verified and 0 otherwise. By noticing that (we omit the arguments of ϕ_s for brevity):

$$\phi_s \cdot (I_{y'_t = y_t} - I_{y'_t \neq y_t}) = \begin{cases} +1 & \text{if } \phi_s = 1 \wedge y'_t = y_t \\ -1 & \text{if } \phi_s = 1 \wedge y'_t \neq y_t \\ 0 & \text{if } \phi_s \text{ is } 0 \end{cases}$$

it is immediate to verify that the rule emphasizes features ϕ_s positively correlated with good classification and de-emphasizes those that are negatively correlated with it.

4.5 Features Design

The features are used to provide discriminative power to the learning system and incorporate domain knowledge. They are evaluated at each time point t in order to compute an informed prediction about the label to be assigned to the event. Formal definitions are reported in Table 1. Before illustrating the features in detail, let us start by summarizing few elements that have been introduced:

1. We distinguish between vertical and horizontal information. The former one is concerned with the portion of tonal harmonic theory that describes rules governing simultaneous sounds, whilst the latter one is concerned with the part of tonal harmony describing successions.
2. We cast the chord recognition problem to a supervised sequential learning problem, that can be solved as a path-finding problem by means of the HMPerceptron algorithm.
3. In particular, given T events and K labels available for each event, we build a T-layered graph. In this setting, labelling a sequence of chords corresponds to finding an optimum (with maximal reward) path throughout the graph, from the leftmost to the rightmost layer.

In the following, we describe the implemented set of features. We denote the current event with $x_{.,t}$, and with y_t the currently predicted label.

Vertical features typically report about the presence (or absence) of some note in a given event, thus providing a proof for (or against) a given label. The features class Asserted-notes is composed of the features CompletelyStatedChord, AssertedAddedNote, ChordRootAssertedInTheNextEvent and AssertedRootNote. For instance, the feature AssertedRootNote reports about whether the root note of label y_t is present in event $x_{.,t}$. This feature can provide precious cues, since the root note is the most salient sound in any chord. CompletelyStatedChord collects information about whether all the notes in chord y_t are present in event $x_{.,t}$, and so forth.

Features named Asserted_v_NotesOfChord are triggered when *exactly* v notes of y_t are present in $x_{.,t}$. In principle, the smaller is v, the lesser evidence exists for y_t. There are, thus, five different realizations of this feature. Some of them (for v= 3 or 4) strengthen current predictions, while those with low v are actually used to vote against the label y_t. In fact, cases where few notes of a predicted label are sounding should be taken as an evidence that the chord prediction is unreliable.

Features using bass information are grouped in the Bass-At-Degree features class, reporting information about which (if any) degree of the chord y_t is the bass of event $x_{.,t}$. The bass is the most relevant pitch in any chord, often providing cues about the root note of current chord y_t. The other degrees of y_t can be present as the bass of the event $x_{.,t}$. However this class of features should enable the system to apprehend that the third degree, the fifth degree and the possible added note are progressively less informative about the appropriate label.

Horizontal features have been arranged in two classes. One reports about how *meter* and *harmonic changes* relate. The other one reports about some transitions relevant in tonal harmony theory. ChordChanges features account for the correlation of label changes and the *beat level* of a neighborhood of $x_{.,t}$. We distinguish two kinds of beat: accented and unaccented (1 and 0, respectively). In analyzing event $x_{.,t}$, we consider also events $x_{.,t-1}$ and $x_{.,t+1}$. We denote with triplets of 1 and 0 the eight resulting metrical patterns. For instance, 010 denotes an accented event surrounded by two weak beats. The associated feature returns 1 if $y_{t-1} \neq y_t$ (i.e., the chord changes) in correspondence with such a pattern.

Human analysts are known to focus at first on a reduced set of transitions frequent in tonal music. Successions features are used to model them, as a way for biasing the system to behave accordingly. We represent a transition as a pattern, like [Maj7, 5, Maj], composed of *i*) a starting mode and added note; *ii*) a distance between root notes, expressed as the number of intervening semitones; *iii*) an ending mode and added note. That is, [Maj7, 5, Maj] refers to a transition from a major chord with added seventh to a major chord, whose roots are five semitones apart (e.g., from FMaj7 to B♭Maj). Both ChordChanges and Successions features can be used to capture musical aspects that are deeply ingrained with musical style. On the one hand, this implies that restrictive constraints are posed to the stylistic homogeneity between training and testing sets. On the other hand, such style sensitivity grasps valuable 'linguistic' aspects that characterize musical language.

It is relevant to point out that both vertical and horizontal features have been devised so as to *generalize* to unseen labels. For instance, CompletelyStated-Chord will fire any time when all the notes of label y_t are present in $x_{.,t}$, regardless of whether y_t was present in the training set or not. Similarly, the transition [Maj7, 5, Maj] will provide useful information also about transitions between labels never met in the training set. This is in stark contrast with common learning systems. For instance, the transition matrix used in most HMMs-based systems provides detailed accounts of the probabilities of transitions which were observed in the training set. When used on a new dataset containing unseen transitions, the transition matrix cannot provide any useful information about it. On the contrary, based on the

Table 1 Boolean features formal definition. Definitions concerning a feature vector $x_{.,.}$ or concerning a label y are denoted by feature-name$[x_{.,.}]$ and by feature-name$[y]$, respectively. The intended meaning of each feature is reported in the *comment* column. Each definition reports the condition that is met when the feature returns 1. Also, each reported feature is actually a feature template which depends on several parameters. The parameters are mentioned in the feature name and explained in the *comment* column.

φ class	φ n.	φ name	Definition	Comments
Asserted-Notes	1	AssertedRootNote	root$[y_l] \in$ notes$[x_{.,t}]$	root$[y_l]$: root note of chord y_l
	2	ChordRootAssertedInTheNextEvent	root$[y_l] \in$ notes$[x_{.,t+1}]$	
	3	AssertedAddedNote	added$[y_l] \in$ notes$[x_{.,t}]$	added$[y_l]$: added note of chord y_l
	4	CompletelyStatedChord	$\forall i \in \{1..4\}$: note$_i[y_l] \in$ notes$[x_{.,t}]$	note$_i[y_l]$: i-th note in chord y_l
	5-9	Asserted_v_NotesOfChord	$\|$notes$[x_{.,t}] \cap$ notes$[y_l]\| = v$	$v \in \{0..4\}$; notes$[y_l]$: set of notes in y_l
Bass-At-Degree	10	BassIsRootNote	bass$[x_{.,t}] =$ root$[y_l]$	bass$[x_{.,t}]$: bass of event $x_{.,t}$
	11	BassIsThirdDegree	bass$[x_{.,t}] =$ third$[y_l]$	third$[y_l]$: third degree of chord y_l
	12	BassIsFifthDegree	bass$[x_{.,t}] =$ fifth$[y_l]$	fifth$[y_l]$: fifth degree of chord y_l
	13	BassIsAddedNote	bass$[x_{.,t}] =$ added$[y_l]$	
ChordChanges	14-21	ChordChangeOnMetricalPattern_$i_1 i_2 i_3$	m$[x_{.,t-1}] = i_1 \wedge$ m$[x_{.,t}] = i_2 \wedge$ m$[x_{.,t+1}] = i_3 \wedge$ $y_{t-1} \neq y_t$	$i_1, i_2, i_3 \in 0, 1$; m$[x_{.,.}]$ is 1 if event $x_{.,.}$ is accented and 0 otherwise
Successions	22-43	ChordDistance_m_1_i_m_2	mode$[y_{t-1}] = m_1 \wedge$ mode$[y_l] = m_2 \wedge$ root$[y_{t-1}] + i \equiv$ root$[y_l]$ (mod 12)	root$[y_l]$: pitch class of root note of chord y_l

generalization power of the features, BREVE can abstract the transition between chords a perfect fifth apart,thereby recognizing transitions between any two chords such as $(FMaj7\text{-}B\flat Maj)$, $(C\sharp Maj7\text{-}F\sharp Maj)$, $(D\flat Maj7\text{-}G\flat Maj)$ as instances of a transition of type $[Maj7,5,Maj]$. To give an intuition of the generalization power of BREVE, we ran a preliminary experimentation. BREVE has been trained on two chorales by J.S. Bach: one in major key, one in minor key. It then has been tested on 58 chorales by the same author. We repeated this test 5 times, each time with two different training sequences. We obtained an average accuracy of 75.55%. This datum is surprisingly good if we consider that the whole dataset contains 102 different labels, while the training sets, each one composed of a pair of sequences only, contained 24.8 labels on average.

4.6   *CarpeDiem Algorithm*

As mentioned, BREVE mainly exploits vertical information, resorting to horizontal cues primarily to resolve ambiguities. In other words, to label a given vertical, we *first* look at information provided by the current event, and *then* use surrounding context to make a decision if still in doubt. In BREVE, this 'least effort principle' is implemented by the `CarpeDiem` algorithm [9]. `CarpeDiem` is an algorithm allowing one to evaluate the best possible sequence of labels given the vertical and horizontal evidence. It solves the same problem solved by the Viterbi algorithm, but saves some computational effort. In particular, by exploiting vertical information `CarpeDiem` is able to reduce the number of labels taken into consideration by the system.

To provide an intuitive description of the algorithm, it is worth recalling that Viterbi algorithm [29] spends most computational resources to compute:

$$\max_{y_t,y_{t-1}} \left[\overbrace{\omega_{y_{t-1}}}^{\text{weight of the best path to } y_{t-1}} + \overbrace{\sum_s w_s \phi_s(X,y_t,y_{t-1},t)}^{\text{weight for transition } y_{t-1},y_t} \right]. \tag{1}$$

If we partition the set $\{1,2,\ldots p\}$ of all feature *indexes* into the two sets Φ^0 and Φ^1, corresponding to indexes of vertical and horizontal features respectively, then the equation (1) can be rewritten into:

$$\max_{y_t} \left[\sum_{s \in \Phi^0} w_s \phi_s(X,y_t,t) + \max_{y_t 1} \left[\omega_{y_{t-1}} + \sum_{s \in \Phi^1} w_s \phi_s(X,y_t,y_{t-1},t) \right] \right]. \tag{2}$$

Equation (2) is equivalent to Equation (1); in addition, it emphasizes how features in Φ^0 need to be evaluated only once per y_t label, and not once for each y_t,y_{t-1} pair. It is then obvious (and in accord with the intuition) that whenever $\Phi^1 = \emptyset$ the cost of the Viterbi algorithm can be reduced to be linear in the number of labels. The core idea underlying `CarpeDiem` is to exploit vertical information to avoid the evaluation of the inner maximization as long as possible.

5 Evaluation of the System

A corpus of 60 four-parts chorales harmonized by J.S. Bach (1675-1750) was anno-tated by a human expert with the bass and the appropriate chord label, as described in Section 4.1. Additional information on metrical salience of the music events has been computed by using the *meter* program by Temperley [28]. The sequences in the dataset are composed, on average, by 94.5 events; on the whole, the corpus contains 5,664 events.

BREVE has been validated using 10-fold cross validation: the estimated gener-alization error is 19.94%, thus providing an accuracy rate of 80.06%. Previous re-sults of a preceding implementation of the system, tested on a less homogeneous dataset (namely, the Kostka-Payne corpus [14]), obtained a 73.8% accuracy. Also, BREVE exploits the state-of-the-art 'decoding' algorithm CarpeDiem, which al-lows BREVE to run in just 7.35% of the time required by an implementation based on the Viterbi algorithm [10].

In the following we elaborate on the *quality* of the output of BREVE and, specifi-cally, *i*) we examine the learnt parameters from a musical perspective; *ii*) we inspect the errors committed in the classification and consider whether the analysis of er-rors can be useful in suggesting any improvement to the set of features or to the classification strategy.

In the next sections we elaborate on the weights learnt by the system and on the errors it commits. Since it is unclear how to aggregate the weights of the ten classi-fiers acquired during the cross validation, we will consider the weights obtained by training an individual classifier on half of the music sequences in the dataset. Also, we will examine errors committed by the same classifier, which has been tested on the second half of the dataset.

5.1 Musical Interpretation of Acquired Classifiers

In Appendix 7 we present the features list with the learnt weights, arranged into the four classes outlined in Section 4.5.

As expected, information about which notes are currently sounding prevails over contextual information. The highest positive weights involve vertical features, namely Asserted_v_NotesOfChord with v=3 and v=2 (features number 7 and 8 in Table 2) and CompletelyStatedChord. In considering Asserted_v_NotesOf-Chord features, we see that the case v=3 received more emphasis with respect to the case v=4. This is likely due to the fact that in the examined corpus, events composed of 4 or more pitches mostly contain passing tones which mislead the As-serted_4_NotesOfChord feature. Furthermore, the lower frequency with which the feature is asserted may have affected the magnitude of the learned weight as well. As regards v=1, the feature has been used by the system as evidence *against* a given label, rather than supporting it. We also observe how the feature with v=1 received a penalty analogous, in magnitude, to the reward obtained by the feature with v=2.

Fig. 6 The fundamental steps involved in cadence

By looking at the AssertedDegrees features class, we observe that the features CompletelyStatedChord (feature number 4), AssertedRootNote (feature number 1), and Asserted_(2|3)_NotesOfChord necessarily fire simultaneously. In such situations, the 'amount of evidence' in favour of y_t is overwhelming and BREVE will probably investigate only few alternative labels.

Let us now consider horizontal features (numbered from 14 to 43). Horizontal features are arranged in two classes, Successions and ChordChanges. The former ones collect information about some recurrent transitions, whilst the latter ones are mainly concerned with the correlation between harmonic change and meter strength. Weights associated to harmony changes in correspondence with weak beats (patterns with shape ?0?) receive the highest penalty among all features. While harmony change is globally discouraged, the system assigns only weak penalties to harmonic changes in correspondence with accented beats (feature 17 over metrical pattern 011 and feature 20, over metrical pattern 110). The preference for chord transitions in correspondence with accented events substantially fits to experimental evidences and analytical strategies known in literature, e.g., the *Strong beat rule* proposed by Lerdahl & Jackendoff [17], and implemented by Temperley as the preference for "chord spans that start on strong beats of the meter" [28]. The speed of harmony changes is deeply with musical style as well as with the musical form under consideration: since these features are a simple –though effective– way of modeling the harmonic rhythm, they provide interesting high-level information about style.

ChordDistance_m_1_i_m_2 features identify roughly three kinds of transitions: 1) those providing strong positive evidence in favor of a given label y_t; 2) those used as refinement criteria; 3) those used to forbid the transitions that have associated large negative numbers. All transitions involved in the *cadence* have been identified among the most relevant horizontal features (see features number 30, 22, 34, 25, 26, 32, 23, 24 in Table 2) and Figure 6.

Having identified the main components of the cadence shows that the system recognizes the relevance of such transitions in the considered corpus. In facts, four-parts chorales where harmonized by starting from melodies (properly, the chorales themselves) composed of phrases. Typically, the end of each phrase is marked by a cadence, so that the harmonization of chorales corresponds to a good extent to harmonizing the cadences, and then to connecting such 'fixed' points with further chords.

5.2   Errors Analysis

As mentioned, BREVE incorrectly classifies chords in about 20% of cases. A closer look at the errors reveals some well-defined error classes. First, in about 30% of errors, labels computed are wrong: in these cases BREVE has been totally misled. Also, determining why the system provided incorrect output is not simple. Somewhere, it has been caught in interpretative nuances that are hardly avoided given the simplified input representation (e.g., the system has no information on double notes, contrapuntal structure, which provide useful cues to human analysts). In other cases, the resistance to chord changes forces the system to inherit by mistake previously attributed labels (this is especially true for chord changes on weak beats).

Among other mistakes, only few errors appear execrable from the viewpoint of harmony theory. For instance, those due to confusing chord modes such as major with minor chords or viceversa. This is an error seldom committed by human analysts, and by BREVE, too. Precisely, only 8.60% (i.e., 1.71% of the labels in the dataset) of the overall errors fall in this class. One may argue that a lager context would be helpful to improve on this issue.

The errors due to confusion between relative keys (e.g., *Amin* instead of *Cmaj*) amount to 14.81% of the total error. Relative keys are intrinsically connected tones, in that they share the same *key signature* while having different roots, and to confuse between them is not a serious error from a musical viewpoint.

Many errors (namely 21.02%) are due to misclassifying the added note (root and mode being correct): in these cases the chord is analyzed in a substantially correct fashion. In most cases such errors are to be imputed to the resistance of the system to chord changes. For instance, this is the case of the succession *FMaj* ⇓-*FMaj* ⇑-*B♭Maj* ⇓ when the seventh occurs in correspondence with the second *–unaccented–* *FMaj* chord. This case can be controversial even for human analysts, in that both *FMaj* and *FMaj7* can be considered correct classifications for the second vertical. A human expert would make a decision inspecting whether the *E♭* can be considered a passing note or not. For example, a clue in favor of *FMaj7* would be provided by the resolution of the seventh a semitone below, on the third degree of the new chord.

Additionally, we observe that many errors occur on unaccented beats. Namely, if we consider that only 34.87% of events in the dataset fall on unaccented beats, the fact that 38.85% of errors involve weak events shows that some improvement in this direction is possible.

On average, BREVE inspects only a fraction of all possible labels.[2] If we consider that the complexity of the Viterbi is $\Theta(TK^2)$ (see Figure 5), this datum help explaining how the implementation of BREVE based on CarpeDiem saves 92.65% of computation time with respect to implementation based on the Viterbi algorithm.

---

[2] We note that the number of inspected nodes reported here is not a direct measure of the time required by the algorithm to run over the dataset. In facts, even for nodes that we consider as visited, not necessarily all incoming edges have been inspected by CarpeDiem.

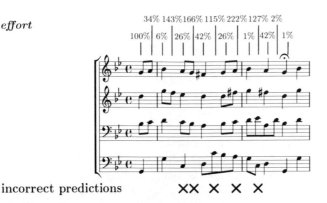

effort

incorrect predictions

Fig. 7 How computational effort, in terms of label evaluations, and incorrect predictions co-occur. Excerpt taken from the four-parts chorale *BWV 6.6* by J.S. Bach.

In particular, we can define an *effort measure* to characterize the amount of problem space explored, and compute such *effort* as the percentage of labels inspected by CarpeDiem with respect to Viterbi to compute the optimal label for current event. Such figure surpasses 100% when CarpeDiem visits some labels in previous layers (of course, this happens if the corresponding vertices had been left previously unvisited). Since on average CarpeDiem does not explore more than 100% of available labels, this fact provides an intuitive argument to corroborate the fact that CarpeDiem is never asymptotically worse than the Viterbi algorithm [9]. Interestingly, we observed that such *effort* correlates (though only mildly) with the classification error. On average, in correspondence with correctly predicted labels BREVE inspected the 65.47% of available labels; on the contrary, in correspondence with incorrectly predicted labels, BREVE inspected the 78.59% of available labels. In Figure 7 we report an excerpt with annotated above the staves the *effort* required by the analysis, along with –below the staves– the events for which BREVE provided incorrect predictions. As outlined above, errors are more frequent in correspondence with weak beats, where harmony quickly changes. For example, the first event in measure 3 only requires considering the 1% of available labels, whilst the subsequent, unaccented, event requires inspecting 127% of labels (thus implying the need to backtrack to previous level of the graph). This is clearly due to a harmonic movement faster than usual, where evidence provided by vertical features (considering the asserted pitches) is contradicted by the horizontal ones, highly penalizing label changes on unaccented events surrounded by accented events (see feature number 19 in Table 2).

In summary, our review of the errors committed reveals that most errors are either venial, or justifiable on musical accounts. Moreover, more difficult passages require increased effort, and BREVE has been proved to spend more computational resources exactly on those events.

6   Conclusions

This chapter described BREVE, a system for chord recognition, a task that is at the ridge of AI, Cognitive Science and Computer Music. The approach implemented by BREVE is to cast the chord recognition problem to a Supervised Sequential Learning approach: the musical flow is mapped onto a sequence of events, each one labelled with a chord. BREVE exploits a corpus of annotated musical pieces in order to 'learn' how to label new excerpts.

Among several possible approaches to solve this kind of learning problems, BREVE exploits the HMPerceptron algorithm: it converges faster than most competitor approaches, meanwhile retaining comparable classification performance. In BREVE few tens of Boolean features encode rich domain knowledge. This is in stark contrast with many recently proposed generative models where the number of parameters to be learnt ranges in the thousands.

The experiments over a corpus of chorales from J.S. Bach show that the system has performances similar to competitor systems. However, the fewer number of parameters involved in BREVE arguably allows for better explanation of the results and a deeper understanding of the system. In facts, the learnt weights turned out to be be musically meaningful; also the errors committed fall in few and clearly identifiable classes, therefore leaving room to future refinements.

Future refinements may address two aspects: from a musical viewpoint, the music representation could be extended so as to allow taking into account the role played by key as a center of gravity of the whole composition, and functional analysis. From a machine learning perspective, the system could be extended to take into account also information about transitions between non-adjacent layers, thereby exploiting higher order Markov assumptions.

Acknowledgements

This research has been partly supported by the postdoctoral research grant Università di Torino – A.A, 200.102 POSTDOC.

References

1. Barthelemy, J., Bonardi, A.: Figured bass and tonality recognition. In: Procs. of the 2nd Annual International Symposium on Music Information Retrieval 2001 (2001)
2. Bent, I. (ed.): Music Analysis in the Nineteenth Century. Cambridge University Press, Cambridge (1994)
3. Bigand, E., Pineau, M.: Global context effects on musical expectancy. Perception and psychophysics 59(7), 1098–1107 (1997)
4. Collins, M.: Discriminative training methods for hidden markov models: Theory and experiments with perceptron algorithms. In: Proceedings of the Conference on Empirical Methods in Natural Language Processing (2002),
   http://citeseer.ist.psu.edu/collins02discriminative.html
5. Cope, D.: A Musical Learning Algorithm. Comput. Music J. 28(3), 12–27 (2004)

6. Demany, L., Semal, C.: Harmonic and melodic octave templates. The Journal of the Acoustical Society of America 88(5), 2126–2135 (1990)
7. Dietterich, T.: Machine Learning for Sequential Data: A Review. In: Caelli, T.M., Amin, A., Duin, R.P.W., Kamel, M.S., de Ridder, D. (eds.) SPR 2002 and SSPR 2002. LNCS, vol. 2396, pp. 15–30. Springer, Heidelberg (2002)
8. Dietterich, T.G., Ashenfelter, A., Bulatov, Y.: Training conditional random fields via gradient tree boosting. In: ICML 2004: Twenty-first international conference on Machine learning. ACM Press, New York (2004),
   http://doi.acm.org/10.1145/1015330.1015428
9. Esposito, R., Radicioni, D.P.: CarpeDiem: an Algorithm for the Fast Evaluation of SSL Classifiers. In: Proceedings of the 24th Annual International Conference on Machine Learning, ICML 2007 (2007)
10. Esposito, R., Radicioni, D.P.: Trip Around the HMPerceptron Algorithm: Empirical Findings and Theoretical Tenets. In: Basili, R., Pazienza, M.T. (eds.) AI*IA 2007. LNCS (LNAI), vol. 4733, pp. 242–253. Springer, Heidelberg (2007)
11. Freeman, J.: Fast generation of audio signatures to describe iTunes libraries. Journal of New Music Research 35(1), 51–61 (2006)
12. Grachten, M., Arcos, J.L., López de Mantaras, R.: A Case Based Approach to Expressivity-Aware Tempo Transformation. Mach. Learn. 65(2-3), 411–437 (2006)
13. Harte, C., Sandler, M., Abdallah, S., Gomez, E.: Symbolic Representation of Musical Chords: A Proposed Syntax for Text Annotation. In: International Conference on Music Information Retrieval, pp. 66–71. Queen Mary, University of London (2005)
14. Kostka, S., Payne, D.: Tonal Harmony. McGraw-Hill, New York (1984)
15. Lafferty, J., Pereira, F.: Conditional random fields: Probabilistic models for segmenting and labeling sequence data. In: Procs. of the 18th International Conference on Machine Learning, pp. 282–289. Morgan Kaufmann, San Francisco (2001)
16. Lee, K., Slaney, M.: Acoustic Chord Transcription and Key Extraction from Audio Using Key-Dependent HMMs Trained on Synthesized Audio. IEEE Transactions on Audio, Speech and Language Processing 16, 291–301 (2008)
17. Lerdahl, F., Jackendoff, R.: A Generative Theory of Tonal Music. MIT Press, Cambridge (1983)
18. López de Mantaras, R., Arcos, J.: AI and Music: From Composition to Expressive Performance. AI Magazine 23, 43–57 (2002)
19. McCallum, A., Freitag, D., Pereira, F.: Maximum entropy Markov models for information extraction and segmentation. In: Proc. 17th International Conf. on Machine Learning, pp. 591–598. Morgan Kaufmann, San Francisco (2000),
   citeseer.ist.psu.edu/mccallum00maximum.html
20. Pardo, B., Birmingham, W.P.: Automated partitioning of tonal music. In: Procs. of the Thirteenth International Florida Artificial Intelligence Research Society Conference, FLAIRS 2000 (2000)
21. Pardo, B., Birmingham, W.P.: Algorithms for chordal analysis. Comput. Music J. 26, 27–49 (2002)
22. Rabiner, L.R.: A tutorial on Hidden Markov Models and Selected Applications in Speech Recognition. Proceedings of the IEEE 77, 267–296 (1989)
23. Raphael, C.: A Hybrid Graphical Model for Rhytmic Parsing. Artif. Intell. 137(1-2), 217–238 (2002)
24. Raphael, C., Stoddard, J.: Functional harmonic analysis using probabilistic models. Computer Music Journal 28(3), 45–52 (2004)
25. Rosenblatt, F.: The perceptron: A probabilistic model for information storage and organization in the brain. Psychological Review (Reprinted in Neurocomputing (MIT Press, 1998)) 65, 386–408 (1958)

26. Schoenberg, A.: Structural Functions in Harmony. Norton, New York (1969)
27. Scholz, R., Ramalho, G.: Cochonut: Recognizing Complex Chords from MIDI Guitar Sequences. In: Procs. of the 2008 International Conference on Music Information Retrieval, pp. 27–32 (2008)
28. Temperley, D.: The Cognition of Basic Musical Structures. MIT Press, Cambridge (2001)
29. Viterbi, A.J.: Error Bounds for Convolutional Codes and an Asymptotically Optimum Decoding Algorithm. IEEE Transactions on Information Theory 13, 260–269 (1967)
30. Widmer, G.: Discovering Simple Rules in Complex Data: A Meta-learning Algorithm and Some Surprising Musical Discoveries. Artif. Intell. 146, 129–148 (2001)
31. Widmer, G.: Guest editorial: Machine learning in and for music. Machine Learning 65, 343–346 (2006)
32. Winograd, T.: Linguistics and computer analysis of tonal harmony. Journal of Music Theory 12, 2–49 (1968)

7   Features Used by BREVE

Table 2 The feature list with the learnt weights. For each feature ϕ we report its class, its number, its name, and the weight learnt by the HMPerceptron.

ϕ class	ϕ number	ϕ name	weight
Asserted-notes	1	AssertedRootNote	2.600
	2	ChordRootAssertedInTheNextEvent	12.000
	3	AssertedAddedNote	-3.200
	4	CompletelyStatedChord	45.400
	5	Asserted_0_NotesOfChord	-70.600
	6	Asserted_1_NotesOfChord	-23.400
	7	Asserted_2_NotesOfChord	22.800
	8	Asserted_3_NotesOfChord	52.000
	9	Asserted_4_NotesOfChord	19.200
Bass-At-Degree	10	BassIsRootNote	17.400
	11	BassIsThirdDegree	13.400
	12	BassIsFifthDegree	0.200
	13	BassIsAddedNote	-0.600
ChordChanges	14	ChordChangeOnMetricalPattern_000	-9.600
	15	ChordChangeOnMetricalPattern_001	-95.200
	16	ChordChangeOnMetricalPattern_010	-23.600
	17	ChordChangeOnMetricalPattern_011	-1.800
	18	ChordChangeOnMetricalPattern_100	-84.600
	19	ChordChangeOnMetricalPattern_101	-552.200
	20	ChordChangeOnMetricalPattern_110	-10.800
	21	ChordChangeOnMetricalPattern_111	-82.200
Successions	22	ChordDistance_M_5_M	20.000
	23	ChordDistance_M_5_m	8.200
	24	ChordDistance_M7_5_m	21.600
	25	ChordDistance_M7_5_M	18.600
	26	ChordDistance_m_5_M	4.000
	27	ChordDistance_m_5_M7	-2.400
	28	ChordDistance_m7_5_M	6.800
	29	ChordDistance_m7_5_M7	0.400
	30	ChordDistance_M_7_M	13.800
	31	ChordDistance_M_7_M7	-13.200
	32	ChordDistance_M_2_M	8.200
	33	ChordDistance_M_2_m	7.800
	34	ChordDistance_M_2_M7	18.800
	35	ChordDistance_m6_2_M	-3.600
	36	ChordDistance_m6_2_M7	1.200
	37	ChordDistance_d_1_M	6.800
	38	ChordDistance_d_1_m	1.400
	39	ChordDistance_m_3_M	-36.600
	40	ChordDistance_m_8_M	-17.400
	41	ChordDistance_M_9_m	15.600
	42	ChordDistance_M_9_m7	-13.000
	43	ChordDistance_M4_0_M7	0.400

Analysis of Chord Progression Data

Brandt Absolu, Tao Li, and Mitsunori Ogihara

Abstract. Harmony is an important component in music. Chord progressions, which represent harmonic changes of music with understandable notations, have been used in popular music and Jazz. This article explores the question of whether a chord progression can be summarized for music retrieval. Various possibilities for chord progression simplification schemes, N-gram construction schemes, and distance functions are explored. Experiments demonstrate that such profiles can be used for artist grouping and for composition retrieval via top-k queries.

1 Introduction

The chord progression is an important component in music. Musicians and listeners speak of novel and influential chord progressions. A well-known example of famous chord progressions is the Tristan Chord of Richard Wagner, the very first two chords in the First Act Prelude of "Tristan und Isolde" and a motif that reappears over and over again in the ensuing four hours of drama. Another example is "Because" by The Beatles, whose main theme runs on an eight-bar chord sequence that is sometimes rumored to have been produced by reversing the chord progression for the main theme of the "Moonlight" Piano Sonata, one of the most famous piano compositions by Ludwig van Beethoven (Sonata Op.27 No.2). Yet another example is "Giant Steps" by John Coltrane, which uses a combination of dominant-to-tonic [V7 - I] cadence and repeatedly raises the key by major third.

Brandt Absolu
Maritime and Science Technology High School, Key Biscayne, FL, USA
e-mail: bla8291@aol.com

Tao Li
School of Computer Science
Florida International University, Miami, FL, USA
e-mail: taoli@cs.fiu.edu

Mitsunori Ogihara
Department of Computer Science
University of Miami, Coral Gables, FL, USA
e-mail: ogihara@cs.miami.edu

Z.W. Raś and A.A. Wieczorkowska (Eds.): Adv. in Music Inform. Retrieval, SCI 274, pp. 165–184.
springerlink.com
© Springer-Verlag Berlin Heidelberg 2010

Among many genres of music the role of chord progressions appears to be the most significant in Jazz. The performance in Jazz takes the form of Theme - Improvisation - Theme, where the middle part is improvisation in which the melody is spontaneously created while the chord progression of the main theme is being played repeatedly. Keeping in mind that the melody has to be created spontaneously, Jazz performers select tunes with chord progressions having certain characteristics.

Many Jazz compositions are based on one of two well-known chord progression forms. One is the 12-bar blues progressions and the other is "I Got Rhythm" by George Gershwin, where the chord progression of a tune is constructed out of the base progression and a new melody is played over the new progression. The abundance of such tunes witnesses the fact that Jazz music is highly improvisational (the tunes themselves might have been composed spontaneously by way of improvisation) and the fact that chord progressions play an extremely important role in that genre.

Quite often in studio recordings and live performances of Jazz, their performance programs consist of many tunes. Sometimes the tunes are compositions by the performers themselves, but more frequently they are compositions by someone else. If one surveys a large collection of Jazz recordings he/she will notice that many of them contain compositions from a small set of famous Jazz composers, such as Duke Ellington, Wayne Shorter, and Thelonius Monk. The high popularity of these composers, along with the fact that Jazz performers select tunes based on chord progression, suggests that there are Jazz composers with a unique chord progression style. Thus we here hypothesize:

there is a group of popular Jazz composers whose composition style can be well represented by chord progressions.

This article explores this hypothesis from the perspectives of clustering (the problem of grouping data according to their similarity) and similarity search (the problem of finding data objects similar to an input data object).

Fundamental to this exploration is a method for assigning a distance value given two chord progressions. An approach for designing a distance measure is sequence alignment, which is often used in melody-based music retrieval systems (see, e.g., [2, 6, 12, 13]). The basis for the sequence-alignment approach is a theory that models transformation of a chord progression to another (see, e.g., [9, 11]). Such a generative theory can offer a highly understandable explanation as to why two progressions are similar or why they aren't, but has a substantial limitation that computing a transformational path might be very difficult for tunes that have musically little to do with each other. It may be possible to deal with this issue by the use of partial alignments, as has been done in other scientific disciplines, but such a solution for chord progression analysis is yet to be established.

Also, the sequence alignment approach has a limitation that the pairwise similarity does not enable calculation of the mean—the chord progression that represents a collection of progressions as a whole—that is an essential component in clustering. Furthermore, computation of pairwise distance via sequence alignment is very expensive, which might limit its practical usage if alignment must be computed on the spot for a large number of chord progression pairs.

This consideration suggests the use of statistics to summarize chord progressions and then to compare chord progressions. The simplest of such statistics will be frequency counts of chords, i.e., how often the chords or chord components are used in a composition. The use of statistics has two major advantages. One, a single scan will be sufficient to compute frequencies of chords in a progression, so such statistics are easy to compute. Two, the instrument a composer uses for composing may result in a certain bias in the statistics. However, the simple statistics are insufficient for our purpose, since the frequencies chords do not provide information about the order in which chords appear.

We address the above issue by using of N-grams—the patterns consisting of N-consecutive chords that appear in a chord progression. The N-gram is a standard tool in natural language understanding (see, e.g., [5]), and has been used in the area of music information retrieval, in particular, in the melodic contour analysis [3, 4, 10]. Notably, the recent work of Mauch et al. [7] use 4-grams of triads to compare the compositions by The Beatles and Jazz tunes. The work used triads because a large portion of the chords in the Beatles compositions are simply triads.

The present article, extending an earlier work by a subset of the authors [8], considers the use of other chord tones in the analysis. Highly prominent in the Jazz harmony are the tone group of the 6th, 7th and major 7th notes and the group of tension notes (the 9th, the 11th, and the 13th notes). The former signifies the functions that chords possess while the latter adds color to triads. Chord progression analysis in terms of triads is likely to enable fundamental understanding of the chord structure. However, deeper understanding perhaps cannot be obtained without examining these non-triad notes, in particular, for comparing Jazz compositions.

While the chord progressions are an important subject in musicology, one might ask how chord progressions can be successfully incorporated into a music information retrieval system. One possible scenario is where tunes are retrieved by fragments of chord progression and accompanying metadata. In such a system, the user provides a chord sequence (either typed or copy-pasted from a sequence on screen) as input and the system retrieves tunes that contain a part with either exactly the same as (with the possibility of allowing transposition of the key) or similar to the input sequence, where the input chord progression is specified using an unambiguous notation system (such as the one in [1]). Also, the accompanying metadata (artist, genre, etc.) is used to narrow the scope of the search.

Another possible scenario is the retrieval of tunes with a certain set of chords as a constraint. In such a system, the user specifies the set of chords he/she can play well on his/her instrument (for example, a guitar) and metadata (again, artist, genre, etc.), and the system retrieves the tunes that meet the criteria. For example, the user may say "I need a Beatles song that uses chords only from { G, EMI, A, C, B7 }" and then the system retrieves "Run For Your Life" (which actually is based on the five chords with BMI in place of B7).

1.1   Contributions of This Article

This article presents a novel concept of N-gram profiles for the purpose of computing numerical representation of a single chord progression as well as a collection of chord progressions. First, the efficacy of the proposed profiles is tested using hierarchical clustering of famous Jazz composers according to their profiles. There are too many possible N-grams because the native chord space is gigantic. Thus, the chord name space has to be reduced using some chord simplification. With respect to two selected N-gram formats the hierarchical representation of Jazz composers reflects very well the historic development of the Jazz compositional idioms. Next, the use of N-gram profiles for similarity search is tested using a larger set of compositions and composers. Here different formats of profiles can be mixed together to represent a composition. The best combination of formats is searched for using a greedy algorithm with the effectiveness of top-K queries as the guide. Finally, the combinatorial search for the top-K queries is applied to the problem of identifying the composer given a composition as input.

1.2   Organization of the Article

This article is organized as follows. The next section discusses in detail how a chord is defined and how N-grams of a chord progression can be computed. Section 3 presents a proof-of-concept analysis of chord progression profiles via hierarchical clustering of Jazz composers. Section 4 presents exploration of best chord progression profiles using top-K query analysis.

2   Chord and N-grams Profiles

2.1   Chord Name Space

The Oxford University Press defines a chord as: "Any simultaneous combination of notes, but usually of not fewer than 3. The use of chords is the basic foundation of harmony." This definition readily accepts as a chord any multiple number of simultaneously played notes and thus a smash of keys on the piano is considered to be a chord. However, since the focus of this article is chord progression analysis that is useful for retrieving tunes, the definition of chords must be narrowed so that all the chords can be presented using a compact and clear notational scheme without specifying a chord as the collection of pitch names that are present in it. Chords presented with such a scheme are instrument-independent in the following manner. Given any instrument or any ensemble of instruments, as long as all the chord notes are presented in the harmony and no others are, we must think that the chord is correctly presented.

Such notational schemes indeed exist. Books of popular music often present chord names in addition to the the melody, lyrics, piano accompaniment chart, and somewhat less frequently present guitar tabs. There even exist "fakebooks" that present only the melody, lyrics, and chords, which are often used by Jazz musicians for free-style interpretation of compositions. While the chord names that appear in these books have their basis in the Western classical music theory, as pointed by Brandt and Roemer in [1], there exists conspicuous, and curious, ambiguity in the notation. The 7th chord of G with the augmented fifth and with the 9th note (that is, the chord consisting of notes G, B, $D^\sharp$, F and A with the G at the bottom), can be written in six different ways: G AUG 9, G+9, G9 ($^\sharp$5), G AUG 7 (9), G+7 (9) and G7 ($^\sharp$5 9). Since the first three are also used for the same chord without the 7th, the coexistence of various chord name scheming is very confusing. To resolve this issue Brandt and Romer [1] proposed a unified chord naming scheme that is both succinct and compact. The chord names considered in this article are all representable using this unified scheme.

In the chord notation scheme by Brandt and Roemer, a chord consists of four major parts: (1) the triad (the root, the 3rd, the 5th), (2) the 6th/7th, (3) the tension notes (the 9th, the 11th, the 13th), and (4) the added bass note. In the proposed chord notation the names start with the root and the triad together (the triad being denoted as empty for the major triad, MI for the minor triad, and SUS for the suspended 4th triad) followed by the 6th/7th note specification (6 for the sixth, 7 for the seventh, and MA7 for the major 7th). This is followed by additional information presented within a pair parentheses, which consists of alterations to the 3rd and the 5th notes and of the tension notes, and then a special keyword "on" and the bass note name if there is an added bass note. Also, the combinations (the 7th and the 9th), (the 7th, the 9th and the 11th), and (the 7th, the 9th, the 11th, and the 13th) are respectively represented by the numbers 9, 11, and 13 attached immediately after the root name for short-hand, with the exception that (a) if the triad is major then the second combination will be used and the third combination does not include the 11th; and (b) the 7th note may be the major 7th note, in which case, the numbers 9, 11, and 13 will be preceded by letters "MA". For example, DMIMA11 is equivalent to DMIMA7 (9 11). (The interested reader is encouraged to consult with the book for more detail.) An implicit restriction here is that not more than one note can be present from each of the four note groups: the 6th/7th notes, the 9th notes, the 11th notes, and the 13th notes.

2.2 Chord N-grams

2.2.1 Formal definition of an N-gram

For a set of symbols, U, for an integer $N \geq 1$, an N-gram over U is an ordered N-tuple $(u_1, \ldots, u_N)$ such that $u_1, \ldots, u_N \in U$. An N-gram $(u_1, \ldots, u_N)$ is said to be *proper* if for all i, $1 \leq i \leq N - 1$, it holds that $u_i \neq u_{i+1}$.

2.3 Chord Simplification

The chord name space defined in Section 2.1 is enormous. There are twelve possible choices for the root (without distinguishing between two notes that refer to the same note in the equal temperament); four for the 3rd (Minor, Major, Suspended 4th, and Omitted 3rd); four for the 5th ($^\flat 5$, $^\natural 5$, $^\sharp 5$, and Omitted 5th); four for the 6th/7th (6th, Minor 7th, Major 7th, and none of the three being used); four for the 9th ($^\flat 9$, $^\natural 9$, $^\sharp 9$, and none of the three being used); four for the 11th ($^\natural 11$, $^\sharp 11$, and neither of the two being used); four for the 13th ($^\flat 13$, $^\natural 13$, and neither of the two being used); and finally 12 for the added bass note. These make the total number of choices more than 320,000. This means that the total number of possible N-grams is more than 100 billions for $N = 2$ and 27 trillions for $N = 3$.

One must, however, be cautioned that although the space of N-grams is enormous, the N-grams that actually appear in a chord progression are very small in quantity. In fact, for a sequence of M chords, there are only $M - N + 1$ positions from which an N-gram can be started, the number of unique N-grams appearing in the sequence is at most $M - N + 1$. Even though the distributions of chords are often very skewed (towards certain keys and towards chords without tension notes), the vastness may make it unlikely for the N-gram profile of a chord progression with highly enriched chords to intersect with the N-gram profile of another chord progression. This problem can be overcome by simplifying chords.

The concept of chord simplification corresponds well with the concept of stemming in document processing, which is the process of removing modifiers of words thereby making words generated from the same root with difference modifiers treated as identical words. The process of simplifying a chord can be divided into two parts: (1) turning a chord with an attached bass note (such as AMI7 on B) into a non-fractional chord and (2) simplifying the tensions and the use of 6th and 7th notes.

There are three options for the first part:

- (B_0) simply removing the bass note (for example, AMI7 on B is changed to AMI7),
- (B_1) reorganizing the chord notes so that the bass note becomes the root (for example, AMI7 on B is changed to B7SUS4 ($^\sharp 5^\flat 9$)), and
- (B_2) incorporating the bass note as a tension (for example, AMI7 on B is changed to AMI9).

There are three options for the second part:

- (T_0) removing entirely the tensions and the 6th/7th note,
- (T_1) removing entirely the tensions but keeping the 6th/7th note, and
- (T_2) replacing the whole tension notes with a single bit of information as to whether the chord has any tension and keeping the 6th/7th note.

Also included in the list of possibilities are the possibility to keep the bass note, which will be denoted by B_3, and the possibility to keep all the tensions intact and keeping the 6th/7th note, which will be denoted by T_3.

The simplification options that are considered here then can be denoted by a pair (B_i, T_j) such that $0 \le i \le 3$ and $0 \le j \le 3$. The most aggressive simplifications are $(B_i, T_0), 0 \le i \le 3$. Each of these simplifications has the effect of reducing any chord to a triad or a chord that is a proper subset of a triad and therefore reduces the number of possibilities for a chord name to 192. For a progression Π and a simplification method τ, we will use $\tau(\Pi)$ to denote the progression Π after applying τ. Table 1 shows an example of how these simplifications work. The listed are the chords generated by simplifying AMI7(11) on F. For T_2 simplification, we will use the symbol of (9) to show that there is a tension. Note that the incorporation

Table 1 Various simplifications of the chord AMI7(11) on F

	T_0	T_1	T_2	T_3
B_0	AMI	AMI7	AMI7(9)	AMI7(11)
B_1	AMI	AMI7	AMI7(9)	AMI7(11^{b}13)
B_2	F	FMA7	FMA7(9)	FMA7(9 11 13)
B_3	AMI on F	AMI7 on F	AMI7(9) on F	AMI7(11) on F

of the bass note as a chord note required in the B_1 simplification may make the accompanying melody inconsistent with the chord. This characteristic is highly more prominent in the T_2 simplification, where all the tension notes are represented by the (9) tension symbol.

2.3.1 Measuring the Length of an N-gram

There are two important issues to consider when defining chord N-grams. The first is whether consecutive repetitions of the same chord should be permitted in a chord N-gram. The second is how to consider the number of beats assigned to each component of an N-gram. The two issues are related to each other and come directly from the fact that an arbitrary number of beats can be allocated to a single chord. For example, considering a 12-bar chord progression with the rhythm signature of 4/4 where the first four measures are [F – F($^\sharp$5) – F6 – F7], the next two are B^{b}7, the next two are F7, and then during the next two bars the chord moves G7 G^{b}7, F7, E7, E^{b}7, D7, D^{b}7, and C7, resolving to F7 in the last two measures. If the progression is scanned with a sliding window of two measures, the chromatic descent in measures 9 and 10 are captured in three windows, while the [I – IV – I] motion that occurs in measures 4 through 7 can never be captured within such a small window. One may suggest to use a double-sized window for the scan, but then the clone of the progression in which each chord has twice as many number of beats as the original creates exactly the same problem.

To resolve these issues, an N-gram here is considered to be proper. This allows an N-gram to have an arbitrarily large number of beats in it. In the above example, the [I – IV – I] motion is captured as a 3-gram [F7 – B^{b}7 – F7] with the weight of 32 beats, and the eight-chord chromatic descent is captured as a collection of six 3-grams [G7 – G^{b}7 – F7], ..., [D7 – D^{b}7 – C7] with the weight of 3 beats each. Also, after simplification, all the consecutive entries whose chord-part are identical to each other should be merged into a single entry. Once this modification has been done, the simplified chord progression has the property that every neighboring pair of chords are different and thus every one of its subsequences is a proper N-gram. Such a chord progression is called a *proper* chord progression. The reader might be cautioned that application of simplification to a proper chord progression without neighbor merging may produce a non-proper N-gram. For example, in the above example, the original sequence with the inclusion of duration can be represented as:

$$[\texttt{F}:4 - \texttt{F}(^{\sharp}5):4 - \texttt{F6}:4 - \texttt{F7}:4 - \texttt{B}^{\flat}\texttt{7}:8 -$$
$$\texttt{F7}:8 - \texttt{G7}:1 - \texttt{G}^{\flat}\texttt{7}:1 - \texttt{F7}:1 - \texttt{E7}:1 - \texttt{E}^{\flat}\texttt{7}:1 -$$
$$\texttt{D7}:1 - \texttt{D}^{\flat}\texttt{7}:1 - \texttt{C7}:1 - \texttt{F7}:8].$$

Applying the T_0-simplification and merging identical neighbors yields

$$[\texttt{F}:16 - \texttt{B}^{\flat}:8 - \texttt{F}:8 - \texttt{G}:1 - \texttt{G}^{\flat}:1 - \texttt{F}:1 -$$
$$\texttt{E}:1 - \texttt{E}^{\flat}:1 - \texttt{D}:1 - \texttt{D}^{\flat}:1 - \texttt{C}:1 - \texttt{F}:8].$$

2.3.2   N-gram Transposition

Since popular songs are transposed to different keys, one might be interested in studying chord changes relative to the first chord of the N-gram. One can thus transpose each N-gram locally, in such a way that each N-gram starts with a code having A as the root. Since A is simply nominal, the Roman numerals I, II, III, and so on, can be used. For example, from a five-chord sequence [FMI7, $\texttt{B}^{\flat}\texttt{7}$, $\texttt{E}^{\flat}\texttt{MA7}$, CMI7, B7], three 3-grams can be obtained, [FMI7 – $\texttt{B}^{\flat}\texttt{7}$ – $\texttt{E}^{\flat}\texttt{MA7}$], [$\texttt{B}^{\flat}\texttt{7}$ – $\texttt{E}^{\flat}\texttt{MA7}$ – CMI7], and [$\texttt{E}^{\flat}\texttt{MA7}$ – CMI7 – B7], which are then transposed respectively to [IMI7 – IV7 – $\texttt{VII}^{\flat}\texttt{MA7}$], [I7 – IVMA7 – IIMI], and [IMA7 – VIMI7 – $\texttt{VI}^{\flat}\texttt{7}$]. This transposition process is called the *A-transpose*.

2.3.3   Chord Sequences and Weight of N-grams

As mentioned earlier, a chord progression is a series of chord names such that each chord name is accompanied by a positive rational that represents the number of beats during which its chord is to be played. For an N-gram of a chord progression, its weight represents the contribution that the N-gram makes to the whole progression. For example, the weight has to be assigned so as to distinguish the contribution of a 4-chord pattern DMI7 – G7 – EMI7 – A7 with one beat assigned to each of the four chords from the contribution of the same 4-chord pattern appearing elsewhere in the same chord progression progress with four beats assigned to each chord. The contribution of an N-chord pattern can be approximated by the total number of beats assigned to the chords. Let $\Pi = [a_1 : \ell_1, \ldots, a_M : \ell_M]$ be a progression; that is, it is a series of M chord names $a_1, \ldots, a_M$ and for each i, $1 \leq i \leq M$, ℓ_i is the number of beats assigned to the chord name a_i. Then, for each i, $1 \leq i \leq M - N + 1$, the contribution of the N-gram at position i, $(a_i, a_{i+1}, \ldots, a_{i+N-1})$, is defined to be $\ell_i + \cdots + \ell_{i+N-1}$.

2.3.4   N-gram Profile of a Chord Progression

Figure 1 shows the melody and the chord progression of "Witch Hunt" composed by a Jazz giant Wayne Shorter. Without any simplification the progression is

$$[\text{CMI7}:32 - \text{E}^b7:16 - \text{CMI7}:16 - \text{G}^b7:4 - \text{F7}:4 - $$
$$\text{E7}:4 - \text{E}^b7:4 - \text{A}^b\text{MI7}(11):4 - \text{A on A}^b:4 - $$
$$\text{A}^b\text{MI7}(11):4 - \text{G7}(^b5):4].$$

With the B_0 (remove bass) simplification and the T_1 (no tension notes) simplification, the progression becomes

$$[\text{CMI7}:32 - \text{E}^b7:16 - \text{CMI7}:16 - \text{G}^b7:4 - \text{F7}:4 - $$
$$\text{E7}:4 - \text{E}^b7:4 - \text{A}^b\text{MI7}:4 - \text{A}:4 - \text{A}^b\text{MI7}:4 - \text{G7}(^b5):4].$$

Without transpose, the progression has the following 3-grams:

- $[\text{CMI7} - \text{E}^b7 - \text{CMI7}]$ (64 beats),
- $[\text{E}^b7 - \text{CMI7} - \text{G}^b7]$ (36 beats),
- $[\text{CMI7} - \text{G}^b7 - \text{F7}]$ (24 beats),
- $[\text{G}^b7 - \text{F7} - \text{E7}]$ (12 beats),
- $[\text{F7} - \text{E7} - \text{E}^b7]$ (12 beats),
- $[\text{E7} - \text{E}^b7 - \text{A}^b\text{MI7}]$ (12 beats),
- $[\text{E}^b7 - \text{A}^b\text{MI7} - \text{A}]$ (12 beats),
- $[\text{A}^b\text{MI7} - \text{A} - \text{A}^b\text{MI7}]$ (12 beats), and
- $[\text{A} - \text{A}^b\text{MI7} - \text{G7}(^b5)]$ (12 beats).

With transpose, the fourth and the fifth ones become identical, so we have

- $[\text{IMI7} - \text{III}^b7 - \text{IMI7}]$ (64 beats),
- $[\text{I}^b7 - \text{VIMI7} - \text{III}^b7]$ (36 beats),
- $[\text{IMI7} - \text{V}^b7 - \text{IV7}]$ (24 beats),
- $[\text{I7} - \text{VII7} - \text{VII}^b7]$ (24 beats),
- $[\text{I7} - \text{VII7} - \text{IIIMI7}]$ (12 beats),
- $[\text{I7} - \text{IVMI7} - \text{V}^b]$ (12 beats),
- $[\text{IMI7} - \text{II}^b - \text{IMI7}]$ (12 beats), and
- $[\text{I} - \text{VIIMI7} - \text{VII}^b7(^b5)]$ (12 beats).

Now we obtain the profile of this composition with respect to the (B_0, T_1)-simplification by dividing the weight in terms of the number of beats by their sum.

2.3.5   An Alternative Weighting Scheme

An alternative to the number-of-beats-based weight is the simple frequency count. Let Π be a proper chord progression generated from a given input progression after a certain simplification. Let $g_1, \ldots, g_k$ be an enumeration of all unique N-grams appearing in Π and for each i, $1 \leq i \leq k$, let m_i be the number of times that the N-grams g_i appears in Π. Then for each i, $1 \leq i \leq k$, we define the weight c_i assigned to g_i to be $m_i/(m_1 + \cdots + m_k)$.

Table 2 shows the weights of the N-grams of "Witch Hunt" in the two schemes.

2.3.6   Mathematical Notation

We view an N-gram profile construction scheme is a triple consisting of the N-gram length N, the choice of whether or not to transpose, and the choice of weighting

Fig. 1 The melody and the chord progression of "Witch Hunt" by Wayne Shorter

Table 2 A comparison between the two weight schemes on "Witch Hunt"

3-gram	Weight Scheme	
	Number of Beats	Frequency
[IMI7 - III$^\flat$7 - IMI7]	0.3265	0.1111
[I$^\flat$7 - VIMI7 - III$^\flat$7]	0.1837	0.1111
[IMI7 - V$^\flat$7 - IV7]	0.1224	0.1111
[I7 - VII7 - VII$^\flat$7]	0.1224	0.2222
[I7 - VII7 - IIIMI7]	0.0662	0.1111
[I7 - IVMI7 - V$^\flat$]	0.0662	0.1111
[IMI7 - II$^\flat$ - IMI7]	0.0662	0.1111
[I - VIIMI7 - VII$^\flat$7 ($^\flat$5)]	0.0662	0.1111

scheme. For an N-gram profile construction v and a (simplified) chord progression θ, $v(\theta)$ represents the chord progression profile created from θ by applying the N-gram construction scheme v. For a collection of some k tunes, $\theta_1, \ldots, \theta_k$, the collective N-gram profile of the collection with respect to v is

$$v(\{\theta_1, \ldots, \theta_k\}) = \frac{1}{k} v(\theta_i).$$

We can view an N-gram profile as the set of all pairs $w : c$, where w is a proper N-gram appearing in Π and c is the total contribution of w (since w may appear at more than one place in Π) scaled by the total contribution of all N-grams appearing in Π. Since the N-gram profile is created using a fixed length for N-grams, the total number of N-grams that can appear is finite. By assuming that the weight is 0 for all N-grams not appearing in Π, $\Theta[\tau,N](\Pi)$ can be viewed as a vector of finite dimension, whose entries are all nonnegative and add up to 1.

2.4   Comparison Using Cosine-Based Similarity Measure

Given two vectors with the same number of dimensions, $u = (u_1,\ldots,u_d)$ and $v = (v_1,\ldots,v_d)$, their mutual distance can be measured using various methods. In particular, we test the cosine distance

$$1 - \frac{u_1 v_1 + \cdots + u_k v_k}{\sqrt{u_1^2 + \cdots + u_k^2}\sqrt{v_1^2 + \cdots + v_k^2}},$$

and the Hellinger distance,

$$\frac{\sum_{i=1}^{d}(\sqrt{u_i} - \sqrt{v_i})^2}{2}.$$

Both distance measures have the value range of $[0,1]$. Also, both have the property that the value is 0 if and only if $u = v$.

Let π and σ be two chord progressions. Let τ be a simplification and let v be an N-gram scheme. Let δ be a distance function. Then the distance between π and σ with respect to the triple $\xi = (\tau,v,\delta)$ is defined to be

$$\text{dist}[(\tau,v,\delta)](\pi,\sigma) = \delta(v(\tau(\pi)),v(\tau(\sigma))). \tag{1}$$

In other words, it is the distance with respect to δ between the vector representation of the N-gram profile constructed from π by applying τ and v and the one from σ by applying τ and v. Given a collection D of such triples, $[(\tau_1,v_1,\delta_1),\ldots,(\tau_m,v_m,\delta_m)]$, the distance between π and σ with respect to the collection is the average of the m distance values, that is,

$$\text{dist}[\Delta](\pi,\sigma) = \frac{1}{m}\sum_{i=1}^{m}\text{dist}[(\tau_i,v_i,\delta_i)](\pi,\sigma) = \frac{1}{m}\delta(v(\tau(\pi)),v(\tau(\sigma))). \tag{2}$$

3   Proof-of-Concept Analysis

A proof-of-concept analysis has been carried out on a data collected from ten composer groups to test the efficacy of N-gram chord progression profiles.

3.1   Data

A data base of 301 chord progressions is constructed from various sources. The data base consists of the following:

- 218 compositions of composers John Coltrane (28 tunes), Chick Corea (25 tunes), Duke Ellington (25 tunes), Herbie Hancock (16 tunes), Freddie Hubbard (17 tunes), Thelonius Monk (27 tunes), Wayne Shorter (47 tunes), and Horace Silver (33 tunes), collected from Jazz fake books (Real Book 1, 2, and 3; New Real Book 1, 2, and 3; Jazz Limited);
- 63 "standard" tunes from Real Book 1, excluding compositions by modern Jazz musicians and Bossa Nova tunes;
- 20 compositions of The Beatles from the Hal Leonard Publishing "Anthology Volume 3".

The Beatles compositions are considered to be something very different from standards or Jazz composer tunes. The data can be obtained at the first author's web page: http://www.cs.miami.edu/~ogihara/chord-sequence-files.zip.

3.2 Comparison of the Simplification Methods

3.2.1   The choice of N and bass note simplification

To determine the value for N and to choose the bass note simplification, we calculate the cosine-based similarity between the standards and D. Ellington with respect to each of the twelve simplification methods and for $N = 1, 2, 3, 4$. Since D. Ellington played the most prominent role in founding the modern Jazz theory and the chord progressions of the Fakebook standard tunes in some sense summarize the chord sequences resulting from Jazz reharmonization, it is anticipated that the two groups are very similar, in particular, when the tension notes are excluded (namely, T_0 simplification). The similarity values are shown in Table 3. It appears that either $N = 3$ or $N = 4$ will be a good choice.

The choice of the bass note simplification (the B-part) does not seem to affect much the similarity measure, while the choice of the tension note simplification (the T-part) makes a substantial difference, in particular, for 3-grams and 4-grams. The phenomenon that the selection on the bass note simplification does not change much the similarity value can be explained by the fact that only a small fraction (less than 5%) of the chords appearing the data had a bass note. This observation leads us to

Table 3 The cosine-based similarity between the standards and D. Ellington with respect to various simplification methods and for $N = 1, 2, 3, 4$.

Method		N				Method		N			
T	B	1	2	3	4	T	B	1	2	3	4
T_0	B_0	0.990	0.950	0.818	0.579	T_2	B_0	0.950	0.798	0.504	0.197
	B_1	0.990	0.950	0.818	0.579		B_1	0.949	0.797	0.500	0.190
	B_2	0.990	0.950	0.818	0.576		B_2	0.947	0.796	0.497	0.187
T_1	B_0	0.954	0.835	0.628	0.319	T_3	B_0	0.952	0.805	0.502	0.194
	B_1	0.953	0.836	0.630	0.320		B_1	0.951	0.804	0.500	0.189
	B_2	0.952	0.834	0.626	0.310		B_2	0.950	0.804	0.500	0.185

choose B_0 (bass note omission) for the bass note simplification, because it is the simplest operation.

3.2.2 Tension Simplification

We next examine how the similarity values vary depending on the choice of the T-part. It is anticipated that the more aggressive the simplification is, the higher the similarity value becomes. This anticipation is clearly confirmed in Table 3, which shows the similarity values between the standards and the D. Ellington tunes. According to the table, there isn't much difference between the T_2 and T_3 simplifications. Since T_2 is more aggressive than T_3, and thus, the resulting chord notation is generally simpler with T_2 than with T_3, we should choose T_2 over T_3.

We then compare T_0 and T_1 using the songs by The Beatles and those by the others. The similarity values are shown in Table 4. There is a substantial difference in the similarity value between T_0 and T_1. Given that The Beatles is in the Pop/Rock genre and the rest are in Jazz, we feel that T_1 is more appropriate than T_0. Since the

Table 4 Comparison between T_0 and T_1

Composer	1-gram		2-gram	
	T_0	T_1	T_0	T_1
CC	0.933	0.594	0.527	0.250
DE	0.993	0.521	0.715	0.239
FH	0.921	0.570	0.456	0.114
HH	0.827	0.354	0.346	0.078
HS	0.962	0.483	0.621	0.178
JC	0.983	0.562	0.790	0.241
TM	0.998	0.551	0.691	0.243
WS	0.950	0.373	0.500	0.164

similarity of The Beatles to these composers seems very high for T_0, we consider using T_1 instead of T_0. These observations narrow our choices down to (B_0, T_1) and (B_0, T_2).

Table 5 shows the comparison of the standards against The Beatles, T. Monk, and H. Hancock with respect to the (B_0, T_1)-simplification and the (B_0, T_3)-simplification. We note that as N increases the similarity of the standards more quickly decays with The Beatles and Herbie Hancock than with Thelonius Monk and the decay with respect to the (B_0, T_1) simplification appears to be more dramatic than the decay with resect to the (B_0, T_2) simplification.

Figure 2 shows the cosine-based similarity of the profiles among the Jazz composers with respect to 3-grams and (B_0, T_2)-simplification. Two composers are connected if the similarity is 0.2500 or higher. The thicker the line is, the higher the similarity value is. Since the similarity is symmetric, the upper right portion of the table is left blank and the two $<$'s appearing in the last line indicate that the similarity value is not more than 0.2500.

Table 5 Cosine-distance-based similarity between the standards and each of The Beatles, T. Monk, and H. Hancock

	(B_0, T_1)-simplification				(B_0, T_3)-simplification		
	Standards Versus				Standards Versus		
N	The Beatles	T. Monk	H. Hancock	N	The Beatles	T. Monk	H. Hancock
1	0.430	0.922	0.875	1	0.414	0.886	0.829
2	0.163	0.716	0.390	2	0.162	0.676	0.185
3	0.040	0.437	0.114	3	0.040	0.378	0.051
4	0.017	0.199	0.038	4	0.018	0.1580	0.010

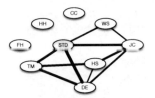

(a) The similarity graph of the Jazz composers.

	STD	DE	HS
DE	0.504		
HS	0.349	0.376	
TM	0.379	0.422	0.363
JC	0.402	0.278	0.349
WS	0.267	<	<

(b) The similarity table.

Fig. 2 The composer similarity

This graph seems to reflect well the relations among the composers from the historical perspective. According to the year of the first recording session as a leader, these composers are ordered as follows: Ellington (1924), Monk (1947), Silver (1955), Coltrane (1957), Shorter (1959), Hubbard (1960), Hancock (1962), and Corea (1966). The graph connects among the first five along with the standards and disconnects the remaining three from every one else.

3.3   Artist Clustering Using Profiles

The observation that the 3-gram similarity with respect to the (B_0, T_2) simplification reflects relations among artists from the Jazz historical perspective appears to be more strongly represented in hierarchical clustering of the composers. Figures 3 shows the hierarchical clusters of the composers generated using 3-grams.

4   Exhaustive Analysis Using Top-K Queries

The analysis presented in the previous section shows that among all possible triples of distance measure, simplification method, and N-gram scheme, there exist some combinations that very well reflect the development of Jazz music when they are

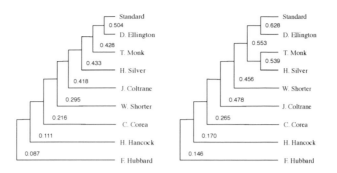

Fig. 3 Hierarchical clustering of the composers. Left panel: with respet to the (B_0, T_1)-simplification. Right panel: with respect to the (B_0, T_2)-simplification.

used for the purpose of comparing different composers. This section considers the problem of distinguishing a composer from the others in a top-k query environment by bringing together more than one triple of distance measure, simplification method, and N-gram scheme.

4.1 Method

Let k be a fixed integer. Suppose that a data base of chord progressions D is given. Let dist$[\Delta]$ be a distance measure. That is, Δ is a series of triples, $\xi_1, \ldots, \xi_r$, where for each i, $1 \le i \le r$, δ_i is a triple of distance function, simplification method, and N-gram scheme. We define the precision of dist$[\Delta]$ on a given data set D to be the proportion of compositions Π in D such that the set of k compositions in the data base $D - \{\Pi\}$ that are the closest to Π with respect to dist$[\Delta]$ have at least one composition by the composer of Π.

Let $w_{max} \ge 1$ be a parameter that bounds from above the N-gram length. Let $r_{max} \ge 1$ be a parameter that bounds from above the number of triples in Δ. Let $c_{max} \ge 1$ be a parameter that specifies the number of elements carried over from a stage to the next in the algorithm below. We search for the best distance measure in terms of the aforementioned precision value, in a greedy manner as follows:

Step 1   Set T to the collection of all N-gram schemes where the N-gram length is at most w_{max}. Set U to the collection of all distance functions of interest. Set $\Delta = []$, $C = \{\Delta\}$, and $W = \emptyset$.

Step 2   For $i = 1$ to r_{max}, do the following:

Step 2a   For each member Δ in C, for each distance function δ in U, for each simplification method $\tau in S$, and for each N-gram scheme $v \in T$, do the following:

Step-2a(i)   Let Δ' be the series constructed from Δ by appending (τ, v, δ).
Step-2a(ii)   Compute the precision of dist$[\Delta']$ with respect to top k-queries.

Step-2a(iii)   If C_0 has less than c_{max} elements, add $[\Delta']$ to C_0; otherwise, if $\text{dist}[\Delta']$ has precision higher than the distance measure in C_0 with the lowest precision among of the group, then replace that distance measure by $[\Delta']$.

Step 2b   Set $C = C_0$. Add to W the element in C_0 having the highest precision value.

Step 3   Output the element in W having the highest precision value.

Note that the members of C at the end of each loop body with respect to i has i components each, so the distance measure with the highest precision value produced by the algorithm has at most r_{max} components. Note also that components in a distance measure may be identical. Since the components are assigned an equal weight, a triple that appears n times receives weight n times as high as the weight a triple appearing only once receives. This in a naive way makes it possible to assign unequal weights to triples.

4.2   Experiments

4.2.1   Data set

A data set consisting of 340 chord progressions is used. The set covers 17 composers and from each composer 20 compositions are selected. The composers are the previous 10 plus seven new: Richie Beirach, Bill Evans (pianist), Keith Jarrett, Pat Metheny, and Steve Swallow; a Brazilian Bossa Nova composer Antonio Carlos Jobim; and an Argentinian "Nuevo Tango" composer Astor Piazzolla.

4.2.2   Parameter Choices

We set r_{max}, the maximum number of rounds, to 10, set c_{max}, the number of distance measures carried over to the next round, to 10, and set w_{max}, the maximum N for the N-gram length N, to 4.

4.3   Results

Figure 4 shows the result of the experiment with respect to top-k queries for $k = 2, \ldots, k = 10$. The precision is the proportion of chord progressions for which at least one of the three closest progressions is composed by the same composer.

In all cases, the precision increases steadily in the first three rounds and then, for a majority of the k-values, the growth tapers off.

Table 6 shows the plotted precision values in a chart. The last row of the table is the baseline precision; that is, the probability that a set of randomly selected pairwise-distinct k compositions from the pool of compositions other than the query contains the composition by the same composer. The query fails when the selected k distinct elements are compositions by someone else. There are 320 compositions composed by someone else and so the number of selections that lead to failure is

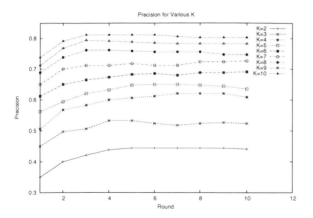

Fig. 4 The accuracy of the best performer in each round for each value of $k, 2 \leq k \leq 10$

Table 6 The precision table

k		2	3	4	5	6	7	8	9	10
Round	1	0.3500	0.4500	0.5059	0.5618	0.6118	0.6500	0.6882	0.7118	0.7382
	2	0.4000	0.4971	0.5676	0.5941	0.6500	0.7000	0.7382	0.7676	0.7912
	3	0.4206	0.5059	0.5824	0.6206	0.6647	0.7118	**0.7618**	**0.7941**	**0.8118**
	4	0.4382	**0.5324**	0.6000	0.6324	0.6735	0.7118	0.7618	0.7912	0.8118
	5	**0.4441**	0.5324	0.6059	0.6471	0.6824	0.7176	0.7588	0.7882	0.8118
	6	0.4441	0.5235	0.6118	**0.6500**	0.6853	0.7118	0.7559	0.7853	0.8118
	7	0.4441	0.5176	**0.6206**	0.6500	0.6794	0.7118	0.7559	0.7824	0.8059
	8	0.4441	0.5235	0.6206	0.6471	0.6882	0.7235	0.7559	0.7824	0.8029
	9	0.4441	0.5265	0.6206	0.6441	0.6882	0.7235	0.7471	0.7824	0.8029
	10	0.4412	0.5235	0.6088	0.6353	**0.6912**	**0.7265**	0.7471	0.7824	0.8029
Best		0.4441	0.5324	0.6206	0.6500	0.6912	0.7265	0.7618	0.7941	0.8118
Baseline		0.1091	0.1593	0.2069	0.2519	0.2944	0.3347	0.3728	0.4088	0.4428
Gap		0.3350	0.3731	0.4137	0.3981	0.3968	0.3918	0.3890	0.3853	0.3690

$\binom{320}{k}$. On the other hand, since there are 339 compositions other than the query itself, the number of possible selections of k distinct elements is $\binom{339}{k}$. Thus, the probability of failure is $\binom{320}{k}/\binom{339}{k}$ and the probability of success is: $1 - \binom{320}{k}/\binom{339}{k}$. Note that the gain from the baseline by the use of chord progression profile ranges from 0.33 to 0.41. This indicates that the chord progression profile can be a highly effective method for identifying compositions by the same composer.

Table 7 shows the summary of precision values of the 17 composers over the ten rounds for top-5 query analysis. At each round, 10 distance measures that achieved the highest precision are selected. For each such measure (there are a total of 100 measures), the precision (or accuracy) is calculated with respect to each artist. The maximum, minimum, the average, and the standard deviation of the 100 values for

Table 7 Composer-wise accuracy distribution for top-5 queries. The composers are presented in the decreasing order of average accuracy.

Composer	Max	Min	Average	StdDev
BEATLES	1.0000	0.8500	0.9335	0.0332
STANDARDS	1.0000	0.7500	0.9025	0.0597
ASTOR PIAZZOLLA	0.9500	0.7500	0.8915	0.0469
THELONIUS MONK	0.8500	0.5000	0.7685	0.0599
WAYNE SHORTER	0.8500	0.2500	0.7360	0.1229
DUKE ELLINGTON	0.8500	0.4000	0.7500	0.0745
BILL EVANS	0.8500	0.2000	0.6595	0.0999
PAT METHENY	0.8000	0.3500	0.6140	0.0645
RICHIE BEIRACH	0.6500	0.4000	0.5930	0.0806
KEITH JARRETT	0.7000	0.2500	0.5625	0.0931
HORACE SILVER	0.6500	0.3500	0.5345	0.0523
ANTONIO CARLOS JOBIM	0.7000	0.3500	0.5275	0.0676
STEVE SWALLOW	0.6000	0.3000	0.4945	0.0509
JOHN COLTRANE	0.7500	0.4000	0.4845	0.0695
HERBIE HANCOCK	0.5000	0.2000	0.4325	0.0694
FREDDIE HUBBARD	0.7500	0.2000	0.3890	0.0780
CHICK COREA	0.5500	0.2000	0.3325	0.0719

each artist are presented. The 10 top composers in this ranking are: The Beatles, Standards, Astor Piazzolla, Thelonius Monk, Wayne Shorter, Duke Ellington, Bill Evans, Pat Metheny, Richie Beirach, and Keith Jarrett. The standard deviation is small for The Beatles, Standards, Astor Piazzolla, Thelonius Monk, Horace Silver, and Steve Swallow. This indicates that for these composers the top-k query makes consistent performance.

Next, for each value of k the distance measure components (that is, the triples of simplification, N-gram scheme, and distance function) are collected from the best performing distance measure. A total of 58 components are collected, which are shown in Table 8. The most frequently occurring simplifications are (B_0, T_1), (B_0, T_2), and (B_2, T_0). They appear 11 times, 9 times, and 7 times, respectively. The N-gram length is 1 for 15 times, 2 for 11 times, 3 for 7 times, and 4 for 15 times. The average length is 2.07.

5 Conclusion

This article explores the question of whether a chord progression can be summarized for music retrieval. Various possibilities for chord progression simplification schemes, N-gram construction schemes, and distance functions are explored. Experiments demonstrate that such profiles can be used for artist grouping and for composition retrieval via top-k queries. The precision of nearly 65% is achieved with top-5 queries involving 17 composers, with a large margin of 40% from the

Table 8 The table of components appearing in the best performing distance measures

Bass	Count	Tension	Count	Transpose	Count	Length	Distance	Count
B_0	29	T_0	4	No	2	4	Cosine Frequency	1
						2	Hellinger Weight	1
				Yes	2	4	Hellinger Frequency	1
						2	Hellinger Weight	1
		T_1	11	No	10	1	Cosine Frequency	1
						1	Hellinger Frequency	2
						1	Hellinger Weight	3
						3	Cosine Weight	1
						4	Cosine Frequency	2
						4	Hellinger Weight	1
						4	Cosine Weight	1
				Yes	1	4	Cosine Weight	1
		T_2	9	No	5	1	Hellinger Frequency	3
						4	Cosine Weight	1
						4	Cosine Frequency	1
				Yes	4	2	Cosine Frequency	4
		T_3	5	No	2	4	Hellinger Weight	1
						4	Cosine Frequency	1
				Yes	3	2	Cosine Frequency	2
						2	Hellinger Frequency	1
B_1	5	T_1	1	No	1	1	Hellinger Frequency	1
		T_2	2	Yes	2	1	Hellinger Frequency	1
						2	Cosine Frequency	1
		T_3	2	Yes	2	1	Cosine Weight	1
						4	Hellinger Weight	1
B_2	12	T_0	7	No	4	3	Hellinger Weight	1
						3	Cosine Frequency	3
				Yes	3	4	Hellinger Weight	1
						4	Cosine Frequency	2
		T_1	4	No	2	3	Hellinger Frequency	1
						4	Cosine Weight	1
				Yes	2	1	Hellinger Frequency	1
						3	Cosine Weight	1
		T_2	1	No	1	4	Cosine Frequency	1
B_3	2	T_1	1	Yes	1	1	Hellinger Frequency	1
		T_1	1	Yes	1	1	Cosine Weight	1

baseline of 25%. This result seems highly promising. An interesting question will be how the performance decays for much larger sets of diverse composers. Another question is whether the N-gram profiles will be effective in identifying composers in terms of composer classification or genre/style classification. Finally, it will be interesting to study extensions of such approaches to include melodic fragments.

References

1. Brandt, C., Roemer, C.: Standardized chord symbol notation: a uniform system for the music profession, 2nd edn. Roerick Music Co., Sherman Oaks (1976)
2. Cahill, M., O'Maidín, D.: Melodic similarity algorithms – using similarity ratings for development and early evaluation. In: Proceedings of the 6th International Conference on Music Information Retrieval, pp. 450–453 (2005)
3. Doraisamy, S.C., Rüger, S.M.: Robust polyphonic music retrieval with n-grams. Journal of Intelligent Information Systems 21(1), 53–70 (2003)
4. Downie, J.S.: Evaluating a simple approach to music information retreival: Conceiving melodic n-grams as text. PhD thesis, University of Western Ontario, London, Ontario, Canada (1999)
5. Jurafsky, D., Martin, J.H.: Speech and Language Processing. Prentice Hall, Upper Saddle River (2000)
6. Kline, R.L., Glinert, E.P.: Approximate matching algorithms for music information retrieval using vocal input. In: Proceedings of the Eleventh ACM International Conference on Multimedia, pp. 130–139 (2003)
7. Mauch, M., Dixon, S., Casey, M., Harte, C., Fields, B.: Discovering chord idioms through Beatles and Real Book songs. In: Proceedings of the International Symposium on Music Information Retrieval, pp. 255–258 (2007)
8. Ogihara, M., Li, T.: N-gram chord profiles for composer style representation. In: Proceedings of 9th International Conference on Music Information Retrieval, pp. 671–676 (2008)
9. Paiement, J.-F., Eck, D., Bengio, S., Barber, D.: A graphical model for chord progressions embedded in a psychoacoustic sapce. In: Proceedings of the 22nd International Conference on Machine Learning, Bonn, Germany (2005)
10. Swanson, R., Chew, E., Gordon, A.: Supporting musical creativity with unsupervised syntactic parsing. In: Creative Intelligent Systems, AAAI Spring Symposium Series (2008)
11. Tojo, S., Oka, Y., Nishida, M.: Analysis of chord progression by HPSG. In: AIA 2006: Proceedings of the 24th IASTED International Conference on Artificial Intelligence and Applications, Anaheim, CA, USA, pp. 305–310. ACTA Press (2006)
12. Uitdenbogerd, A.L., Zobel, J.: Matchng techniques for large music databases. In: Proceedings of the Seventh ACM International Conference on Multimedia, pp. 57–66 (1999)
13. Volk, A., Garbers, J., van Kranenborg, P., Wiering, F., Veltkamp, R.C., Grijp, L.P.: Applying rhythmic similarity based on inner metric analysis to folksong research. In: Proceedings of the Eighth International Symposium on Music Information Retrieval, pp. 293–300 (2007)

Part III
Content-Based Identification and Retrieval of Musical Information

Statistical Music Modeling Aimed at Identification and Alignment

Riccardo Miotto, Nicola Montecchio, and Nicola Orio

Abstract. This paper describes a methodology for the statistical modeling of music works. Starting from either the representation of the symbolic score or the audio recording of a performance, a hidden Markov model is built to represent the corresponding music work. The model can be used to identify unknown recordings and to align them with the corresponding score. Experimental evaluation using a collection of classical music recordings showed that this approach is effective in terms of both identification and alignment. The methodology can be exploited as the core component for a set of tools aimed at accessing and actively listening to a music collection.

1 Introduction

The act of performing a music work, which has been coded in a music score by a composer, can be considered as a process that converts score symbols into acoustic features. To this aim, performers allow composers to communicate with the audience by transforming a sequence of symbols into something that can be perceived: the sound. While playing the role of intermediaries, musicians can also add their own interpretation to the music work, because music is both a *composing* and a *performing* art. The degree of freedom allowed to the performers is mostly genre-dependent. For instance, Western art music – also called tonal Western music, or

Riccardo Miotto
Department of Information Engineering, University of Padova
e-mail: miottori@dei.unipd.it

Nicola Montecchio
Department of Information Engineering, University of Padova
e-mail: montecc2@dei.unipd.it

Nicola Orio
Department of Information Engineering, University of Padova
e-mail: orio@dei.unipd.it

Z.W. Raś and A.A. Wieczorkowska (Eds.): Adv. in Music Inform. Retrieval, SCI 274, pp. 187–212.
springerlink.com   © Springer-Verlag Berlin Heidelberg 2010

more generally classical music – imposes a strict adherence to the score, which generally prescribes all the notes that have to be played by each instrument (in most cases it also indicates which are the instruments associated to each part) and gives indications about timing, articulation, and dynamics. Other genres, such as jazz or fusion, grant performers the freedom to make substantial changes to the main music dimensions – melody, rhythm, and even harmony – and music scores usually do not even represent articulation or dynamics. In between these two extremes, genres such as pop and rock let performers change the arrangement and the orchestration, but usually the main melody and the chord progression are subject to minor modifications.

While listening, the audience can associate a performance to a given composition according to different strategies, which depend on the degree of personal interpretation that is expected by the performers for a given music genre. Similar strategies are exploited when the listeners try to follow an ongoing performance along a symbolic score. In both cases the *expected* acoustic parameters, that are inferred from the symbolic representation of the music work, are compared with the *perceived* acoustic parameters, and a number of hypotheses are formulated considering the probability that differences are the result of the personal interpretation by the performers. This process of identification and alignment is probably related to the central role that symbolic representation plays in Western art music. Although a music score is only an approximate representation of a music work, because it cannot express all the possible nuances of a music performance [21], the score is often considered as the ideal version of a music work, to which performances are only approximations [16].

Given these considerations, it can be assumed that there is a statistical dependence between the symbolic representation and the acoustic performance of a given music work. The degree of correlation between symbols and acoustic parameters is clearly connected to the freedom of interpretation granted to performers. Moreover, it can be assumed that a statistical dependence exists between the acoustic parameters of two different performances of the same music work.

This paper presents an approach to statistical music modeling based on an application of Hidden Markov Models (HMMs). To this end, we provide a unified methodology that allows us to generate a HMM, which is the abstraction of a music work and models the possible differences of its performances, starting from either a symbolic representation of the score or an acoustic recording. Once a HMM is created to represent a music score, it can be used to simulate the listener's behavior both in identifying a music work given a performance and in following a performance along the corresponding score.

As an initial step towards the definition of the statistical dependence between different representations of a music work, we focus on Western art music that, as previously mentioned, has a clear definition of the musical parameters that can or cannot be modified by performers. This genre is also particularly suitable for the main application domain that we envisage, which is the access to music cultural heritage in an educational context.

This paper is structured as follows. In Sect. 2 we provide a short review of the main problems addressed through HMMs and the description of different approaches to music identification and alignment that have been presented in the literature. The feature extraction steps are described in Sect. 3, while Sect. 4 presents the methodology to automatically build HMMs from either a symbolic representation or a digital recording of a music score. The two main applications of HMMs proposed in this paper are described in Sect. 5 and evaluated in Sect. 6. The last section draws some conclusions about the proposed approach and discusses directions for future work.

2 Background

There is an increasingly large literature addressing music access and retrieval. In this review we focus only on the aspects that are directly connected to the proposed approach. Thus, after reviewing HMM-based recognition and alignment, we describe related work on these two topics in the music domain.

2.1 Review of Hidden Markov Models

HMMs are a powerful statistical tool that has been applied to several different tasks, ranging from speech recognition [19] and music information retrieval [30] to biological sequence analysis [8]. The tutorial written by Rabiner [27] in the late eighties is still one of the most complete introductions to HMMs.

HMMs are stochastic finite-state automata where transitions between states are ruled by probability functions. At each transition, the new state emits a random vector with a given probability density function. A HMM λ is completely defined by a set of N states $Q = \{q_1, \ldots, q_N\}$, a probability distribution for state transitions, which defines the probability to go from state q_i to state q_j $\forall i, j \in \{1 \ldots N\}$, and a probability distribution for observations, which defines the probability to observe a particular feature vector r when in state q_j $\forall j \in \{1 \ldots N\}$, for each possible feature vector r. Rabiner described the three main problems that can be addressed using HMMs: recognition, decoding, and training. Given the aims of this paper, we focus on the first two problems.

The recognition, or identification, problem applied to the music domain can be stated as: given an unknown audio recording, described by a sequence of audio features $R = \{r(1), \cdots, r(T)\}$, and given a set $\{\lambda_i\}$ of competing models, find the model that most likely generates R. This can be described by the simple maximization:

$$\overline{\lambda} = \arg\max_i P(R|\lambda_i) \tag{1}$$

which can be computed efficiently using *forward variables* [27]. In particular, forward variables allow the computation of the probability that model λ generates the sequence R in $\mathbf{O}(N^2 T)$, where N is the number of states of λ and T is the length of the observation vector R.

The decoding problem applied to the music domain can be stated as: given an audio recording, described by a sequence of audio features $R = \{r(1), \cdots, r(T)\}$, and given the model λ that generated it, find the state sequence $W = \{q(1), \cdots, q(T)\}$ that most likely corresponds to the generation of R. Also in this case, the definition does not impose a particular optimality criterion, although in general it is assumed that the state sequence should be globally optimal, thus it can be described by the maximization

$$\hat{W} = \arg\max_{W} P(W|R, \lambda) \tag{2}$$

which can be computed using the Viterbi algorithm [27] in $\mathbf{O}(N^2 T)$, where as usual N is the number of states of model λ and T is the length of vector R.

Clearly, the optimal path can be computed only after all observations are available, thus this strategy cannot be exploited in a real time task. Moreover, a globally optimal path may be less robust to local mismatches, for instance due to additional noise in the recordings or to large variations in the performance. In this case, a local criterion of optimality can be introduced,

$$\overline{q}(t) = \arg\max_{i \in \{1...N\}} P(q_i(t)|R, \lambda) \tag{3}$$

which can be computed using the forward probabilities in $\mathbf{O}(N^2 T)$, obtaining the final state sequence $\overline{W} = \{\overline{q}(1), \ldots, \overline{q}(T)\}$. It is important to note that $\overline{W}$ may not correspond to a real state sequence, that is the transition probability between states $\overline{q}(t)$ and $\overline{q}(t+1)$ may be zero. This characteristic allows the approach to recover faster from local mismatches, because the computed paths do not need to be feasible.

Although HMMs are the state of the art in speech recognition, alternative approaches are more common for music identification, as described in the following section.

2.2 Music Identification

In literature different methodologies have been proposed for music identification. One of the most common feature set applied to an identification task is *chroma vectors*, which was introduced initially in [9]. The concept behind chroma is that octaves play a peculiar role in music perception and composition [1]: the perceived quality of a given chord – i.e. major, minor, diminished – depends only marginally on the actual octaves where it spans, while it is strictly related to the pitch classes of its notes. This characteristic has been exploited in a number of identification tasks related to harmonic features, such as chord estimation [10, 26] and detection of harmonic changes [12].

Following the considerations about the freedom of interpretation typical of different genres, chroma features can be applied to an identification task when the harmonic structure and the chord progressions are expected to be only marginally altered by a performer. This is typical of pop and rock music (and classical music of course) where the sequence of chroma features of the song to be identified can be aligned and compared with the songs in the database either using Dynamic Time Warping (DTW), as described in [13, 14] for cover identification task in pop music, or using linear warping of the music features in [15] for a classical music matching. It is interesting to note that in [13] identification is carried out using a collection of MIDI songs that are synthesized automatically in order to compute the chroma features, while in [14, 15] features are directly extracted from audio recordings. In Sect. 4 we present a methodology that can be applied both to score-based and performance-based identification.

An alternative representation of the spectral content of the music signal has been proposed in [22, 24], where the first harmonics of the notes to be modeled are represented by a set of rectangular filters, while the application of HMMs to a music identification task instead of DTW has already been proposed in [17]. This paper partially builds from these two contributions.

2.3 Music Alignment

Approaches to music alignment reported in the literature are usually based on the assumption that a digital recording has to be aligned with a symbolic representation of the corresponding score. Yet, audio to score alignment must deal with the fact that at least one of the two forms has to be transformed in order to compute a match with the other. For example, the information in the score can be used to create a set of filterbanks, where each filter is centered on the expected harmonics of the signal, and the local match can be computed by measuring the energy that is output by the filterbank [24]. Alternatively, the concept of filterbank can be substituted by modeling the main statistical parameters of the expected harmonics [28]. The score can also be used to create an artificial performance, that is then matched against the real performance [32].

One of the main applications of automatic audio to score alignment is score following. In this case a local alignment is computed in real time between an ongoing performance, which is digitized and processed in real time, and a digital score stored in the system. The goal is to perform an automatic accompaniment capable of following the time deviations of the performance and possibly resynchronize in case of errors made by the musicians. Applications range from tools for instrumental practising to complete systems for public performances.

Early score following systems were based on dynamic programming approaches, such as [5], using MIDI format to represent both the score to be followed and the ongoing performance. It is interesting to notice that, already in 1993, an information retrieval approach was applied to a score following task [31], where DTW was used with MIDI signals. While early approaches focused on the alignment of a monophonic solo instrument, and were based on external pitch trackers, more recent approaches directly deal with the audio signal, using statistical models [11], hidden Markov models [2, 20, 23, 28], and hierarchical hidden Markov models [4] (a statistical model derived from HMMs, in which each state is considered to be a self contained probabilistic model).

Audio to audio alignment has been less investigated, probably because fewer applications can be based on such technology. In particular, the idea of exploiting the alignment to provide additional information to the listeners assumes that the symbolic score contains metadata – e.g., the instruments that are playing, the individual themes played by any single voice, composer's indications, performers's annotations – which can increase the quality of the listening experience. To this end, the alignment of two acoustic performances can be used only if one of the recordings has already been annotated. For this reason, although the presented model is general, the experimental evaluation presented in this paper does not include audio to audio alignment. Alternative applications of the alignment between two acoustic recordings have been the comparison of the style of different performances of classical music [6], or the development of aiding tools for musicological analysis of electroacoustic music [25]. In both cases, DTW has been used to align the spectral features of two performances.

3   Description of the Music Works

The main idea presented in this paper is that the most relevant acoustic features of a music performance can be modeled by a HMM. As discussed in Sect. 1, the process of converting a music work into an acoustic performance is stochastic because of the freedom of interpretation granted to the performers. Yet, the knowledge of a music work that can be obtained either from the score or from a performance can be exploited to create a statistical model alternative performances. To this end, both sources have to be processed to highlight the parameters that better describe a music work and their main features.

3.1   Segmentation in Events

A music work can be considered as a sequence of music events. In this context, an event can be a single note, a rest, or a chord, which corresponds either to a group of symbols explicitly represented in a score or to a segment of the signal representing a performance. The methodologies to segment a music work into its events clearly depend on the kind of media.

3.1.1   Parsing Acoustic Recordings

The audio recording of a music performance is a continuous flow of acoustic features, which depend on the characteristics of the music notes – pitch, amplitude, and timbre – that vary with time according to the music score and to the choices of the musicians. In order for these features to be structured, the audio information must undergo a *segmentation* process where the goal is to segment music signals into subsequences bounded by (consecutive) music events. An event occurs whenever the current pattern of a music work is modified; in particular, such modifications can be due to one or more new notes being played or stopped. This approach to segmentation is motivated by the central role that pitch plays in music language: in fact the segmentation may be considered as the process of highlighting audio excerpts described by a stable pitch. The granularity of the segmentation can vary considerably, ranging from being very fine (at individual notes level), or very general (at the level of music themes). Alternative segmentation strategies can be considered, for example, according to timbre or instrumentation criteria.

In the methodology described in [17], the first step of the algorithm is the computation of the *similarity* between audio frames. This is computed as the cosine of the angle between the frequency representations of two audio frames. Thus, given X and Y as column vectors containing the magnitude of the Fourier transforms for two frames, their similarity is defined as

$$sim(X,Y) = \frac{X^T Y}{|X|\,|Y|} \qquad (4)$$

A high similarity value is expected for frames where the same notes are playing, while a drop is related to a change in the active notes. A similarity matrix S can be defined as $(s_{ij}) = sim(X_i, Y_j)$.

Pure similarity values based on this measure may not be completely reliable for a segmentation task, as it has been shown for text segmentation, because changes in the local correlation could be more relevant than its absolute value. For instance, the value of local correlation of a note sung with vibrato is expected to be lower than in case of a steady tone played with a keyboard. Yet, in both case it is expected a decrease in local correlation when the note changes.

For this reason, segmentation has been carried out according to the methodology proposed in [3] for text segmentation. The basic idea is that in non-parametric statistical analysis one compares the rank of data sets when qualitative behavior is similar but the absolute quantities are unreliable. Thus, for each couple (X,Y) of frames that represents an element of the similarity matrix, the similarity value is substituted by its *rank*, which is defined as the number of neighbor elements whose similarity is smaller than $sim(X,Y)$. That is

$$r(X,Y) = ||\{(A,B) : sim(A,B) < sim(X,Y), (A,B) \in N(X,Y)\}|| \qquad (5)$$

where $N(X,Y)$ denotes the set of neighbors of (X,Y) in the similarity matrix.

Once the rank is computed for each couple of frames, hierarchical clustering on the similarity matrix is used to divide a sequence of features into coherent passages. The clustering step computes the location of boundaries using Reynar's maximization algorithm [29], a method to find the segmentation that maximizes the inside density of the segments. A preliminary analysis of the segmentation step allows us to set a threshold for the optimal termination of the hierarchical clustering. It is interesting to note that it is possible to tune the termination of the clustering step to obtain different levels of granularity.

3.1.2 Parsing Symbolic Scores

Score parsing is carried out automatically starting from the information stored in the symbolic format. Due to their large availability, MIDI files can be used as an approximate representation of the music score, even though more expressive formats such as Lilypond (http://lilypond.org) could be more appropriate for this task. In the case of a monophonic score, each note (or rest) corresponds to an event, while in a polyphonic score events are bounded by the different onsets and durations of all the notes being played by the various instruments/voices. Fig. 1 reports an example of score parsing.

Fig. 1 A polyphonic excerpt and the corresponding event sequence; long notes can be divided into multiple shorter events depending on other voices (for example the first note, a G5, is divided in five parts and assigned to the first five events)

3.2 Feature Extraction from Events

The main goal of this step is to define a uniform set of features, which can be computed from both symbolic and audio formats. Unfortunately, there is a small overlap between the kind of features that can be reliably extracted from the two formats. On the one hand, symbolic notation directly represents the pitch of all the voices, which is particularly relevant for the application to classical music, where melody plays a central role, while there is often no information about timbre (and little about dynamics). On the other hand, audio processing techniques are still far from reliably identifying the pitches in a polyphonic source. We propose a description of the events of a music work through an intermediate representation of the melodic content of the events, representing the expected spectral content corresponding to the notes that are contained in each event.

3.2.1 Extraction from Acoustic Recordings

In order to obtain a general representation of a music performance, each segment needs to be described by a compact set of features that are automatically extracted. Similarly to the segmentation approach, parameter extraction is based on the idea that pitch information is the most relevant for a music identification task. Since pitch is related to the presence of peaks in the frequency representation of an audio frame, the parameter extraction step is based on the computation of local maxima in the magnitude of the Fourier transform of each segment, averaged over all the frames in the segment.

The positions of local maxima are likely to be related to the positions along the frequency axis of the fundamental frequency and the first harmonics of the notes that are played in each frame. Considering differences in performing styles, timbre, room acoustics, recording equipment, and audio post processing among different versions, a general assumption is that alternative performances will have at least similar local maxima in the frequency representations, that is the dominant pitches will be in close positions.

In order to deal with noise due to signal windowing, imprecise tuning and different reference frequency, features are computed by averaging the FFT values of all the frames in a segment, selecting the positions of the local maxima, and associating to each maximum a frequency interval of the size of a semitone (the frequency of the maximum plus/minus a quarter tone). Fig. 2 exemplifies the approach: the light lines depict the average FFT of a segment, while the darker rectangles show the selected intervals.

The number of intervals is computed automatically, by requiring that the sum of the energy components that fall within the selected intervals is above a given threshold. The threshold is computed as a fraction of the overall energy of the frame. Peaks are taken starting from the highest and continuing in decreasing order. Fig. 2 depicts two possible sets of relevant intervals, depending on the percentage of the overall energy required: 70% on the left and 95% on the right. It can be noted that a low threshold may exclude some of the peaks, which are thus not used as content descriptors.

3.2.2 Extraction from Symbolic Scores

The extraction of the event pitches from a score is straightforward, because the information is readily available from symbolic formats. Since our goal is to obtain the same statistical modeling from both acoustic and symbolic sources, pitch information is used to compute a set of relevant frequency intervals in the form of a bank of bandpass filters. Each filter is centered on the harmonic frequencies of the notes forming the event, with a bandwidth that should deal with possible differences in tuning.

Different settings for the number and type of harmonics and for the width of each filter play a significant role in the accuracy of the alignment system. Typically, the

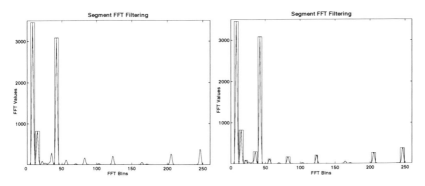

Fig. 2 Parameters extraction considering the peaks that carry, on the whole, 70% (left) and 95% (right) of the overall energy

first four harmonics are used, each filter being a semitone wide; the use of only odd harmonics can be useful when following a single instrument such as a clarinet, characterized by the absence of even harmonics. It can be noted that it is also possible to adjust the width of the filters to find a tradeoff between the robustness to intonation differences, the number of bins in the frequency domain, and the effect of leakage due to signal windowing. For instance, in case of low frequency notes, the effect of windowing may become more relevant than imprecise intonation, requiring to modify the size of a bandpass filters accordingly.

A simple approach, in the case of a single-note event with fundamental frequency f_0 is to create the filterbanks through equation

$$H_v(f) = \begin{cases} 1 & pf_0 c^{-1} < f < pf_0 c \\ 0 & \text{otherwise} \end{cases} \tag{6}$$

where $c = \sqrt[24]{2}$ is the quarter tone interval and $p = 1, 2, \ldots, P$ is the number of modeled harmonics. Polyphonic events can be modeled through the superposition of individual filterbanks.

4   Generation of the Hidden Markov Models

HMMs are a particularly suitable tool for identifying and aligning music. In fact, an ideal representation of a music performance can be considered to be a *hidden process*, because the evolution of the music itself and the actual position on the music work cannot be directly observed. As in any application of HMMs, what is observed is only the result of this process, because a listener can hear the sound produced by the musician but can only guess which are the notes in the score that the musician is actually looking at. Moreover, given the sequential nature of a music performance, it can be reasonably assumed that the process is *Markovian*, because a position along the music work summarizes all the information about the expected acoustic features that are observed and the expected future position. Each music work is modeled by a HMM providing that states are labeled with music events, transition probabilities model the temporal evolution of music events, and observations model the audio features that are related to each event.

4.1   Topology of the Model

The sequence of events is converted into a graph, where states correspond to music events and edges are associated to their interconnection (i.e. their adjacency in the score). Two levels of abstraction can be distinguished for the graph, namely a *higher level*, corresponding to the sequence of the music events, and a *lower level*, used to model each event. This separation reflects the origin of two different sources of possible mismatch: the higher level addresses discrepancies between the "ideal" music work and its representation in a score or a performance – or possible errors in the two representations – while the lower level models the duration and the acoustical features of each event.

4.1.1 Higher Level

In its simplest form, the topology of the higher level graph resembles the idea of a score as a succession of events: the states, each corresponding to a single music event, form a linear chain, as shown in Fig. 3.

Fig. 3 The simplest possible structure for the higher level of the score graph

This approach however is not robust enough for modeling complex music performances, because there is no explicit model for local differences between the representation and the actual performance that has to be identified or aligned. This limitation becomes particularly relevant for the alignment task: for instance, a skipped event, which should create only a local mismatch, can extend its effect also when subsequent correct events are played resulting in larger differences in the alignment; in the worst case, this could result in a completely wrong alignment.

A solution to this problem is the introduction of a special type of states, namely *ghost states* – as opposed to *event states*, which correspond to real events in the music work. In the proposed approach, each event state is linked to an associated ghost state, which in turn is linked to subsequent event states, forming a parallel series of states as shown in Fig. 4.

Fig. 4 Improved structure for the higher level of the score graph: the chain of *event states* is supported by a parallel chain of *ghost states*

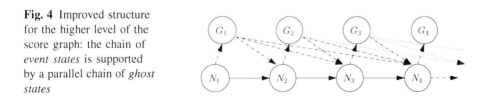

This approach allows us to model the possibility of wrong, additional or missing events: in case of a correct performance, the most probable path obtained via decoding will pass through the lower chain of Fig. 4 (the chain of events), while in case of an error it will pass through one or more ghost states during the mismatch and realign on the lower chain when the performance corresponds again to the representation of the music piece.

The transition probabilities from event states to corresponding ghost states are typically fixed, whereas the transition probabilities from a ghost state to subsequent event states follow a decreasing function of distance: this resembles the idea of *locality* of a mismatch due to an error. Extensive testing showed that this approach gives a considerable improvement over setting a constant value. Ghost states were initially proposed, although in a different form, in [23]. Their introduction has been motivated by their positive effect on real time alignment when complex orchestral pieces are to be followed, while experiments showed that ghost states do not play a significant role when aligning monophonic performances.

4.1.2   Lower Level

The lower level models the expected features of the incoming audio signal. Each state of the higher level is modeled as a chain of *sustain states* of the lower level, which can be follower by a *rest state*. Each sustain state has a self-loop probability p, as shown in Fig. 5. Sustain states model the features of the sustain part of an event, while rest states model the possible presence of silence at the end of each event that can be due to effects such as staccato playing style. Event attacks and decays are not explicitly represented. As regards attack, initial experiments showed that the modeling of the initial part of an event did not improve the alignment, while sometimes degraded the performances in terms of identification. As regards decay, we preferred to introduce a simple representation which is robust to dynamics rather than create a complex representation.

The number of states in the model is proportional to the number of events in the performance. In particular, experimentation has been carried out using a fixed number of n states for each segment, where states can either perform a self-transition or a forward transition. As described in [28], if all the states in a given event have the same self-transition probability p, the probability of having a segment duration d is modeled by a negative binomial distribution

$$P(d) = \binom{d-1}{n-1} p^{d-n}(1-p)^n \tag{7}$$

Parameters n and p are calculated in such a way to get the desired expected value, which corresponds to the event duration, and the variance for the distribution, which models the possible differences due to interpretation. A simple approach is to set the value of n and to compute p accordingly. Because small values of n have a positive impact on the computational complexity, experiments have been carried out using a fixed value of $n = 4$.

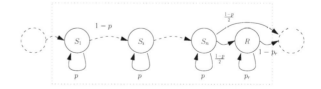

Fig. 5 Lower level model of an event state

4.2 Modeling Observations

States of the lower level emit the acoustic features of the incoming signal. Ideally, the most useful information would be the pitch of the notes simultaneously played during each event. Because polyphonic pitch detection is still unreliable, the signal spectrum is directly compared to the expected features of the HMM emissions. A detailed description of the observation probability computation strategy follows for each type of state.

4.2.1 Sustain states

As described in Sect. 3.2, each state is associated to a bank of bandpass filters. At each audio frame a Fourier analysis is carried out on the incoming signal, and the spectrum is compared to the spectra corresponding to each sustain state. A graphical example is presented in Fig. 6.

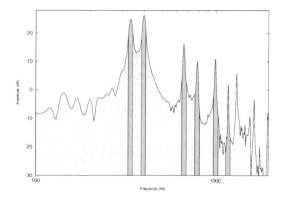

Fig. 6 Spectrum of a two-notes piano chord (E4, G4); the darker shaded regions represent the frequency bands of the filters, each of which is centered on a different harmonic frequency of the notes and is characterized by a width corresponding to a semitone

The emission probability b_i for the i-th sustain state is computed as

$$b_i^{(s)} = F\left(\frac{E_i}{E_{tot}}\right) \qquad (8)$$

where E_{tot} is the energy of the incoming signal, E_i is the energy of the incoming signal filtered by the i-th state's associated filterbank, and $F(\cdot)$ is a continuous probability density function.

To this end, two different distributions, namely *unilateral exponential* and *reversed Rayleigh*, are used as probability density functions for sustain states emission observation.

The unilateral exponential distribution, shown in Fig. 7(a) and initially presented for an alignment task in [23], is a variant of the traditional exponential distribution, from which it differs in the domain (the interval $[0, 1]$), in the fact that it is mirrored and translated and that a scale factor is present in order to obtain $\int_0^1 f(x)dx = 1$. The density function for the unilateral exponential distribution is

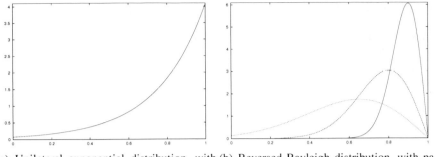

(a) Unilateral exponential distribution, with parameter $\lambda = 4$

(b) Reversed Rayleigh distribution, with parameters $\sigma = 0.1$ (the sharpest one), $\sigma = 0.2$ and $\sigma = 0.35$ (the smoothest one)

Fig. 7 Probability distribution functions for the observation modeling of sustain states

$$f(x) = \frac{e^\lambda}{e^\lambda - 1} \lambda \cdot e^{\lambda(x-1)} \qquad 0 \leq x \leq 1, \ \lambda > 0 \tag{9}$$

and it is characterized by an expected value $\mu = \frac{(\lambda-1)e^\lambda + 1}{\lambda(e^\lambda - 1)}$.

The reversed Rayleigh distribution's density function is

$$f(x) = \frac{-(x-1)}{\sigma^2} e^{\frac{-(x-1)^2}{2\sigma^2}} \cdot C(\sigma) \qquad 0 \leq x \leq 1, \ 0 < \sigma < \sqrt{\frac{2}{\pi}} \tag{10}$$

and is depicted in Fig. 7(b). The function, in order to be a proper distribution, contains a scaling factor $C(\sigma)$, which is approximately 1 for typical values of σ and can be automatically computed from σ. Its expected value is $\mu = 1 - \sigma\sqrt{\frac{\pi}{2}}$.

The reversed Rayleigh distribution has been introduced in [20] to overcome a problem posed by the unilateral exponential distribution. For example, suppose that a dense chord made of many notes is followed by a single note which was part of the former chord; using a unilateral exponential distribution, the state associated to the chord would always be more probable than the one associated to the single note, because this distribution is always increasing in $[0,1]$ (thus having more harmonics in the filter spectrum increases the observation probability). The reversed Rayleigh distribution is instead decreasing after a certain point, to allow less dense chords to become more probable than denser ones.

It is important to note that both distributions are governed by a single parameter: this was an explicit choice, since having a single parameter eases model training; moreover, they have another peculiarity: the higher their expected value is, the more selective they become. This is in agreement with the intuitive hypothesis that a very high observation probability corresponds to a high confidence about the observed features; having a lower expected value means that uncertainty is higher, thus giving a lower observation probability even in case of reaching the maximum value of the function.

4.2.2 Ghost States

According to the modeling strategy introduced in [20], the observation probability of the i-th ghost state is computed as

$$b_i^{(g)} = \sum_{j=i}^{i+k} w_i(j) b_j^{(s)}$$

(11)

that is a weighted average of the sustain observation probabilities of the following event states. The weighting function $w_i(\cdot)$ is tipically a decreasing discrete distribution function (such as a geometric distribution); its presence is motivated by the fact that, intuitively, in case of wrong or skipped notes, the notes played instead would probably be the nearest (in the score) to the expected one. This also makes sense in case of errors in the reference file, which are likely to happen in case of MIDI files, because the weighting function induces the system to quickly realign on near notes. This strategy turned out to work particularly well in case of complex polyphonic performances.

4.2.3 Rest States

The observation probability for the i-th rest state is computed as a decreasing function of the ratio of the current audio frame energy E_{tot} over a reference threshold E_{max} representing the maximum signal energy

$$b_i^{(r)} = F\left(\frac{E_{tot}}{E_{max}}\right)$$

(12)

5 Identification and Alignment

After generating the HMMs, identification and alignment tasks can be carried out using classical techniques. As described in Sect. 2.1, identification and decoding are typical problems of HMMs, that can be solved using dynamic programming approaches. In our approach both tasks are carried out locally: music works are divided into overlapping parts of about 20 seconds, and identification (and subsequent alignment) is carried out on these excerpts. This approach reflects the fact that local information about the music structure is normally used to define whether or not a given recording corresponds to a music work.

Recalling the definitions presented in Sect. 2.1, identification is carried out using Eq. 1, while alignment is carried out using Eq. 3.

Although based on the same modeling of the music works, the identification task is based on a simpler topology of the HMM graph. In particular, extensive testing showed that the use of ghost states at the higher level and rest states at the lower level did not significantly improve the identification rate, while decreasing efficiency. The simplified version of the HMM topology used for music identification is shown in Fig. 8.

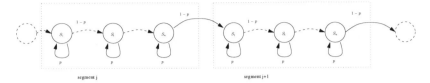

Fig. 8 Graphical representation for two segments of an HMM in the simplified representation used for the identification task

5.1 Regression Factor

The computation of Eq. 1 is based on the summation over all the states of the values given by Eq. 3. This means that an alignment between the events and the acoustic features is available while computing the probability that a model generated an observation sequence. This information can be useful to compute an additional parameter, which measures the distance between the computed path and an estimated linear path. In fact, in case of a potential correct match, the alignment of the query through the model is likely to have a linear trend. Such linear path can be estimated by considering a regression analysis of the computed alignment points. In fact, by computing the best fit line among the forward values and the distance of the points from the line, it is possible to have a parameter that measures how the path evolves through the model.

The best fit line associated with the n points $(x_1, y_1), (x_2, y_2), \ldots, (x_n, y_n)$, where x_i represents the generic state i of the model and y_i is the value of the forward variable, has the form

$$y = ax + b \tag{13}$$

where coefficients a and b are computed following the Ordinary Least Squares (OLS) method. The deviation Γ from the best fit line is computed as

$$\Gamma = \sum_i (y_i - (ax_i + b))^2 \tag{14}$$

In case of high values of Γ, a forward path can be considered unfeasible and then discarded in the final rank. Moreover, the forward path can be discarded when the slope of the line is outside a certain interval of values, because the evolution across the states of the HMM is likely to be either too slow or too fast to correspond to a realistic performance (a typical situation being when the optimal path is based on a long sequence of self-transitions).

It has to be noted that the best fit line represents an abstraction of a correct alignment. In reality, tempo fluctuations, changes in articulations, the presence of accelerando or rallentando, never let the correct alignment being on a simple line. For this reason, information about the the best fit line cannot be used to adjust the alignment but can give only a general description of the alignment trend.

5.2 Computational Complexity

The computation of either the identification or the alignment task can be carried out using a dynamic programming approach. As already mentioned in Sect. 2.1, the computational complexity of both identification and alignment tasks is $\mathbf{O}(N^2 T)$, where N is the number of states and T is the length of the observation sequence.

Yet, the proposed topology defines a limited number of transitions for each state. In particular, states can perform either a self-transition or at most two forward-transition, depending on the presence of a rest state (see Figure 5); if ghost states are used, the most distant transition is further, but still limited. It can be shown that the computational complexity becomes $\mathbf{O}(NT)$, thus linear also with the number of states in the model.

The task of music identification requires that each model in the collection is compared against the unknown performance, thus implying a linear search in the number of models. This approach can create scalability problems when large collections are used. For instance, with our experimental setup the identification of a single recording from a collection of one thousand models can be carried out in about two seconds, implying efficiency issues. To this end it is proposed to perform a clustering of the collection, as described in [18], in order to carry out linear search only on a limited number of models. Experimental results show that a cluster of about one hundred models is sufficient for achieving scalability without a loss in effectiveness.

6 Experimental Results

The proposed approach to HMM-based music modeling has been tested using a collection of classical music. The effectiveness of the identification task has been evaluated by using the audio excerpt of a performance as the query – that is the music piece that has to be recognized – and a collection of both MIDI and digital audio files to create the HMM. We did not consider the identification of symbolic as a relevant task, because it can be done using either metadata contained in the MIDI format or music retrieval techniques such as the one described in [7, 33] for polyphonic music. The effectiveness of the alignment task has been tested matching an audio recording with the corresponding symbolic score.

We did not evaluate the audio to audio alignment, because we believe that the most useful application for alignment is to annotate – and present to the user – a performance with the information carried in the score, or added by performers, musicologists, and music teachers.

6.1 Music Identification

The music identification methodology has been evaluated with real acoustic data from original recordings taken from the personal collections of the authors. Orchestral music repertoire has been used as a testbed because of the high number of instruments that play simultaneously, making the identification a particularly difficult task. The audio performances to be identified were all recordings of well known composers of Baroque, Classical, and Romantic periods. All the audio files have a sampling rate of 44.1 kHz, and they have been divided into frames of 2048 samples, applying a Hamming window, with an overlap of 1024 samples. With these parameters, observations are computed every 23.2 ms. It is worth mentioning that with this resolution a quarter note played at a fast tempo (Presto at 200 bpm) will last for more than 10 observations.

The aim of both tests was to measure the effectiveness of the methodology in terms of identification.

6.1.1 Audio to Midi Identification

In this experiment the collection was composed of 115 MIDI files representing the scores of different orchestral music works. All of these works were modeled by HMMs following the process described in Sect. 4. The audio files to identify were 49 audio recordings of some of the works stored in the collection. Fig. 9 reports the precision obtained with this dataset.

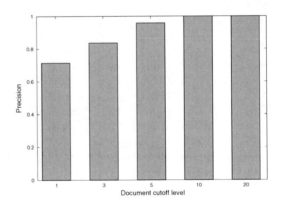

Fig. 9 Precision at different cutoff levels for the audio to score identification

As it can be seen all the queries were correctly identified within the first 20 positions, and 71.4% of them were correctly identified at the first position. The mean average precision was 81.2%. The results are satisfactory, proving the validity of the model even if the size of the collection is not large enough to provide a reliable evaluation.

It is interesting to note that the results were consistent among the repertories. In particular, we did not find a correlation between the repertoire – e.g., Baroque, Classic, Romantic – and the effectiveness of the retrieval. Precision depended more on the adherence of the performance to the score, either because of an imprecise notation in the MIDI files or because the score did not include all the information, as in the case of the music works with basso continuo.

6.1.2 Audio to Audio Identification

In this second test, the identification methodology has been evaluated with a larger testbed made up of a database of 1000 recordings and a query set of 50 different performances of a subset of the works in the database. The query set the same as in the previous experiment. Fig. 10 shows the precision without using the regression factor described in Sect. 5.1.

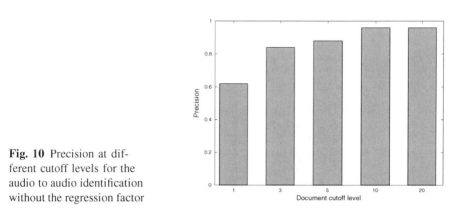

Fig. 10 Precision at different cutoff levels for the audio to audio identification without the regression factor

After removing the models with an incorrect alignment according to the regression factor, the results improve and almost reach the ones presented in Sect. 6.1.1, as it can be seen in Fig. 11. In particular, the regression factor allowed us to increase the precision of the query correctly identified in the first position. The mean average precision was 77.1%.

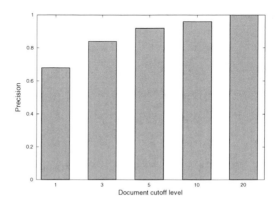

Fig. 11 Precision at different cutoff levels for the audio to audio identification using the regression factor

This result is very important, especially considering that it has been achieved with a collection almost ten times bigger than the MIDI collection. Although the parsing of a symbolic representation should not introduce errors, it could be argued that the score does not carry all the information required for an identification task. For instance, in most cases dynamics was not reported in the files. Moreover, MIDI files were usually provided by non-expert users and thus may be an approximate representation of the music work, because not all the voices are represented and many transcription errors often occur. Instead, audio recordings were all taken from high quality commercial CDs.

6.2   Music Alignment

An objective evaluation of an audio to score alignment system is a particularly difficult task, because of the lack of a manually annotated test collection, i.e. a collection of recordings in which the onset times relative to the events in the score are manually tagged by an expert.

In order to deal with this issue, a simple evaluation methodology is presented and used to measure the effectiveness of the proposed alignment system on two different test collections, the former made up of single-instrument, monophonic pieces and the latter comprising excerpts from complex orchestral polyphonic music. The audio files are professionally recorded (monophonic pieces) or extracted from commercial CDs (orchestral music), while the scores are parsed from MIDI files.

6.2.1   Evaluation Methodology

The output of the alignment system for a single performance/score couple is a list of value pairs in the form *(audiotime,miditime)*. Once all the performances in a collection are aligned to their corresponding score, these alignments are analyzed to extract a measure of precision based on the average deviation from the best fit line.

This measure is based on the hypothesis that a performer plays more or less "a tempo": while the tempo of the performance might be different from the MIDI tempo marking, the deviations are supposed to be negligible. Under this assumption, it is clear that a graphic representation of the alignment should follow a straight line, similarly to the situation described in Sect. 5.1. While this is clearly a strong and potentially incorrect assumption, the suitability of the particular performances in the test collections was verified by the authors.

The best fit line associated with the alignment data is assumed to be the correct alignment; once its slope a and intercept b are computed, the average deviation Δ_{avg} from the n data points is computed as

$$\Delta_{avg} = \frac{\sum_{i=1}^{n} |y_i - (ax_i + b)|}{n} \tag{15}$$

Clearly, under the assumption of a performance characterized by a steady tempo, the lower Δ_{avg} is, the higher is the adherence of the alignment data points to the best fit line and hence the alignment accuracy. An example of such situation is pictured in Fig. 12.

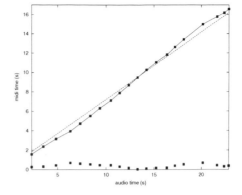

Fig. 12 Alignment data (continuous-dotted), corresponding best fit line (dashed) and absolute difference (dotted). In this example, where the alignment is correct and the tempo is quite steady, $\Delta_{avg} = 0.38$s

An alignment is then classified as "wrong" when at least one of the following situations occurs:

- the slope value, that is the ratio between the performance and the MIDI tempo markings, is outside the interval $[0.5, 2]$ (half/double speed);
- not enough alignment data pairs are present, which usually happens when the system is not able to perform properly;
- Δ_{avg} is larger than a fixed threshold (when $\Delta_{avg} > 3$s the resulting alignment makes no sense at a visual inspection);

A typical situation is presented in Fig. 13(a) where glitches are observed in the time alignment. Those are usually caused by sudden "jumps" of the forward variables due to the effect of ghost states (as can be seen in Fig. 13(b)) and can be effectively corrected using simple heuristics, which have not been applied to the results presented below.

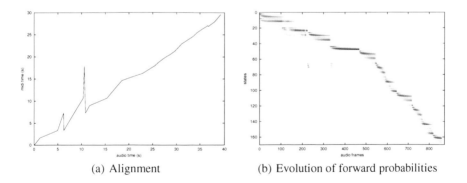

(a) Alignment   (b) Evolution of forward probabilities

Fig. 13 Alignment of the first movement from Haydn's Symphony n. 104

6.2.2   Experimental Results for Monophonic Recordings

The evaluation with monophonic recordings has been carried out using 28 phrases
(typically characterized by a length of about 10s) played by different monophonic
instruments. The compositions where these phrases are taken from include:

- J. S. Bach - Flute Sonata BWV 1013, Goldberg Variations (performed on a cello);
- J. Brahms - Clarinet Sonata Op. 120 No. 1;
- M. Mussorgsky - Pictures at an Exhibition, orchestrated by M. Ravel (excerpts
  of flute, trumpet and violin parts);

The system performs well with solo instruments excerpts, as can be seen from the
histogram in Fig. 14.

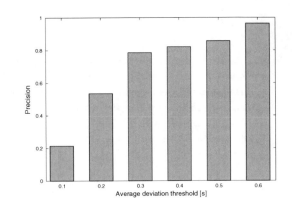

Fig. 14 Percentage of files
for which Δ_{avg} is inferior
to given thresholds, for the
monophonic collection

In one case an alignment is considered "wrong": a closer analysis revealed that a
single score event was misaligned, but this difference was large enough to make the
slope of the best fit line fall outside the $[0.5, 2]$ range.

6.2.3   Experimental Results for Polyphonic Recordings

The collection comprises the same 49 recordings used for testing the identification
effectiveness described in Sect.6.1.1. All the excerpts have a length of 20 seconds
and were taken from:

- L. v. Beethoven - Symphonies No. 3, 7, 9 and Egmont Overture;
- F. J. Haydn - Symphony No. 104 "London";
- F. Mendelssohn - Symphony No. 4;
- W. A. Mozart - Divertimento K136, Horn Concerto K412, Eine Kleine Nacht-
  musik K525, Symphony No. 40 K550;
- F. Schubert - Quartettsatz D703;
- A. Vivaldi - The Four Seasons;

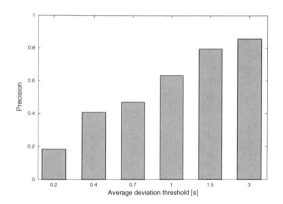

Fig. 15 Percentage of files for which Δ_{avg} is inferior to given thresholds, for the polyphonic collection

Experimental results with this collection are presented in Fig. 15. The system was not able to properly align 7 out of 49 files; those files are the ones whose scores are particularly different from an accurate transcription of the recording: a notable example is Vivaldi's "Winter", where the basso continuo, played by a harpsichord, is not transcribed. It is worth mentioning that, historically, the basso continuo had a different notation that can be interpreted by trained performers.

This provides additional evidence to the fact that the efficacy of the alignment is highly dependent on the accuracy and completeness of the score file, tough the proposed system proves to be generally robust with the provided scores, which were downloaded from generic Web sites and not modified by the authors.

7 Conclusions

This paper presents a unified approach to the statistical modeling of a music work, starting from either a symbolic representation of the score or from a digital recording of a performance. Statistical modeling is based on an application of HMMs, that are generated automatically starting from a segmentation of the music work into events. The approach has been applied to identify and align pieces of the classical music repertoire.

The proposed methodology can have several applications, in particular in the field of music education and dissemination. The first application is the identification of unknown recordings, in order to retrieve relevant metadata to be presented to the user. It should be noted that the identification can be carried out locally, by comparing an audio excerpt with fragments of the music works of the collection. In this way it is possible to identify at the same time the music work and the position in the corresponding file stored in the collection, which can be either in MIDI or in audio formats.

Once the recording has been identified, it is also possible to exploit alignment techniques to provide a more active listening experience. In fact, there is a gap between the access to music content by musicians and musicologists, who normally take most of the information about a music work directly from a structured music score, and musically untrained listeners, who are not able to read a music score. The alignment of the recording of interest with the score can allow the user to: read annotations made by the composer, be aware of changes in tempo and tonality, identify which are the instruments that are playing at any moment, and selectively listen to them. Visual cues can be added to particular events in a score and synchronously presented to the listeners each time they are listening to a different performance.

The approach presented in this paper, although experimental results show that there is still room from improving its effectiveness, can be exploited to build the core components of music teaching tools. To this end, the choice of classical music as the targeted genre becomes particularly suitable, in order to promote music cultural heritage through the use of new technologies.

References

1. Bartsch, M.A., Wakefield, G.H.: Audio thumbnailing of popular music using chroma-based representations. IEEE Transactions on Multimedia 7(1), 96–104 (2005)
2. Cano, P., Loscos, A., Bonada, J.: Score-performance matching using HMMs. In: Proceedings of the International Computer Music Conference, pp. 441–444 (1999)
3. Choi, F.Y.Y.: Advances in domain independent linear text segmentation. In: Proceedings of the Conference on North American chapter of the Association for Computational Linguistics, pp. 26–33 (2000)
4. Cont, A.: Realtime audio to score alignment for polyphonic music instruments using sparse non-negative constraints and hierarchical HMMs. In: IEEE International Conference in Acoustics and Speech Signal Processing, pp. V245–V248 (2006)
5. Dannenberg, R.B., Mukaino, H.: New techniques for enhanced quality of computer accompaniment. In: Proceedings of the International Computer Music Conference, pp. 243–249 (1988)
6. Dixon, S., Widmer, G.: MATCH: a music alignment tool chest. In: Proceedings of the International Conference of Music Information Retrieval, pp. 492–497 (2005)
7. Doraisamy, S., Rüger, S.: A polyphonic music retrieval system using N-grams. In: Proceedings of the International Conference on Music Information Retrieval, pp. 204–209 (2004)
8. Durbin, R., Eddy, S., Krogh, A., Mitchison, G.: Biological Sequence Analysis. Cambridge University Press, Cambridge (2000)
9. Fujishima, T.: Realtime chord recognition of musical sound: a system using common Lisp music. In: Proceedings of the International Computer Music Conference, pp. 464–467 (1999)
10. Gómez, E., Herrera, P.: Estimating the tonality of polyphonic audio files: Cognitive versus machine learning modelling strategies. In: Proceedings of the International Conference on Music Information Retrieval, pp. 92–95 (2004)
11. Grubb, L., Dannenberg, R.B.: A stochastic method of tracking a vocal performer. In: Proceedings of the International Computer Music Conference, pp. 301–308 (1997)
12. Harte, C., Sandler, M., Abdallah, S., Gómez, E.: Symbolic representation of musical chords: a proposed syntax for text annotations. In: Proceedings of the International Conference on Music Information Retrieval, pp. 66–71 (2005)

13. Hu, N., Dannenberg, R.B., Tzanetakis, G.: Polyphonic audio matching and alignment for music retrieval. In: Proceedings of the IEEE Workshop on Applications of Signal Processing to Audio and Acoustics, pp. 185–188 (2003)
14. Herrera, P., Serrá, J., Gómez, E., Serra, X.: Chroma binary similarity and local alignment applied to cover song identification. IEEE Transactions on Audio, Speech, and Language Processing 16(6), 1138–1151 (2008)
15. Kurth, F., Müller, M.: Efficient index-based audio matching. IEEE Transactions on Audio, Speech, and Language Processing 16(2), 382–395 (2008)
16. Middleton, R.: Studying Popular Music. Open University Press, Philadelphia (2002)
17. Miotto, R., Orio, N.: Automatic identification of music works through audio matching. In: Proceedings of 11th European Conference on Digital Libraries, pp. 124–135 (2007)
18. Miotto, R., Orio, N.: A music identification system based on chroma indexing and statistical modeling. In: Proceedings of the International Conference on Music Information Retrieval, pp. 301–306 (2008)
19. Mohri, M.: Finite-state transducers in language and speech processing. Computational Linguistics 23(2), 269–311 (1997)
20. Montecchio, N., Orio, N.: Automatic alignment of music performances with scores aimed at educational applications. In: AXMEDIS 2008: Proceedings of the 2008 International Conference on Automated solutions for Cross Media Content and Multi-channel Distribution, Washington, DC, USA, pp. 17–24. IEEE Computer Society, Los Alamitos (2008)
21. Nattiez, J.-J.: Musicologie générale et sémiologie. Christian Bourgois éditeur, Paris, FR (1987)
22. Orio, N.: Alignment of performances with scores aimed at content-based music access and retrieval. In: Proceedings of European Conference on Digital Libraries, pp. 479–492 (2002)
23. Orio, N., Déchelle, F.: Score following using spectral analysis and hidden Markov models. In: Proceedings of the International Computer Music Conference, pp. 125–129 (2001)
24. Orio, N., Schwarz, D.: Alignment of monophonic and polyphonic music to a score. In: Proceedings of the International Computer Music Conference, pp. 129–132 (2001)
25. Orio, N., Zattra, L.: Audio matching for the philological analysis of electroacoustic music. In: Proceedings of the International Computer Music Conference, pp. 157–164 (2007)
26. Peeters, G.: Chroma-based estimation of musical key from audio-signal analysis. In: Proceedings of the International Conference of Music Information Retrieval, pp. 115–120 (2006)
27. Rabiner, L.R.: A tutorial on hidden Markov models and selected application. Proceedings of the IEEE 77(2), 257–286 (1989)
28. Raphael, C.: Automatic segmentation of acoustic musical signals using hidden Markov models. IEEE Transactions on Pattern Analysis and Machine Intelligence 21(4), 360–370 (1999)
29. Reynar, J.C.: Topic Segmentations: Algorithms and Applications. PhD Thesis, Computer and Information Science, University of Pennsylvania, USA (1998)
30. Shifrin, J., Pardo, B., Meek, C., Birmingham, W.: HMM-based musical query retrieval. In: Proceedings of the ACM/IEEE Joint Conference on Digital Libraries, pp. 295–300 (2002)

31. Stammen, D.R., Pennycook, B.: Real-time recognition of melodic fragments using the dynamic timewarp algorithm. In: Proceedings of the International Computer Music Conference, pp. 232–235 (1993)
32. Turetsky, R.J., Ellis, D.P.W.: Ground-truth transcriptions of real music from force-aligned MIDI syntheses. In: Proceedings of the International Conference of Music Information Retrieval, pp. 135–141 (2003)
33. Typke, R., Wiering, F., Veltkamp, R.C.: A search method for notated polyphonic music with pitch and tempo fluctuations. In: Proceedings of the International Conference of Music Information Retrieval, pp. 281–288 (2004)

Harmonic and Percussive Sound Separation and Its Application to MIR-Related Tasks

Nobutaka Ono, Kenichi Miyamoto, Hirokazu Kameoka, Jonathan Le Roux,
Yuuki Uchiyama, Emiru Tsunoo, Takuya Nishimoto, and Shigeki Sagayama

Abstract. In this chapter, we present a simple and fast method to separate a monaural audio signal into harmonic and percussive components, which leads to a useful pre-processing for MIR-related tasks. Exploiting the anisotropies of the power spectrograms of harmonic and percussive components, we define objective functions based on spectrogram gradients, and, applying to them the auxiliary function approach, we derive simple and fast update equations which guarantee the decrease of the objective function at each iteration. We show experimental results for sound separation on popular and jazz music pieces, and also present the application of the proposed technique to automatic chord recognition and rhythm-pattern extraction.

1 Introduction

Recently, music signal has become an important target in the signal processing field. In the Music Information Retrieval Evaluation eXchange (MIREX), various tasks related to music information retrieval (MIR) have been discussed such as audio onset detection, multiple fundamental frequency estimation, audio chord detection, and so on [1]. Since music signals consist of various kinds of tones due to different instruments and different expressions, these tasks are difficult and challenging.

The many tones which form a music piece can be broadly classified into two components: a harmonic one and a percussive one. Their simultaneous presence makes some tasks much harder because of their very different spectral structures. For instance, most of the multi-pitch analysis methods are disturbed by percussive tones,

Nobutaka Ono · Kenichi Miyamoto · Hirokazu Kameoka · Jonathan Le Roux · Yuuki Uchiyama · Emiru Tsunoo · Takuya Nishimoto · Shigeki Sagayama
Department of Information Physics and Computing, Graduate School of Information Science and Technology, The University of Tokyo, 7-3-1 Hongo Bunkyo-ku, Tokyo, 113-8656, Japan
e-mail: {onono,miyamoto,kameoka,leroux,uchiyama,tsunoo,nishi,
sagayama}@hil.t.u-tokyo.ac.jp
http://hil.t.u-tokyo.ac.jp/index-e.html

Z.W. Raś and A.A. Wieczorkowska (Eds.): Adv. in Music Inform. Retrieval, SCI 274, pp. 213–236.
springerlink.com   © Springer-Verlag Berlin Heidelberg 2010

while the suppression of harmonic components would facilitate drum detection or rhythm analysis.

The separation of a monaural audio signal into harmonic and percussive components has been discussed in several pilot works. Uhle *et al.* applied Independent Component Analysis (ICA) to the magnitude spectrogram, and classified the extracted independent components into a harmonic group and a percussive group based on several features like percussiveness or noise-likeness [2]. Helen *et al.* utilized Non-negative Matrix Factorization (NMF) for decomposing the spectrogram into elementary patterns and classified them by pre-trained Support Vector Machine (SVM) [3]. Through the modeling of harmonic and inharmonic tones at the spectrogram level, Itoyama *et al.* aimed at developing an instrument equalizer and proposed a method for separating an audio signal into single-instrument tracks based on MIDI information synchronized to the input audio signal [4].

For pre-processing of MIR-related tasks, it is often not practical to exploit a priori knowledge of the score or of the included instruments of the input audio signals as such information is in general not available, and a simple and fast algorithm which does not require such knowledge is preferable. Aiming towards such an algorithm, we have developed the so-called harmonic and percussive sound separation (HPSS) technique, which relies solely on the difference between the structures of the spectrograms of harmonic and percussive components and does not need any pre-learning.

While the approaches mentioned above are pattern recognition oriented, our approach is closer to a "sinusoid plus transient model". Because many audio signals including speech and music consist of steady-state parts and transient parts, this modeling can be widely applied for signal enhancement, time stretching, pitch conversion, coding, information retrieval, etc. Daudet, in his very good review paper, has classified recent algorithms for the separation of sinusoid and transient components into three categories [5]:

1. Linear prediction: provides a decomposition of the sound into its excitation signal and a resonating filter.
2. Tonal extraction: does not define transients directly, but rather extract from the signal its "tonal" part (also called sinusoidal part). The residual is then assumed to contain mostly transients. Adaptive phase vocoder, sinusoidal model and subspace methods are classified into this category.
3. Sines + Transients + Noise Models: based on some explicit model for sinusoids and transients, decomposes the sound into a sinusoidal part, a transient part, and a residual noise part. Sequential estimation in orthonormal bases, adapted time-frequency tiles, matching pursuit, etc., are classified into this category.

Because HPSS has explicit models for both harmonic (sinusoid) and percussive (transient) components, it should be classified into the third category. The remarkable features of HPSS are the following.

- The model is very simple: the harmonic and the percussive components should be smooth horizontally and vertically, respectively, in the spectrogram domain.
- No dictionaries are used, and the method does not need pre- learning.

- The two separated components are obtained in the spectrogram domain, and can thus be directly used to perform some feature extraction without having to go via the time domain.
- The iterative calculation can be performed sequentially by sliding-block analysis (described in section 4), which enables us to implement it in real time.

In the following, we present the formulation of the separation as an optimization problem, derive a fast iterative solution to that problem through the auxiliary function approach, and evaluate the separation performance by experiments on popular and jazz music pieces. As applications of the proposed technique to MIR-related tasks, automatic chord detection and rhythm pattern extraction are also described.

2 Formulation of Harmonic/Percussive Separation

2.1 Anisotropy of Harmonic and Percussive Spectrograms

Let $F_{\omega,\tau}$ be the short-time Fourier transform (STFT) of a monaural audio signal $f(t)$ and $W_{\omega,\tau} = |F_{\omega,\tau}|^2$ be its power spectrogram, where ω and τ represent the angular frequency bin and time frame indices, respectively.

A typical spectrogram of a popular music piece is shown in Fig. 1, where the vertical and horizontal structures are clearly observed. The harmonic component usually has stable pitch and forms parallel horizontal ridges with smooth temporal envelopes on the spectrogram, while the energy of the percussive tone is concentrated in a short time frame, which forms vertical ridges with wide-band spectral envelopes. Hence, at the power spectrogram level, the harmonic component $H_{\omega,\tau}$ and the percussive component $P_{\omega,\tau}$ should have the following properties:

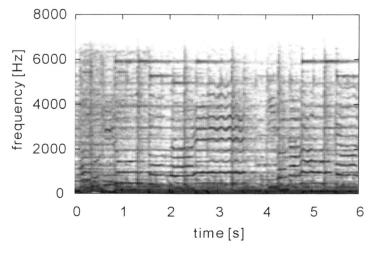

Fig. 1 A typical spectrogram of a popular music piece

- $H_{\omega,\tau}$ is horizontally smooth.
- $P_{\omega,\tau}$ is vertically smooth.
- The sum of $H_{\omega,\tau}$ and $P_{\omega,\tau}$ is close to the original power spectrogram $W_{\omega,\tau}$, although additivity does not rigorously hold in the power spectrogram domain.
- $H_{\omega,\tau}$ and $P_{\omega,\tau}$ are non-negative since they represent power.

The problem here is to find $H_{\omega,\tau}$ and $P_{\omega,\tau}$ satisfying these properties from the observed power spectrogram $W_{\omega,\tau}$. It can be formulated as an optimization problem, that of minimizing

$$J(H,P) = \sum_{\omega,\tau} D(W_{\omega,\tau},\ H_{\omega,\tau} + P_{\omega,\tau})$$

$$+ \frac{1}{2\sigma_H^2}\sum_{\omega,\tau}(H_{\omega,\tau-1}^{\gamma} - H_{\omega,\tau}^{\gamma})^2 + \frac{1}{2\sigma_P^2}\sum_{\omega,\tau}(P_{\omega-1,\tau}^{\gamma} - P_{\omega,\tau}^{\gamma})^2 \quad (1)$$

under the conditions that $H_{\omega,\tau} \geq 0$ and $P_{\omega,\tau} \geq 0$, where $H = (H_{\omega,\tau})_{\omega,\tau}$ and $P = (P_{\omega,\tau})_{\omega,\tau}$ represent the sets of all $H_{\omega,\tau}$ and $P_{\omega,\tau}$, respectively. The first term of the objective function measures the distance between $W_{\omega,\tau}$ and $H_{\omega,\tau} + P_{\omega,\tau}$, while the second and third terms are cost functions on the smoothness of $H_{\omega,\tau}$ and $P_{\omega,\tau}$, respectively, with σ_H and σ_P determining their weights. A range-compression factor γ is introduced for balancing the first term and the second and the third terms as described later. There are several possibilities for selecting the distance measure function $D(\cdot,\cdot)$ and the balance parameter γ. Among them, we shall investigate in particular two different objective functions in the following sections.

2.2   Method 1: I-Divergence-Based Method

As power spectrograms can be considered more generally as non-negative distributions, a measure of the difference between two power spectrograms $A_{\omega,\tau}$ and $B_{\omega,\tau}$ can be obtained through their I-divergence [6] defined by

$$I(A_{\omega,\tau}, B_{\omega,\tau}) = \left(A_{\omega,\tau}\log\frac{A_{\omega,\tau}}{B_{\omega,\tau}} - A_{\omega,\tau} + B_{\omega,\tau}\right). \quad (2)$$

Strictly speaking, it is not a distance but a divergence since it is not symmetric in $A_{\omega,\tau}$ and $B_{\omega,\tau}$. But due to its logarithmic nature which fits well auditory perception and the fact that it is easy to handle mathematically, the I-divergence has been used in several power-spectrogram-based signal processing methods such as NMF and Harmonic Temporal Clustering (HTC) [7, 8, 9, 10] and can also be used in our problem [11].

When using I-divergence as a distance measure in eq. (1), a desirable property is to keep the balance between the distance measure (the first term) and the smoothness cost (the second and third terms) under scale change. Since $I(aA_{\omega,\tau}, aB_{\omega,\tau}) = aI(A_{\omega,\tau}, B_{\omega,\tau})$ for any non-negative scale parameter a, the range-compression factor

γ should be set to 0.5 to ensure scale invariance. The objective function can then be written as

$$J_1(H,P) = \sum_{\omega,\tau} \left\{ W_{\omega,\tau} \log \frac{W_{\omega,\tau}}{H_{\omega,\tau} + P_{\omega,\tau}} - W_{\omega,\tau} + H_{\omega,\tau} + P_{\omega,\tau} \right.$$
$$\left. + \frac{1}{2\sigma_H^2} (\sqrt{H_{\omega,\tau-1}} - \sqrt{H_{\omega,\tau}})^2 + \frac{1}{2\sigma_P^2} (\sqrt{P_{\omega-1,\tau}} - \sqrt{P_{\omega,\tau}})^2 \right\}. \quad (3)$$

2.3   Method 2: L_2-Norm-Based Method

Another approach aims at obtaining a simpler formulation [12]. Since the intersection of the horizontal and vertical ridges is small, it can be assumed that they are approximately disjoint. If either $W_{\omega,\tau} = H_{\omega,\tau}$ or $W_{\omega,\tau} = P_{\omega,\tau}$ are almost satisfied at each time-frequency bin (ω, τ), we can assume that

$$\tilde{W}_{\omega,\tau} = \tilde{H}_{\omega,\tau} + \tilde{P}_{\omega,\tau}, \quad (4)$$

for any γ where

$$\tilde{W}_{\omega,\tau} = W_{\omega,\tau}^{\gamma}, \quad \tilde{H}_{\omega,\tau} = H_{\omega,\tau}^{\gamma}, \quad \tilde{P}_{\omega,\tau} = P_{\omega,\tau}^{\gamma}. \quad (5)$$

Although eq. (4) is a rather rough assumption, it leads to a simple formulation. Under the constraint of eq. (4), the distance term in eq. (1) vanishes and the objective function is given by

$$J_2(\tilde{H},\tilde{P}) = \frac{1}{2\sigma_H^2} \sum_{\omega,\tau} (\tilde{H}_{\omega,\tau-1} - \tilde{H}_{\omega,\tau})^2 + \frac{1}{2\sigma_P^2} \sum_{\omega,\tau} (\tilde{P}_{\omega-1,\tau} - \tilde{P}_{\omega,\tau})^2, \quad (6)$$

where $\tilde{H} = (\tilde{H}_{\omega,\tau})_{\omega,\tau}$ and $\tilde{P} = (\tilde{P}_{\omega,\tau})_{\omega,\tau}$ represent the sets of all $\tilde{H}_{\omega,\tau}$ and $\tilde{P}_{\omega,\tau}$, respectively. As eq. (6) is a quadratic form in $\tilde{H}_{\omega,\tau}$ and $\tilde{P}_{\omega,\tau}$, it has a single global minimum. Hence, when using this objective function, we do not need to worry about local minimum problems.

3   Derivation of Update Equations through the Auxiliary Function Approach

3.1   Auxiliary Function Approach

Minimizing eq. (3) is a nonlinear optimization problem for which there is no closed-form solution. Although eq. (6) is a quadratic form in $\tilde{H}_{\omega,\tau}$ and $\tilde{P}_{\omega,\tau}$ with a linear constraint and has a closed-form solution, its computation requires that of the inverse of a matrix with a very large number of variables, which is equal to the number of time-frequency bins. In order to avoid this computation and derive an effective iterative algorithm, we apply the auxiliary function approach, which is an extension of EM algorithm and has been recently applied to solve optimization problems in the signal processing field [7, 8, 9, 10].

In order to introduce the auxiliary function approach, let us here consider the general optimization problem of the minimization of an objective function $J(\theta)$ where θ represents a parameter vector. The problem is to find $\theta = \theta^*$ satisfying

$$\theta^* = \mathrm{argmin}_\theta\, J(\theta). \tag{7}$$

A simple way to solve this problem is, under sufficient smoothness conditions, to find a solution of the following equation:

$$\frac{\partial J(\theta)}{\partial \theta} = 0. \tag{8}$$

But, in many cases, this equation has no closed-form solution.

In the auxiliary function approach, a function $Q(\theta, \bar{\theta})$ is designed such that it satisfies

$$J(\theta) = \min_{\bar{\theta}}\, Q(\theta, \bar{\theta}). \tag{9}$$

$Q(\theta, \bar{\theta})$ is called an auxiliary function for $J(\theta)$, and $\bar{\theta}$ are called auxiliary variables. Then, instead of directly minimizing the objective function $J(\theta)$, the auxiliary function $Q(\theta, \bar{\theta})$ is minimized in terms of θ and $\bar{\theta}$, alternatively, the variables being iteratively updated as

$$\bar{\theta}^{(l+1)} = \mathrm{argmin}_{\bar{\theta}}\, Q(\theta^{(l)}, \bar{\theta}), \tag{10}$$

$$\theta^{(l+1)} = \mathrm{argmin}_\theta\, Q(\theta, \bar{\theta}^{(l+1)}), \tag{11}$$

where l denotes the iteration index.

The principle of the auxiliary function method is based on the fact that $J(\theta)$ is non-increasing under the above updates, as can be seen in the following simple proof:

1. $Q(\theta^{(l)}, \bar{\theta}^{(l+1)}) = J(\theta^{(l)})$ from eq. (9) and eq. (10),
2. $Q(\theta^{(l+1)}, \bar{\theta}^{(l+1)}) \leq Q(\theta^{(l)}, \bar{\theta}^{(l+1)})$ from eq. (11),
3. $J(\theta^{(l+1)}) \leq Q(\theta^{(l+1)}, \bar{\theta}^{(l+1)})$ from eq. (9),

thus,

$$J(\theta^{(l+1)}) \leq J(\theta^{(l)}), \tag{12}$$

which guarantees that the objective function is non-increasing. Note that even if eq. (7) has no closed-form solutions, in some cases we can design an auxiliary function $Q(\theta, \bar{\theta})$ satisfying eq. (9) such that both eq. (10) and eq. (11) have closed-form solutions. In such situations, the auxiliary function approach gives us efficient iterative update rules.

3.2   *Derivation of Update Rules for Method 1*

One way to design an auxiliary function for a given objective function is to exploit an inequality. Here, we would like to find an auxiliary function for J_1 defined by eq. (3). Focusing on the term $\log W_{\omega,\tau}/(H_{\omega,\tau}+P_{\omega,\tau})$ included in eq. (3), which is the reason why eq. (3) cannot be directly minimized in terms of $H_{\omega,\tau}$ and $P_{\omega,\tau}$, we can consider applying Jensen's inequality:

$$\lambda_1 f(x_1) + \lambda_2 f(x_2) \geq f(\lambda_1 x_1 + \lambda_2 x_2), \tag{13}$$

which holds for any convex function f (a function f is said to be convex if $f(\beta x_1 + (1-\beta)x_2) \leq \beta f(x_1) + (1-\beta)f(x_2)$ for any x_1 and x_2 and any $0 \leq \beta \leq 1$, which simply means that the graph of $y = f(x)$ is convex upward), and non-negative weights λ_1, λ_2 such that $\lambda_1 + \lambda_2 = 1$.

Let $f(x) = -\log x$ in eq. (13). Replacing x_1 and x_2 by x_1/λ_1 and x_2/λ_2, respectively, we have

$$\lambda_1 \log \frac{\lambda_1}{x_1} + \lambda_2 \log \frac{\lambda_2}{x_2} \geq \log \frac{1}{x_1 + x_2} \tag{14}$$

for non-negative λ_1 and λ_2 under $\lambda_1 + \lambda_2 = 1$, where the equality is satisfied when $\lambda_1 = x_1/(x_1 + x_2)$ and $\lambda_2 = x_2/(x_1 + x_2)$. Let us now look more closely at the term $\log 1/(x_1 + x_2)$ in the right-hand side in eq. (14). Letting $x_1 = H_{\omega,\tau}/W_{\omega,\tau}$ and $x_2 = P_{\omega,\tau}/W_{\omega,\tau}$ in eq. (14), we have

$$\lambda_1 \log \frac{\lambda_1 W_{\omega,\tau}}{H_{\omega,\tau}} + \lambda_2 \log \frac{\lambda_2 W_{\omega,\tau}}{P_{\omega,\tau}} \geq \log \frac{W_{\omega,\tau}}{H_{\omega,\tau} + P_{\omega,\tau}}. \tag{15}$$

Consequently, it is clear that the following function:

$$Q_1(H,P,m_H,m_P) =$$
$$\sum_{\omega,\tau} \left\{ m_{H\,\omega,\tau} W_{\omega,\tau} \log \left(\frac{m_{H\,\omega,\tau} W_{\omega,\tau}}{H_{\omega,\tau}} \right) + m_{P\,\omega,\tau} W_{\omega,\tau} \log \left(\frac{m_{P\,\omega,\tau} W_{\omega,\tau}}{P_{\omega,\tau}} \right) \right.$$
$$- (W_{\omega,\tau} - H_{\omega,\tau} - P_{\omega,\tau})$$
$$\left. + \frac{1}{2\sigma_H^2} (\sqrt{H_{\omega,\tau-1}} - \sqrt{H_{\omega,\tau}})^2 + \frac{1}{2\sigma_P^2} (\sqrt{P_{\omega-1,\tau}} - \sqrt{P_{\omega,\tau}})^2 \right\} \tag{16}$$

verifies

$$J_1(H,P) \leq Q_1(H,P,m_H,m_P), \tag{17}$$

for any H, P, m_H, and m_P under the condition that

$$m_{H\,\omega,\tau} + m_{P\,\omega,\tau} = 1. \tag{18}$$

It means that $Q_1(H,P,m_H,m_P)$ is an auxiliary function for $J_1(H,P)$ where $m_{H\,\omega,\tau}$ and $m_{P\,\omega,\tau}$ are auxiliary variables and m_H and m_P denote the sets of all $m_{H\,\omega,\tau}$ and $m_{P\,\omega,\tau}$, respectively. The equality in eq. (17) is satisfied for

$$m_{H\omega,\tau} = \frac{H_{\omega,\tau}}{H_{\omega,\tau} + P_{\omega,\tau}}, \tag{19}$$

$$m_{P\omega,\tau} = \frac{P_{\omega,\tau}}{H_{\omega,\tau} + P_{\omega,\tau}}. \tag{20}$$

Since eq. (19) and eq. (20) correspond to eq. (10), the auxiliary function Q_1 always decreases through the update of its auxiliary variables and then reaches J_1.

Meanwhile, the update rules corresponding to eq. (11) are obtained by solving $\partial Q_1/\partial H_{\omega,\tau} = 0$ and $\partial Q_1/\partial P_{\omega,\tau} = 0$. First, differentiating Q_1 with respect to $H_{\omega,\tau}$, we have

$$\frac{\partial Q_1}{\partial H_{\omega,\tau}} = -\frac{m_{H\omega,\tau}W_{\omega,\tau}}{H_{\omega,\tau}} + 1 + \frac{1}{2\sigma_H^2\sqrt{H_{\omega,\tau}}}(\sqrt{H_{\omega,\tau}} - \sqrt{H_{\omega,\tau+1}})$$
$$- \frac{1}{2\sigma_H^2\sqrt{H_{\omega,\tau}}}(\sqrt{H_{\omega,\tau-1}} - \sqrt{H_{\omega,\tau}}). \tag{21}$$

Setting the above derivative to zero and multiplying by $2H_{\omega,\tau}$, we get

$$-2m_{H\omega,\tau}W_{\omega,\tau} + 2H_{\omega,\tau} + \frac{\sqrt{H_{\omega,\tau}}}{\sigma_H^2}(\sqrt{H_{\omega,\tau}} - \sqrt{H_{\omega,\tau+1}})$$
$$- \frac{\sqrt{H_{\omega,\tau}}}{\sigma_H^2}(\sqrt{H_{\omega,\tau-1}} - \sqrt{H_{\omega,\tau}}) = 0. \tag{22}$$

It can be rewritten as a simple quadratic equation in $\sqrt{H_{\omega,\tau}}$, in the following form:

$$a_{H\omega,\tau}(\sqrt{H_{\omega,\tau}})^2 - b_{H\omega,\tau}(\sqrt{H_{\omega,\tau}}) - c_{H\omega,\tau} = 0, \tag{23}$$

whose well-known closed-form solution gives the update rule for $H_{\omega,\tau}$. The update rule for $P_{\omega,\tau}$ is obtained in a similar way. Altogether, the update rules are summarized as follows.

Update of the auxiliary variables

$$m_{H\omega,\tau} \leftarrow \frac{H_{\omega,\tau}}{H_{\omega,\tau} + P_{\omega,\tau}} \tag{24}$$

$$m_{P\omega,\tau} \leftarrow \frac{P_{\omega,\tau}}{H_{\omega,\tau} + P_{\omega,\tau}} \tag{25}$$

Update of the parameters

$$H_{\omega,\tau} \leftarrow \left(\frac{b_{H\omega,\tau} + \sqrt{b_{H\omega,\tau}^2 + 4a_{H\omega,\tau}c_{H\omega,\tau}}}{2a_{H\omega,\tau}}\right)^2 \tag{26}$$

$$P_{\omega,\tau} \leftarrow \left(\frac{b_{P\omega,\tau} + \sqrt{b_{P\omega,\tau}^2 + 4a_{P\omega,\tau}c_{P\omega,\tau}}}{2a_{P\omega,\tau}} \right)^2 \tag{27}$$

where

$$a_{H\omega,\tau} = \frac{2}{\sigma_H^2} + 2, \tag{28}$$

$$b_{H\omega,\tau} = \frac{\left(\sqrt{H_{\omega,\tau-1}} + \sqrt{H_{\omega,\tau+1}}\right)}{\sigma_H^2}, \tag{29}$$

$$c_{H\omega,\tau} = 2m_{H\omega,\tau}W_{\omega,\tau}, \tag{30}$$

$$a_{P\omega,\tau} = \frac{2}{\sigma_P^2} + 2, \tag{31}$$

$$b_{P\omega,\tau} = \frac{\left(\sqrt{P_{\omega-1,\tau}} + \sqrt{P_{\omega+1,\tau}}\right)}{\sigma_P^2}, \tag{32}$$

$$c_{P\omega,\tau} = 2m_{P\omega,\tau}W_{\omega,\tau}. \tag{33}$$

Since the update rules for $H_{\omega,\tau}$ and $P_{\omega,\tau}$ include the neighboring parameters ($H_{\omega,\tau-1}$, $P_{\omega+1,\tau}$, etc.) in eq. (29) and eq. (32), they should be applied sequentially, leading to the following algorithm: 1) update the auxiliary variables $m_{H\omega,\tau}$ and $m_{P\omega,\tau}$ at a certain time-frequency bin (ω, τ), 2) update the parameters $H_{\omega,\tau}$ and $P_{\omega,\tau}$ at the same bin (ω, τ), 3) move to the next time-frequency bin and go back to 1), iteratively. Note that their update rules keep $H_{\omega,\tau}$ and $P_{\omega,\tau}$ non-negative.

3.3   Derivation of Update Rules for Method 2

As eq. (6) is a quadratic form, sequential update rules are easily obtained from the differentiation of eq. (6) with respect to $\tilde{H}_{\omega,\tau}$ and $\tilde{P}_{\omega,\tau}$ with Lagrange multiplier terms for the constraints of eq. (4). Here, however, through the auxiliary function approach, we shall derive a parallel update scheme for each time-frequency bin, which we call complementary diffusion due to its interesting resemblance with a physical diffusion process.

Let us begin by noticing that the inequality

$$(A - B)^2 \leq 2(A - X)^2 + 2(B - X)^2 \tag{34}$$

holds for any A, B, and X, since

$$2(A - X)^2 + 2(B - X)^2 - (A - B)^2 = 4\left(X - \frac{A+B}{2}\right)^2 \tag{35}$$

is obviously non-negative and equal to zero when $X = (A + B)/2$. Applying the above inequality to eq. (6), we can show that the following function is an auxiliary function for J_2:

$$Q_2(\tilde{H}, \tilde{P}, U, V) = \frac{1}{\sigma_H^2} \sum_{\omega, \tau} \left\{ (\tilde{H}_{\omega, \tau-1} - U_{\omega, \tau})^2 + (\tilde{H}_{\omega, \tau} - U_{\omega, \tau})^2 \right\}$$

$$+ \frac{1}{\sigma_P^2} \sum_{\omega, \tau} \left\{ (\tilde{P}_{\omega-1, \tau} - V_{\omega, \tau})^2 + (\tilde{P}_{\omega, \tau} - V_{\omega, \tau})^2 \right\}, \quad (36)$$

where $U_{\omega, \tau}$ and $V_{\omega, \tau}$ are auxiliary variables, and U and V are the sets of all $U_{\omega, \tau}$ and $V_{\omega, \tau}$, respectively. Equality in the inequality $J_2(\tilde{H}, \tilde{P}) \leq Q_2(\tilde{H}, \tilde{P}, U, V)$ is obtained for

$$U_{\omega, \tau} = \frac{\tilde{H}_{\omega, \tau-1} + \tilde{H}_{\omega, \tau}}{2}, \quad (37)$$

$$V_{\omega, \tau} = \frac{\tilde{P}_{\omega-1, \tau} + \tilde{P}_{\omega, \tau}}{2}, \quad (38)$$

which are the update rules for the auxiliary variables corresponding to eq. (10).

To obtain the update rules for the parameters, corresponding to eq. (11), under the constraint of eq. (4), we introduce Lagrange multipliers $\lambda_{\omega, \tau}$ and consider

$$Q'_2(\tilde{H}, \tilde{P}, U, V) = Q_2(\tilde{H}, \tilde{P}, U, V) + \sum_{\omega, \tau} \lambda_{\omega, \tau}(\tilde{H}_{\omega, \tau} + \tilde{P}_{\omega, \tau} - W_{\omega, \tau}). \quad (39)$$

Setting to zero the derivatives of the above function with respect to $\tilde{H}_{\omega, \tau}$, $\tilde{P}_{\omega, \tau}$ and $\lambda_{\omega, \tau}$ yields

$$\frac{2}{\sigma_H^2} (2\tilde{H}_{\omega, \tau} - U_{\omega, \tau+1} - U_{\omega, \tau}) + \lambda_{\omega, \tau} = 0, \quad (40)$$

$$\frac{2}{\sigma_P^2} (2\tilde{P}_{\omega, \tau} - V_{\omega+1, \tau} - V_{\omega, \tau}) + \lambda_{\omega, \tau} = 0, \quad (41)$$

$$\tilde{H}_{\omega, \tau} + \tilde{P}_{\omega, \tau} - W_{\omega, \tau} = 0. \quad (42)$$

Solving the above system of equations, we obtain

$$\tilde{H}_{\omega, \tau} = \frac{\alpha}{2} (U_{\omega, \tau+1} + U_{\omega, \tau}) + \frac{(1-\alpha)}{2} (2W_{\omega, \tau} - V_{\omega+1, \tau} - V_{\omega, \tau}), \quad (43)$$

$$\tilde{P}_{\omega, \tau} = \frac{(1-\alpha)}{2} (V_{\omega+1, \tau} + V_{\omega, \tau}) + \frac{\alpha}{2} (2W_{\omega, \tau} - U_{\omega, \tau+1} - U_{\omega, \tau}), \quad (44)$$

where

$$\alpha = \frac{\sigma_P^2}{\sigma_H^2 + \sigma_P^2}. \quad (45)$$

By substituting eq. (37) and eq. (38) into the right-hand sides of eq. (43) and eq. (44) we can remove the auxiliary parameters $U_{h,i}$ and $V_{h,i}$ from the update rules, leading to the following simple expressions:

$$\tilde{H}_{\omega,\tau} \leftarrow \tilde{H}_{\omega,\tau} + \Delta_{\omega,\tau}, \tag{46}$$

$$\tilde{P}_{\omega,\tau} \leftarrow \tilde{P}_{\omega,\tau} - \Delta_{\omega,\tau}, \tag{47}$$

where

$$\Delta_{\omega,\tau} = \alpha \left(\frac{\tilde{H}_{\omega,\tau-1} - 2\tilde{H}_{\omega,\tau} + \tilde{H}_{\omega,\tau+1}}{4} \right) - (1-\alpha) \left(\frac{\tilde{P}_{\omega-1,\tau} - 2\tilde{P}_{\omega,\tau} + \tilde{P}_{\omega+1,\tau}}{4} \right) \tag{48}$$

However, we have to note that as a result of these updates, $\tilde{H}_{\omega,\tau}$ and $\tilde{P}_{\omega,\tau}$ may become negative. In order to keep $\tilde{H}_{\omega,\tau}$ and $\tilde{P}_{\omega,\tau}$ non-negative, the update rules are applied as follows.

Update of the auxiliary variables

$$\Delta_{\omega,\tau} \leftarrow \alpha \left(\frac{\tilde{H}_{\omega,\tau-1} - 2\tilde{H}_{\omega,\tau} + \tilde{H}_{\omega,\tau+1}}{4} \right) - (1-\alpha) \left(\frac{\tilde{P}_{\omega-1,\tau} - 2\tilde{P}_{\omega,\tau} + \tilde{P}_{\omega+1,\tau}}{4} \right) \tag{49}$$

Update of the parameters

$$\tilde{H}_{\omega,\tau} \leftarrow \begin{cases} 0 & (\tilde{H}_{\omega,\tau} + \Delta_{\omega,\tau} < 0) \\ \tilde{W}_{\omega,\tau} & (\tilde{P}_{\omega,\tau} - \Delta_{\omega,\tau} < 0) \\ \tilde{H}_{\omega,\tau} + \Delta_{\omega,\tau} & \text{(otherwise)} \end{cases} \tag{50}$$

$$\tilde{P}_{\omega,\tau} \leftarrow \begin{cases} \tilde{W}_{\omega,\tau} & (\tilde{H}_{\omega,\tau} + \Delta_{\omega,\tau} < 0) \\ 0 & (\tilde{P}_{\omega,\tau} - \Delta_{\omega,\tau} < 0) \\ \tilde{P}_{\omega,\tau} - \Delta_{\omega,\tau} & \text{(otherwise)} \end{cases} \tag{51}$$

These update rules can be applied in parallel. The update procedure is summarized as follows.

1. Update the auxiliary variables $\Delta_{\omega,\tau}$ for all the time-frequency bins.
2. Update the parameters $\tilde{H}_{\omega,\tau}$ and $\tilde{P}_{\omega,\tau}$ for all the time-frequency bins.
3. Go to Step 1.

Since the auxiliary update variable $\Delta_{\omega,\tau}$ consists of the discrete second-order derivative of $\tilde{H}_{\omega,\tau}$ and $\tilde{P}_{\omega,\tau}$, the updates of eq. (50) and eq. (51) have basically the same form as the diffusion equation:

$$\frac{df}{dt} = C\frac{d^2 f}{dx^2}, \tag{52}$$

which represents the dynamics of diffusion phenomena. For example, in the case of heat conduction process, f, t and x represent temperature, time and spatial coordinate, respectively, corresponding in our case to power spectrogram, iteration index and time or frequency, respectively. But unlike the physical diffusion process, in our case, each update of $\tilde{H}_{\omega,\tau}$ and $\tilde{P}_{\omega,\tau}$ includes a negative diffusion term derived from the other. With iterative calculations, the energy distribution of $\tilde{H}_{\omega,\tau}$ on the spectrogram diffuses horizontally and concentrates vertically, while $\tilde{P}_{\omega,\tau}$

follows the inverse way, the two always satisfying $\tilde{H}_{\omega,\tau} + \tilde{P}_{\omega,\tau} = \tilde{W}_{\omega,\tau}$. We denote this diffusion-like process of two energy distributions with a balance as *complementary diffusion*. The balance parameter α ($0 < \alpha < 1$) controls the strength of the diffusion along the vertical and the horizontal directions. Moreover, in Method 2, any value is allowable for the parameter γ, which can thus be tuned. According to our experiments, setting γ to about 0.3 gives a good performance.

4   Real-Time Processing by Sliding-Block Analysis

Although the objective functions eq. (3) and eq. (6) should include all time-frequency bins, the iterative updates for the whole bins are much time- and memory-consuming. In order to obtain an approximate solution in real time and make the computation feasible for signals of any length we propose a sliding-update algorithm. Based on the assumption that the separation of a certain time-frequency bin is weakly affected by bins far away, we limit the processed frames to $n \leq \tau \leq n+B-1$, where B is the size of the analysis block, and make n slide iteratively. The real-time version of Method 1 is summarized as follows.

1. Set the new frame as $H_{\omega,n+B-1} = P_{\omega,n+B-1} = W_{\omega,n+B-1}/2$.
2. Update variables by eq. (24), eq. (25), eq. (26) and eq. (27) sequentially for $n \leq \tau \leq n+B-1$.
3. Convert the nth frame to a waveform by inverse STFT.
4. Increment n to slide the analysis block.

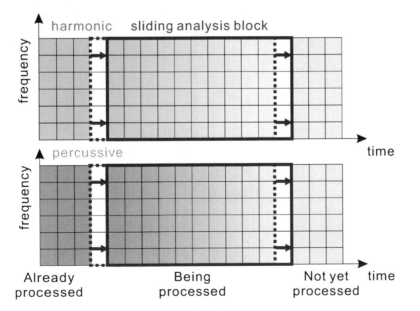

Fig. 2 Illustration of the sliding-block analysis process

The real-time version of Method 2 is processed in the same way. In step 3, the original phase is used to convert the STFT frame back to the time domain. Note that the overlap due to frame shift should actually be considered for the conversion.

Each time-frequency bin is updated only once at each step 2. Altogether, it is totally updated B times after passing through the analysis block shown in Fig. 2. Although larger block sizes B show better performance, the processing time from step 1 to step 4 must be less than the length of the frame shift for real-time processing.

5   Implementation and Evaluations

We have implemented our algorithms in several environments. An implementation for the MS Windows environment with graphical user interface (GUI) is shown in Fig. 3. After clicking the start button, the separation process begins. The processing steps are as follows.

1. Load the next frame-shift long fragment of the input audio signal from the WAV-format file.
2. Calculate the FFT of the new frame.
3. Update stored frames as described in the previous section.
4. Calculate the inverse FFT of the oldest frame.
5. Overlap-add the waveform and play the frame-shift long output fragment.
6. Go to Step 1.

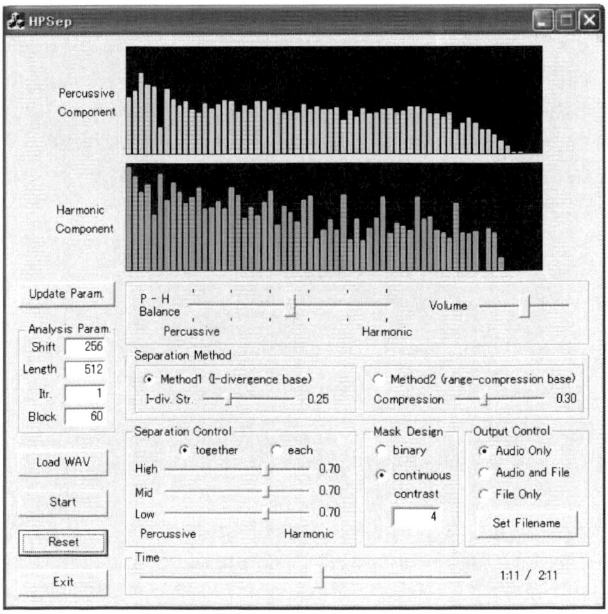

Fig. 3 GUI of the HPSS implementation on MS Windows

The two bar graphs shown in Fig. 3 represent the power spectra of the separated harmonic and percussive components. The sliding bar named "P-H Balance" enables the user to change the volume balance between the harmonic and percussive components while playing. The evolution of the separated spectrograms throughout the course of the sliding-block analysis is illustrated on an example in Fig. 4, where Method 1 was used. We can see that the input power spectrogram is sequentially separated as it passes through the analysis block. Through auditory evaluation, we could make the following observations.

- The pitched instrument tracks and the percussion tracks are well separated by both Method 1 and 2.
- For the same analysis block size, Method 1 gives a slightly better performance than Method 2.
- Method 1 requires about $1.5 \sim 2$ times the computational time of Method 2 because of square root calculations. Thus, Method 2 allows for larger block size.
- The separation results depend on several parameters such as σ_H, σ_P, the frame length, and the frame shift. However, the dependency is not so strong.

Another implementation with character user interface (CUI) on the Linux environment has been developed for fast execution, which is suitable to perform preprocessing for MIR-related task. The current fastest implementation separates a WAV-formated audio signal into harmonic and percussive components about thirty times faster than real time on a PC with 3GHz Pentium CPU and 2GB memory. It means that a song with 3-minute length can be separated into the two components in about 6 seconds.

In order to quantitatively evaluate the performance of the harmonic/percussive separation and its relationship to the block size, we prepared data for each track of two music pieces (RWC-MDB-J-2001 No.16 and RWC-MDB-P-2001 No.18 in [13]) by MIDI-to-WAV conversion and used the summation of all the tracks as input to our algorithms. The experimental conditions are given in Table 1. As a criterion of the performance, the energy ratio of the harmonic component $h(t)$ and the percussive component $p(t)$ included in each track was calculated as

$$r_h = \frac{E_h}{E_h + E_p}, \quad r_p = \frac{E_p}{E_h + E_p}, \tag{53}$$

with

$$E_h = < f_i(t), h(t) >^2, \quad E_p = < f_i(t), p(t) >^2, \tag{54}$$

where $<,>$ represents the cross-correlation operation and $f_i(t)$ the normalized signal of each track. The results are shown in Fig. 5. The pitched instrument tracks and the percussion tracks are represented by solid and dotted lines, respectively. We can see that the separation was generally well performed. Only the bass drum track has a tendency to belong to the harmonic component, which can be considered as a consequence of the long duration of the bass drum sounds. Fig. 5 also shows that a large block size is not required and that the separation performance converges for block sizes of 30 or 40 frames in this condition.

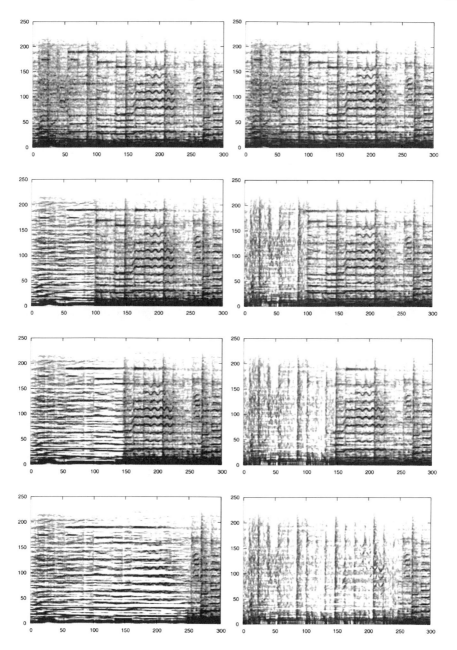

Fig. 4 Example of evolution of the spectrograms of the separated harmonic component (left) and percussive component (right) through sliding-block analysis. The first frame of the analysis block is 0, 50, 100, and 200 from top to bottom, respectively.

Table 1 Experimental conditions

Signal length	10 s
Sampling rate	16 kHz
Frame size	512
Frame shift	256
Range-compression factor (Method 1)	$\gamma = 0.5$
Range-compression factor (Method 2)	$\gamma = 0.3$
Gradient variance	$\sigma_P = \sigma_H = 0.3$

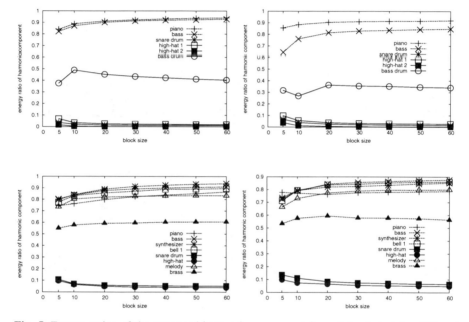

Fig. 5 Energy ratios of the separated harmonic component in each track (r_h) for different block sizes. Top left: Results for Method 1 on RWC-MDB-J-2001 No.16; Top right: Results for Method 2 on RWC-MDB-J-2001 No.16; Bottom left: Results for Method 1 on RWC-MDB-P-2001 No.18; Bottom right: Results for Method 2 on RWC-MDB-P-2001 No.18.

6 Application 1: Automatic Chord Detection

Automatic chord detection is one of the most important tasks in content-based analysis of music, with many applications such as music information retrieval, music identification and automatic music transcription. A chord detection task from audio data has been included in MIREX since 2008 [1].

Kawakami *et al.* [14] applied Hidden Markov Models (HMM) to chord-sequence modeling back in 1999, modeling the hidden chord progression as Markovian state

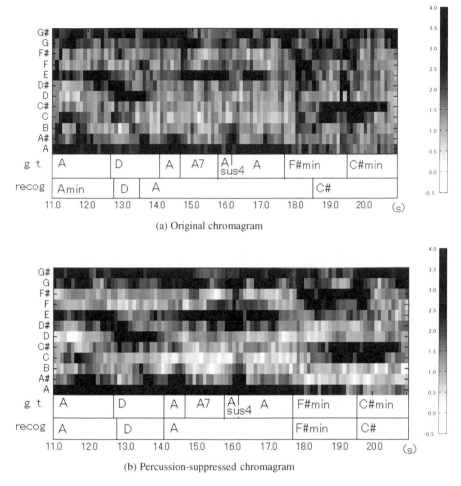

(a) Original chromagram

(b) Percussion-suppressed chromagram

Fig. 6 Comparison of chromagrams before and after applying HPSS for "I Need You" by the Beatles

transitions and considering melody as a stochastic output from the hidden states. This formulation was used to find the optimal chord sequence for an input melody represented in a symbolic (MIDI) form by the Viterbi algorithm. To deal with multi-arrangements of chords, Fujishima used Pitch Class Profiles (PCPs) [15] and a similar idea was proposed by Bartsch with chroma vectors [16]. Both PCPs and chroma vectors are calculated by accumulating the power spectrogram of each chromatic note over octaves to roughly represent the intensities of the twelve chromatic notes. Fujishima exploited this feature to perform symbolic (MIDI) chord detection, while Sheh and Ellis [17] applied the chroma vectors to audio chord detection with HMM.

Since then, the framework using HMM and chroma vectors has become mainstream in audio chord detection [18, 19]. However, popular and jazz music usually contain not only melodic and chordal sounds but also percussive sounds. Since strong percussive tones overwrite the information of the intensities of the chromatic notes in the chroma vectors, they are very harmful in this modeling.

The HPSS technique works as a useful pre-processing to reduce the percussive sounds. Fig. 6 shows the effect of HPSS on a chromagram, where vertical and horizontal axes respectively represent the components of chroma vector and time, and the intensity of the chromatic notes is represented in gray-scale. The upper and lower figures respectively represent the original chromagram and the percussion-suppressed chromagram computed on the power spectrogram obtained by applying HPSS, for the song "I Need You" by the Beatles. The input audio data was first down-sampled to 11.025kHz, then a constant-Q Gabor filter bank was applied, where the Q value was 60 and the center frequencies of the 60 filters were ranged from 55.0Hz (A1) to 1661.2Hz (G6). Finally, the chromagram was computed by accumulating the subband energies into twelve elements corresponding to the twelve pitch classes every 0.1 s intervals. To obtain the percussion-suppressed chromagram, the harmonic component was extracted from the original audio signal by the I-divergence-based HPSS with 2048-point frame length and 1024-point frame shift, and the chromagram was calculated from it in the same way as the original chromagram. In both figures, the ground-truth chords and the recognition results obtained by our HMM-based automatic chord detection system described in [21] are also shown.

The original chromagram looks blurred and the dominant chroma element is not clear in several frames due to the presence of strong percussive tones. On the other hand, in the percussion-suppressed chromagram, the contrast between elements becomes much stronger, which leads to the correction of the original errors on chords A and F#min.

To evaluate our audio chord recognition system, composed of percussion suppression by HPSS and an HMM-based recognizer, the following experiment was conducted. The data set consisted in 180 songs from 12 albums of The Beatles. In this experiment, 25 chord categories (major and minor triads and no chord) were used. For the HMM, each category was modeled by a single hidden state, and the output probability distribution was modeled by a single Gaussian with a diagonal covariance matrix. In two-fold cross-validation, the recognition rate without HPSS was 45.73%, while it increased to 72.48% by suppressing the percussive components with HPSS.

Our system was also evaluated in the audio chord detection task (train test) in MIREX 2008, where submitted programs were trained on 2/3 of the Beatles dataset including 176 songs and evaluated on 1/3. The performance was evaluated by overlap score, which was calculated as the ratio between the overlap of the ground truth and the detected chords and ground truth duration. As shown in Table 2, our system showed the best performance in MIREX 2008.

Table 2 Results of the audio chord detection task (train test) in MIREX 2008 [20]

Rank	Participant	Overlap Score
1	Uchiyama, Miyamoto, Ono & Sagayama	0.72
2	Ellis, D.	0.69
3	Weil & Durrieu (2)	0.62
4	Weil, J. (1)	0.60
5	Lee, K. (withtrain)	0.58
6	Khadkevich & Omologo	0.55
7	Zhang & Lash	0.36

7   Application 2: Extraction of Unit Rhythmic Patterns and Rhythmic Structure

Rhythm is a fundamental element of music and a very significant cue to characterize it. Generally, a music piece is constituted of multiple fundamental rhythmic patterns. These unit rhythmic patterns segment the music piece into measures and determine its structure at the beat and measure levels, while the global structure of the piece can be represented by the way the multiple rhythmic patterns appear in the whole music piece. Being able to properly extract multiple unit rhythmic patterns from a music piece and analyze its musical structure in terms of the use of these unit rhythmic patterns would thus be very helpful in music genre identification and music information retrieval.

The most fundamental aspect of rhythm analysis is beat tracking [23], which is a technique to detect the temporal positions of the beats in a piece of music. To extract the periodicity of the beats, Tzanetakis and Cook proposed a beat histogram in [24], which is obtained by collecting the dominant peaks of the enhanced auto-correlation of temporal envelopes on subbands. Instead of the histogram, Peeters applied the Discrete Fourier Transform (DFT) and the Auto-Correlation Function (ACF) to the onset-energy function in [25]. Extracting rhythmic patterns from audio signals is also an important issue. Paulus and Klapuri represented rhythmic patterns by feature vectors containing loudness, spectral centroid, and MFCCs for measuring similarity [26]. Dixon *et al.* automatically extracted a dominant rhythmic pattern in an audio signal by bar detection in the amplitude envelope and k-means clustering [27]. The effectiveness of temporal and rhythmic features in genre classification has also been investigated [24, 25, 27].

To extract multiple rhythmic patterns in a piece of music, it is essential to accurately measure the similarity between the patterns. However, the simultaneous presence of melodic and chordal sounds makes this problem much harder. Even if percussive sounds are to a certain extent extracted, the automatic segmentation of a piece into multiple patterns is not an easy task. Although the segmentation and the extraction of rhythmic patterns have been independently handled in conventional approaches [26, 27], there is inherently a mutual connection between them, as in a "chicken-or-egg" problem: a fundamental bar-length rhythmic patterns may be determined only after unit boundaries in the music piece are given, while unit

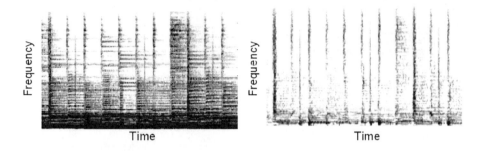

Fig. 7 Original spectrogram (left) and percussion-emphasized spectrogram (right) of a piece of popular music (RWC-MDB-G-2001 No.6) [30]

boundaries can be determined only after unit patterns are given. Another problem is tempo fluctuation, which makes unit rhythmic patterns stretch or shrink. Altogether, the problems to be solved for the extraction of unit rhythmic patterns from an input audio signal can be summarized as the following four points:

(i) an input acoustic signal may contain not only percussive sounds but also melodic and chordal sounds,
(ii) there may be fluctuations in tempo and in pattern itself made by the performer,
(iii) unit segmentation is unknown, and
(iv) unit rhythmic patterns are unknown.

Applying the HPSS technique and extracting the percussive component strongly emphasizes the percussive rhythmic patterns as shown in Fig. 7, giving a very satisfying solution to problem (i).

To solve problems (ii) and (iii), we focus on the analogy with speech recognition. If the set of unit rhythmic patterns is given as templates, problem (iii) can be considered as parallel to the continuous speech recognition problem where the One-Pass DP (Dynamic Programming) algorithm [28] can be employed to find the sequence of uttered words. Accordingly, One-Pass DP can be used to divide a music piece into segments, each optimally corresponding to a template pattern. Moreover, because of One-Pass DP's flexibility in time alignment, problem (ii) is solved simultaneously.

Conversely to problem (iii), problem (iv), the estimation of the unit rhythmic patterns, can be solved by a clustering algorithm if the segmentation is given. Hence, to solve both problems (iii) and (iv), it is necessary to estimate segmentation and unit patterns simultaneously. While this kind of unsupervised training problems have been solved in various ways, the k-means clustering algorithm is employed in combination with the One-Pass DP algorithm in our system, where the unit rhythmic patterns and the music structure are trained iteratively. The details are described in [29, 30].

Fig. 8 shows an experimental result for a piece of dance music, RWC-MDB-G-2001 No. 16 from the RWC music database [31] down-sampled to 22.05 kHz. We obtained the percussive spectral patterns using the I-divergence-based HPSS

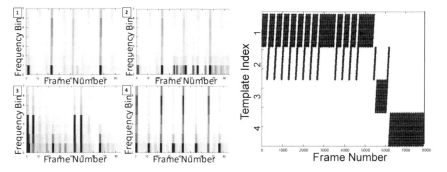

Fig. 8 Four extracted unit rhythmic pattern spectrograms from a piece of dance music (RWC-MDB-G-2001 No. 16) (left) and the corresponding alignment, or "Rhythm Map" (right)

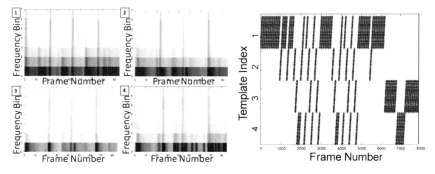

Fig. 9 Four extracted unit pattern spectrograms from a piece of dance music (RWC-MDB-G-2001 No. 16) (left) and the corresponding alignment (right) without HPSS

with 1024-point frame length and 512-point frame shift. Then, to reduce the dimensions of the patterns, the spectrum of each frame was summed up to eight subbands. The number of unit patterns was determined by the Bayesian information criterion (BIC), which was four in the case of the piece. The estimated unit rhythmic patterns are shown in the left part of Fig. 8 and the estimated alignment in the right part. By listening to the music, we were able to tell that pattern 1 was repeatedly played and once in four measures, pattern 2 was played. Following such fundamental rhythms, an interval rhythmic pattern was played (pattern 3), followed by a pattern in the climax part (pattern 4). This can be clearly seen in the right part of Fig. 8, which depicts the music structure in the form of a map of rhythmic patterns, and which we named "Rhythm Map". For comparison, extracted unit patterns and a map of patterns without the pre-processing to emphasize percussive component by HPSS are shown in Fig. 9. Due to strong melodic and chordal sounds, the extracted patterns do not represent characteristic rhythm patterns and the map of patterns are also very different from the rhythm structure of this song. Another example of Rhythm Map on a piece of popular music (RWC-MDB-G2001 No. 6) is illustrated in the right part of Fig. 10, while the left part of Fig. 10 shows the three corresponding

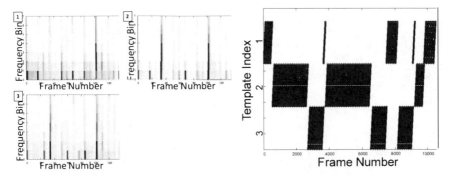

Fig. 10 Three extracted unit rhythmic pattern spectrograms from a piece of popular music (RWC-MDB-G-2001 No. 6) (left) and the corresponding Rhythm Map (right)

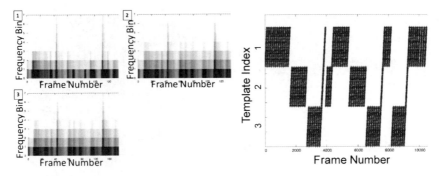

Fig. 11 Three extracted unit pattern spectrograms from a piece of popular music (RWC-MDB-G-2001 No. 6) (left) and the corresponding alignment (right) without HPSS

rhythmic patterns. We are currently investigating the exploitation of the extracted patterns as features for genre classification [32, 33].

8 Conclusion

In this paper, we presented the HPSS technique, which is a simple and fast separation algorithm of a monaural audio signal into harmonic and percussive components that does not require any a priori knowledge of the score or of the included instruments. Its principle is based on the anisotropies of the power spectrograms of the harmonic and percussive components. The update rules derived through auxiliary functions enable us to make the processing faster than real time. We also presented examples of applications of HPSS to automatic chord detection and rhythmic-pattern extraction, where HPSS works as an effective pre-processing for enhancing harmonic or percussive components. Applying it further to other MIR-related tasks such as audio onset detection, melody extraction and multi-pitch analysis is our current concern.

Acknowledgements. This research was supported by Grant-in-Aid for Scientific Research (A) 00303321 and CREST Muse project.

References

1. http://www.music-ir.org/mirex2008/index.php
2. Uhle, C., Dittmar, C., Sporer, T.: Extraction of drum tracks from polyphonic music using independent subspace analysis. In: Proc. ICA, April 2003, pp. 843–847 (2003)
3. Helen, M., Virtanen, T.: Separation of drums from polyphonic music using non-negative matrix factorization and support vector machine. In: Proc. EUSIPCO (September 2005)
4. Itoyama, K., Goto, M., Komatani, K., Ogata, T., Okuno, H.: Integration and Adaptation of Harmonic and Inharmonic Models for Separating Polyphonic Musical Signals. In: Proc. ICASSP, April 2007, pp. 57–60 (2007)
5. Daudet, L.: A Review on Techniques for the Extraction of Transients in Musical Signals. In: Kronland-Martinet, R., Voinier, T., Ystad, S. (eds.) CMMR 2005. LNCS, vol. 3902, pp. 219–232. Springer, Heidelberg (2006)
6. Csiszár, I.: I-Divergence Geometry of Probability Distributions and Minimization Problems. The Annals of Probability 3(1), 146–158 (1975)
7. Lee, D.D., Seung, H.S.: Algorithms for Non-Negative Matrix Factorization. In: Proc. NIPS, pp. 556–562 (2000)
8. Kameoka, H., Nishimoto, T., Sagayama, S.: A Multipitch Analyzer Based on Harmonic Temporal Structured Clustering. IEEE Trans. ASLP 15(3), 982–994 (2007)
9. Le Roux, J., Kameoka, H., Ono, N., de Cheveigne, A., Sagayama, S.: Single and Multiple F0 Contour Estimation Through Parametric Spectrogram Modeling of Speech in Noisy Environments. IEEE Trans. ASLP 15(4), 1135–1145 (2007)
10. Kameoka, H., Ono, N., Sagayama, S.: Auxiliary Function Approach to Parameter Estimation of Constrained Sinusoidal Model. In: Proc. ICASSP, April 2008, pp. 29–32 (2008)
11. Ono, N., Miyamoto, K., Kameoka, H., Sagayama, S.: A Real-time Equalizer of Harmonic and Percussive Components in Music Signals. In: Proc. ISMIR, Sepember 2008, pp. 139–144 (2008)
12. Ono, N., Miyamoto, K., Le Roux, J., Kameoka, H., Sagayama, S.: Separation of a Monaural Audio Signal into Harmonic/Percussive Components by Complementary Diffusion on Spectrogram. In: Proc. EUSIPCO (August 2008)
13. Goto, M., Hashiguchi, H., Nishimura, T., Oka, R.: RWC music database: Popular, classical, and jazz music databases. In: Proc. ISMIR, October 2002, pp. 287–288 (2002)
14. Kawakami, T., Nakai, M., Shimodaira, H., Sagayama, S.: Harmonization for melody using HMM. In: Proc. JHES, F-61, p. 361 (1999) (in Japanese)
15. Fujishima, T.: Real-time chord recognition of musical sound: A system using common lisp music. In: Proc. ICMC, pp. 464–467 (1999)
16. Bartsch, M.A., Wakefield, G.H.: To catch a chorus: Using chroma-based representations for audio thumbnailing. In: Proc. WASPAA, pp. 15–18 (2001)
17. Sheh, A., Ellis, D.P.W.: Chord segmentation and recognition using EM-trained hidden Markov models. In: Proc. ISMIR, pp. 183–189 (2003)

18. Bello, J.P., Pickens, J.: A robust mid-level representation for harmonic content in music signal. In: Proc. ISMIR, pp. 304–311 (2005)
19. Lee, K., Slaney, M.: Acoustic chord transcription and key extraction from audio using key-dependent HMMs trained on synthesized audio. IEEE Trans. on Audio Speech and Language Processing 16(2), 291–301 (2008)
20. http://www.music-ir.org/mirex/2008/results/ MIREX2008_overview_A0.pdf
21. Uchiyama, Y., Miyamoto, K., Nishimoto, T., Ono, N., Sagayama, S.: Automatic Chord Detection Using Harmonic Sound Emphasized Chroma from Musical Acoustic Signal. In: Proc. ASJ Spring Meeting, March 2008, pp. 901–902 (2008) (in Japanese)
22. Uchiyama, Y., Nishimoto, T., Ono, N., Sagayama, S.: HMM-based Audio Chord Detection Using Harmonic Emphasizing and Fourier-Transformed Chroma. IEEE Trans. on Audio Speech and Language Processing (submitted)
23. Goto, M.: An Audio-based Real-time Beat Tracking System for Music With or Without Drum-sounds. Journal of New Music Research 30(2), 159–171 (2001)
24. Tzanetakis, G., Cook, P.: Musical genre classification of audio signals. IEEE Trans. Speech and Audio Processing 10(5), 293–302 (2002)
25. Peeters, G.: Rhythm Classification Using Spectral Rhythm Patterns. In: Proc. ISMIR, September 2005, pp. 644–647 (2005)
26. Paulus, J., Klapuri, A.: Measuring the Similarity of Rhythmic Patterns. In: Proc. ISMIR, pp. 150–156 (2002)
27. Dixon, S., Guyon, F., Widmer, G.: Towards Characterization of Music via Rhythmic Patterns. In: Proc. ISMIR, pp. 509–516 (2004)
28. Ney, H.: The Use of a One-stage Dynamic Programming Algorithm for Connected Word Recognition. In: Proc. ICASSP, pp. 263–271 (1984)
29. Tsunoo, E., Miyamoto, K., Ono, N., Sagayama, S.: Rhythmic Features Extraction from Music Acoustic Signals using Harmonic/Non-Harmonic Sound Separation. In: Proc. ASJ Spring Meeting, March 2008, pp. 905–906 (2008) (in Japanese)
30. Tsunoo, E., Ono, N., Sagayama, S.: Rhythm Map: Extraction of Unit Rhythmic Patterns and Analysis of Rhythmic Structure from Music Acoustic Signals. In: Proc. ICASSP (April 2009)
31. Goto, M., Hashiguchi, H., Nishimura, T., Oka, R.: RWC Music Database: Music Genre Database and Musical Instrument Sound Database. In: Proc. ISMIR, October 2003, pp. 229–230 (2003)
32. Tsunoo, E., Tzanetakis, G., Ono, N., Sagayama, S.: Audio Genre Classification by Clustering Percussive Patterns. In: Proc. ASJ Spring Meeting, March 2009, pp. 877–878 (2009)
33. Tsunoo, E., Tzanetakis, G., Ono, N., Sagayama, S.: Audio Genre Classification Using Percussive Pattern Clustering Combined with Timbral Features. In: Proc. ICME (June 2009)

Violin Sound Quality: Expert Judgements and Objective Measurements

Piotr Wrzeciono and Krzysztof Marasek

Abstract. Searching for objective and subjective parameters is very important in automatic classification of multimedia databases containing recordings of musical instruments sounds. This paper describes these parameters and methods of obtaining them for a violin sound. The objective parameters are violin modes with their properties: frequency and mutual energy factor. The subjective parameter is evaluation of sound quality done by experts. Based on violin modes parameters, expert judgements and harmony perception, a sound quality classifier was created. The estimated value of sound quality evaluation is consistent with expert judgements for 75.5% of instruments from AMATI multimedia database containing recordings of violins from 10th International Henryk Wieniawski Violin Maker Competition.

1 Introduction

1.1 Multimedia Databases of Musical Instruments

There are many kinds of multimedia databases. They contain recordings of songs and various pieces of music, samples of natural sounds, etc. Special kinds of such databases are used for storing sounds of musical instruments. They are created mainly to provide data for psychoacoustic research and for search of specific parameters of instrument sounds. Most multimedia databases of sound contain recordings of a variety of instruments [4, 17, 19]. Other focus on one type of instrument, e.g. a

Piotr Wrzeciono
Polish Japanese Institute of Information Technology, ul. Koszykowa 86, 02-008 Warszawa, Poland
e-mail: `suigan@op.pl`

Krzysztof Marasek
Polish Japanese Institute of Information Technology, ul. Koszykowa 86, 02-008 Warszawa, Poland
e-mail: `kmarasek@pjwstk.edu.pl`

Z.W. Raś and A.A. Wieczorkowska (Eds.): Adv. in Music Inform. Retrieval, SCI 274, pp. 237–260.
springerlink.com   © Springer-Verlag Berlin Heidelberg 2010

violin [14, 15]. These databases also contain some additional information on instruments such as an evaluation of sound quality made by experts. All the data make it possible to analyze the correlation between subjective and objective properties of sound.

Another aspect of multimedia databases of musical instruments is their usefulness both for musicians and instrument makers. Reliable data sets enable experts to find the relations between musicians' opinions and physical properties of the sound.

1.2 Automatic Classification in Databases

The most of sound parameters used in instruments quality evaluation have a subjective character. When creating multimedia databases, one can ask experts to exploit their preferences, but if the database is very large, this method proves ineffective. Experts judgments depend on numerous factors, e.g. on tiredness, current mood, etc. This is the main reason for creating automatic classifiers which can estimate jurors' evaluations. Another approach involves calculation of just the objective parameters [11], but it is very difficult to find the relation between these properties and the physical characteristics of instruments. Therefore, the best way to create reliable classifiers is to search for the relation between jurors' opinions and the physical properties of the instrument.

1.3 Violin Sound Description and Classification

Musicians at present play a variety of music, from the Middle Ages to the modern period. Each kind of music requires a different set of instruments and interpretation [2, 7, 12]. This is also true for the violins. Both soloists and string ensembles want to choose the best instruments for a particular musical performance. This means that the proper description will depend on a reference point. This is especially important in evaluation of dependencies between subjective and objective properties of sound. For instance, spectral analysis of violin sound returns objective parameters, but musicians specialized in baroque music and those who specialize in romantic music will differ in their opinions on quality of the instrument [2, 7]. Therefore, the reference point has to depend on the historical period and the age of the analyzed instrument. Besides, the dependencies found between the objective parameters and the subjective properties cannot be directly applied to the violin from another period.

This is why musicians and violin makers have worked out two main manners of describing the violin sound. First of these is based on concepts derived from subjective terms which describe a sound [1, 24]: classical mean (soprano), bright, noble (soft), nasal, tight and constricted. The other is the evaluation using a point scale of violin sound quality [14, 15] and is typically used during violin makers competitions. Categories used in the latter one are following: sound timbre, loudness of a string, the ease with which a sound is generated by the string, the levelling between the strings and the individual properties of the violin sound [14, 15]. The sound timbre category is similar to the former description, but in fact, it describes the quality of the violin sound, not the kind of sound. Those concepts can be used as subjective classifiers in multimedia databases, although evaluation on the point scale are probably much better than the verbal terms for a sound, as two violins with different kinds of timbre can produce high quality sound [7, 12].

2 AMATI Multimedia Database

This section describes the main source of analyzed violin sounds, the AMATI multimedia database.

2.1 Instruments

The International Henryk Wieniawski Violin Maker Competition is organized in Poznań every five years. Its first edition was held in 1935 in Warsaw, and in 1952 the contest was relocated to Poznań [8]. It was in 2001, during the 10th competition, that the idea of multimedia database of violins used in the event was first offered. Dr Ewa Łukasik is the author of this idea and the Henryk Wieniawski Musical Society of Poznań agreed to create such the database which contains recordings of violin sounds, evaluations made by jury and violins' photos [14, 15].

All the instruments were made especially for this event. The violins came from many countries, e.g. Poland, Russia, Japan, South Korea, Italy, USA and have been designed for romantic and modern music. Seventy instruments were recorded, fifty four of which were included in the multimedia database. One of the violins, an instrument built in the 19th century by Dallenz, was not used in the competition, but was recorded to compare modern and old violins. The instruments in the AMATI multimedia database are marked by randomly assigned numbers.

2.2 Recordings

The violin maker competition was held at two places: the Marcin Groblicz Chamber in the Museum of Musical Instruments and the Concert Hall of Adam Mickiewicz University, both in Poznań, Poland. The recordings, however, were done only at the Marcin Groblicz Chamber to enable a correlation between jurors' opinions and the objective properties of sound. The conditions during the recording session were the same as those during the second stage of the competition. The violinist used the same bow during the entire session. The following sounds were recorded: whole strings (G,D,A,E) in both directions of a bow, pizzicatos on each string, chromatic scales on each string, diatonic scale and Sarabanda from Partita d-minor BWV 1004 by Johann Sebastian Bach. The same repertoire was performed during the second stage of the competition in the Marcin Groblicz Chamber. The scales and whole strings were played without tremolo and vibrato . The recordings were made in the near and far field. The following equipment was used: BRÜEL & KJÆR DPA4011 microphone with a cardioid characteristics for the near field, TONSIL MC358 microphone with a supercardioid characteristic for the far field, SONOSAX SX-M2 microphone preamplifier, and PHILIPS CDR880 CD-recorder. The sampling frequency was 44.1 kHz. The PCM signal had a 16-bit resolution. The recordings were made by Ewa Łukasik and Piotr Wrzeciono. First channel contains near field signal, while the second the far field signal. The near field recordings were made at about 1-m distance from the instrument. The far field recordings were made at about 10-m distance from the violin (Fig. 1).

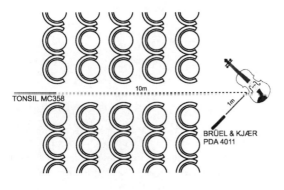

Fig. 1 Positioning of microphones in the Marcin Groblicz Chamber

2.3 Jury

Four independently working musicians formed the jury during the second stage of 10th Henryk Wieniawski International Violin Maker Competition. They judged according to the following categories:

- loudness of sound (per string – from 4 to 20 points),
- timbre of violin sound (per string – from 4 to 20 points),
- ease of sound generation (from 1 to 20 points),
- levelling between strings (from 1 to 15 points),
- correctness of the instrument's assembly (from 1 to 10 points)
- individual properties of the violin sound (from 1 to 15 points).

During each performance the jurors knew only the number assigned to the violin. The evaluation of the violin's assembly was made after the hearings. The members of the jury came from different countries. The final score of instrument was calculated as the arithmetic mean value of four evaluations given by jurors.

The AMATI database contains the complete evaluation from all four jurors.

3 Violin Modes Determination

Violin modes are used as the main parameter in description of instruments and are also very important for violin maker. This section describes properties of modes, methods of searching for them, and a violinist's reaction for their presence in energy spectrum.

3.1   The Violins Modes – Definition

The main parts of a violin body are its belly (table), back and sound post [1, 12]. The belly is the upper wooden plate under the strings, and the back is the lower wooden plate. The sound post is a small rod between the belly and the back. The role of the sound post is to transmit vibrations from the belly to the back. Both plates act as membranes. On these surfaces, as well as inside the violin body, standing waves appear [1,9,12]. These standing waves, which come from the belly, the back and the inside, are called violin modes [1,9,12]. Violin makers distinguish six modes [1,9]: C1, A0, C2, T1, C3 and C4. The C1 mode, which is determined by the shape of plates, has a frequency of 198 Hz. The frequency of the A0 mode is about 270 Hz and comes from the violin body. The T1 mode is named the main wood resonance. Typically, the relation between frequencies of the T1 mode and the A0 mode is 1.5 [1,9]. The C2 and C3 modes are close to the T1 mode. The C2 mode frequency is lower than that of the T1 mode, whereas the C3 frequency is higher than that of the T1 mode. The C4 mode lies in the range from 650 Hz to 810 Hz. A violin may have many more modes, but the presented modes have the highest energy in the spectrum of violin sound.

3.2   Impulse Response of Violin Body

The impulse response of the violin body is an instrument's reaction to an impact by a normalized weight [1, 24]. The normalized weight is dropped onto the bridge of a violin. Next, the vibrations of the instrument are measured by accelerometers and spectrally analyzed. The main assumption of this method is the linearity of the violin body [1, 24]. In reality, the violin is a nonlinear system [18]. Because of this, the analysis of an impulse response cannot be used to predict the instrument's timbre, but it is useful for searching of the violin modes. This method makes possible to find the frequencies of modes but is ineffective when searching for energetic relations between modes or for nonlinear analyses [18].

3.3   Chromatic Scales Played on Violin and Their Spectra

The analysis of the violin sound generated in a normal way (using the bow) is necessary when searching for significant correlations. For proper analysis of violin sound also normally generated sound must be used, e.g. recordings of musical scales. They contain sounds with different fundamental frequencies and harmonics of scale tones. There are numerous musical scales, e.g. major, minor, etc. One of them, the chromatic scale, is very special. It contains all the sounds that can be played on an instrument [6], making the recordings of chromatic scales very useful for the research. In Fig. 2 four energy spectra of chromatic scales played on the violin are shown. These spectra were calculated by DTFT with a 1 Hz frequency resolution. The scale of normalized energy is linear, whereas the scale of frequency is logarithmic.

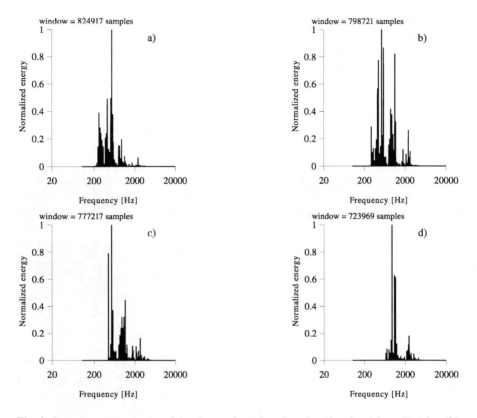

Fig. 2 Sample energy spectra of the chromatic scales played on G string (**a**), on D string (**b**), on A string (**c**) and on E string (**d**) (violin no. 30)

The energy spectra of chromatic scales have a very interesting property. All of them have the most prominent maximum for the same frequency and also the greater energy is concentrated in the neighborhood of these extrema. The frequencies of the maxima are the same as the modes frequencies, e.g. on Figures 2a, 2b, and 2c, the global maximum has the same frequency as the T1 mode. The modes are also present in the violin sound spectrum [20, 21, 22, 23].

3.4 Violin Player and His Reaction to Mode

It is possible to find modes in the energy spectrum of the violin sound (Fig.2). The question, however, is: can the violinist hear them? And how this affects his play? The hearing process is very complex, and so is the violin sound. In an attempt to find an answer to this question analysis of spectral neighborhood of modes was done.

First a statistical analysis of the violinist's precision when playing a chromatic scale was done [23]. In this research, statistical analysis of the fundamental frequencies was used. It was observed that the distributions of fundamental frequency of the tones for frequencies close to the modes were not Gaussian, while the frequency distributions of another tones of chromatic scales were Gaussian. Because of this, we hypothesize that the violinist hears the modes. The musician reacts to the mode and modifies his own style of performance. For this reason, the distribution of the fundamental frequency for tones close the violin modes was not the Gaussian distribution.

In the second analysis a comparison of energy spectra of the chromatic scale tones which are close to the violin modes frequencies was done. It was observed that the resonance curve of the string was modified when the relation between the energy of the fundamental frequency and the mode is less than 10 dB. If this difference was bigger than 10 dB, an unmodified resonance curve was observed. This effect was observed for the most of the instruments from the AMATI database. Fig.3 illustrates this effect for the C3 mode.

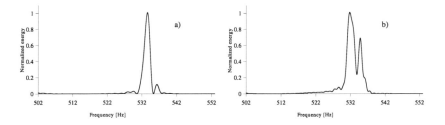

Fig. 3 **a.** The violinist did not hear the mode. The fundamental frequency: 533.68 Hz, the frequency of the mode: 536.38 Hz (violin no. 118) **b.** The violinist heard the mode and modified his own performance. The fundamental frequency: 531.58 Hz, the frequency of the mode: 534.58 Hz (violin no. 85).

All the instruments from the AMATI multimedia database were analyzed in this research. For all the modes the result was the same – when the difference between the energy of the fundamental tone and the energy of the mode is greater than ca 10 dB, the violinist does not hear the mode. In other cases, the violinist always modifies the resonance curve of the string.

Very similar results were obtained by Hafke [5] and Yoshida, Hasegawa and Kasuga [25].

4 Methods of Searching for Violin Modes

This section presents a method of searching for violin modes in energy spectra of chromatic scale. Description contains also a mathematical model of violin sound, and chromatic scales played on this instrument.

4.1 Mathematical Model of Violin Sound

As follows from our analysis the modes are present in the energy spectrum of violin sound and are also recognized by musicians. This can be used in an automatic search of modes [21, 22]. Violin sound consists of a fundamental frequency tone from the string, harmonics of the fundamental tone and the modes. A frequency and amplitude of the fundamental tone and its harmonics are stochastic processes. The amplitude of the modes are also stochastic, but the frequencies of all the modes are deterministic. Therefore, the violin sound can be described as a random process in the time domain:

$$U(t) = \sum_{k=1}^{N} F_k(t)A_k \cos(k\Omega t + \Phi_k) + \sum_{l=1}^{M} G_l(t)B_l \cos(\omega_l t + \Theta_l) \qquad (1)$$

where $U(t)$ is a random process which describes the violin sound. The first part describes the signal which comes from a string, the second describes the signal from the violin modes. Ω is the fundamental pulsation of the sound, N is the number of the harmonics, Φ_k is the phase of k-th harmonic. A_k is the amplitude of k-th harmonic, $F_k(t)$ is the envelope of k-th harmonic. M is the number of the violin modes, ω_l is the pulsation of the l-th violin mode, Θ_l is the phase of the l-th violin mode, B_l is the amplitude of the l-th mode and $G_l(t)$ is the envelope function of the l-th mode. The $F_k(t)$ and $G_l(t)$ are also random processes. A_k, Ω, Φ_k, B_l and Θ_l are the random variables. Only the pulsations of the modes are deterministic. The envelope functions $F_k(t)$ and $G_l(t)$, the amplitudes and the phases depend on many factors e.g. the violinist's performance. The Formula (1) shows the main properties of the violin sound – the deterministic modes of the belly, the back and the whole violin body and the random character of the other components of the sound signal. This formula does not contain any components which come from the non-linear effects of the beat frequencies in the spectral neighborhood of the modes. This effect is clearly visible in the energy spectra of the violin sound, but it does not have any influence on the algorithm of searching for violin modes. Typically, the energy of beat frequencies is lower by at least 10 dB than the energy of the mode or the fundamental frequency. In Figure 4 an energy relations between the C4 mode, the fundamental frequency of the string (the C5 tone) and the beat frequencies are shown.

4.2 Mathematical Model of Chromatic Scale

Formula (1) describes a single violin tone. The chromatic scale have many tones. There are four chromatic scales containing all the possible tones which can be played on the violin [2, 6, 7, 12]. The table below describes all of them with the assumption that the A4 tone has been tuned to 443 Hz. The instruments recorded in the AMATI multimedia database were tuned to this reference frequency. The frequencies in Table 1 are the fundamental frequencies of the tones.

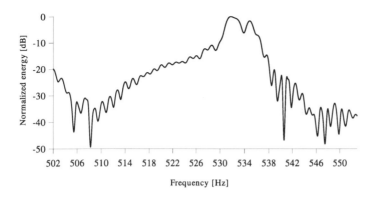

Fig. 4 The beat frequencies in the neighborhood of the C3 mode and the C5 tone played on violin. The global maximum is the fundamental frequency of the tone, the second maximum is the violin mode. (violin no. 85)

Table 1 The chromatic scales on violin

The G string		The D string		The A string		The E string	
Tones	Frequencies	Tones	Frequencies	Tones	Frequencies	Tones	Frequencies
G3	197.33 Hz	D4	295.67 Hz	A4	443.00 Hz	E5	663.75 Hz
$\div$	$\div$	$\div$	$\div$	$\div$	$\div$	$\div$	$\div$
G5	789.34 Hz	D6	1182.67 Hz	A6	1772.00 Hz	E7	2655.00 Hz

Using this information, we can write a formula which describes the chromatic scale played on the violin. This is also a random process.

$$U_{scale}(t) = U_1(t) + U_2(t - \tilde{t}_{sound_1}) + U_3(t - (\tilde{t}_{sound_1} + \tilde{t}_{sound_2})) + ... + \\ + U_{25}(t - (\tilde{t}_{sound_1} + \tilde{t}_{sound_2} + ... + \tilde{t}_{sound_{24}})) \qquad (2)$$

$U_{scale}(t)$ is the random process for the one chromatic scale. $U_1(t), U_2(t), ..., U_{25}(t)$ are the random processes as described by Formula (1). Indexes represent the tones of the chromatic scale. Number 1 is the first, 2 is second, etc. $\tilde{t}_{sound_1}, \tilde{t}_{sound_2}, ..., \tilde{t}_{sound_{24}}$ are duration times of the tones which are also random variables. Using (1) in substitution of $U(t)$:

$$U_{scale}(t) = \sum_{l=1}^{M} \check{G}_l(t) \cos(\omega_l t + \tilde{\theta}_l) + \sum_{k_1=1}^{N_1} \check{F}_{k_1}(t) \cos(k_1 \tilde{\omega}_1 t + \tilde{\phi}_{k_1}) + \\ + \sum_{k_2=1}^{N_2} \check{F}_{k_2}(t - \tilde{t}_{sound_1}) \cos(k_2 \tilde{\omega}_2(t - \tilde{t}_{sound_1}) + \tilde{\phi}_{k_2}) + ... + \qquad (3) \\ + \sum_{k_{25}=1}^{N_{25}} \check{F}_{k_{25}}(t - \sum_{r=1}^{24} \tilde{t}_{sound_r}) \cos(k_{25} \tilde{\omega}_{25}(t - \sum_{r=1}^{24} \tilde{t}_{sound_r}) + \tilde{\phi}_{k_{25}})$$

This formula describes the single chromatic scale, where $N_1, N_2, ..., N_{25}$ are the numbers of the harmonics for each tone of the chromatic scale. The tones are indexed as in Formula (2). $\check{F}_{k_1}(t), \check{F}_{k_2}(t), ..., \check{F}_{k_{25}}(t)$ where $\check{F}_{k_m}(t) = F_{k_m}(t)A_{k_m}$ are the envelope of the harmonics for the tones, $\tilde{\omega}_1, \tilde{\omega}_2, ..., \tilde{\omega}_{25}$ are the fundamental pulsations, $\tilde{\Phi}_{k_1}, \tilde{\Phi}_{k_2}, ..., \tilde{\Phi}_{k_{25}}$ are the phases of k-th harmonic for the tone of the chromatic scale marked by number in index of k, M is the number of the violin modes, $\check{G}_l(t)$ where $\check{G}_l(t) = G_l(t)B_l$ is the envelope function for the l-th mode. The remaining symbols in Formula (3) are the same as in Formula (1).

4.3   Correlation between Different Chromatic Scales

Four chromatic scales were played on each instrument, with one scale on each string. In the time domain, these scales are described by Formula (3), which was used to search for an instrument's modes within its sound.

Mode frequencies are the only deterministic components in Formula (3), while the other elements of the formula are either random variables or stochastic processes. Accordingly, the use of a statistical analysis in the stochastic process (3) made it possible to isolate an instrument's modes from a recorded sound. The autocorrelation function and the cross correlation function were used in the statistical analysis below.

The autocorrelation function is used to isolate the most important temporal elements, whereas the cross correlation function helps to search for common elements in at least two stochastic processes. Both functions were used to study the recordings of chromatic scales.

The power density function, also called the energy spectrum, is a frequency equivalent of the autocorrelation function, the product of Fourier transforms of autocorrelation functions corresponds to the cross correlation function between them in the time domain.

In the energy spectra of chromatic scales energy visibly concentrates around frequencies where modes are expected to be present. What could also be seen there are components derived from the fundamental frequencies of a chromatic scale. Thus, the application of the autocorrelation function alone to establish mode frequencies proved insufficient. However, since four recordings of chromatic scales for different strings and frequencies were available, it was possible to calculate the cross correlation function between autocorrelation functions calculated for each of the four scales.

Just like in the case of the autocorrelation function analysis, the calculations were made in the frequency domain. Prior to that, however, it was necessary to make sure if in this kind of analysis mode-related components would not be accompanied by component from the sounds of a chromatic scale.

A violin is a non-fretted instrument, and so the accuracy of the fundamental frequency of the sound generated by the instrument depends on the precision with which a violinist handles it. If the probability of a violinist hitting the right pitch is very high, the analysis of sound based on the cross correlation function will prove useless for the isolation of mode components from those produced by a string. To see whether spectrum components derived from the sounds of a chromatic scale would also be present in the spectrum of the cross correlation function, the study into the precision of a violinist's performance was conducted [23]. The recordings used in the analysis came from the AMATI database.

The study involved the searching for the fundamental frequency of a scale sound, testing the type of distribution that contains a set of fundamental frequencies for each sound of a chromatic scale, calculating the Q-factor of a string [23], and estimating the probability with which a sound from the same chromatic scale will be played with the accuracy defined by the Q-factor of a string [23].

It was found that in the 197 Hz – 470 Hz band the probability of a violinist hitting the right pitch is considerably high, i.e. 11.73% – 66.6% [23]. Beyond this band, the probability of hitting the right pitch shrinks to 0. Therefore, a spectrum of the correlation between the autocorrelation functions of chromatic scales in the 197 Hz – 470 Hz band will contain mode-related components of the instrument as well as those related to the sounds of a chromatic scale. Consequently, it was necessary to use recordings which did not contain frequencies related to the sounds of the scale within the 197 Hz – 470 Hz band. The recording of a chromatic scale played on E string, where the first sound of the scale (E5) has the frequency of 663.75 Hz (in relation to A4 tuned to 443 Hz), possesses this particular property. In this frequency band, the spectrum of a chromatic scale played on E string contains only those components that are of the mode origin. However, these components have low energy values and are considerably distorted by noise. The application of the cross correlation function in this case led to the isolation of violin modes and the attenuation of the components derived from a vibrating string.

To sum up, the whole analysis is made in the frequency domain, and it is the calculation of the energy spectra of chromatic scales. In the 197 Hz – 650 Hz band, the recordings of chromatic scales on G and E strings are used, whereas in the 650 Hz – 3200 Hz band, all the recordings are used.

The reason why the recordings of chromatic scales on G and E strings were selected is that in the frequency band of 197 Hz – 650 Hz the recording of a chromatic scale played on G string has frequencies in the neighborhood of the C1, A0, C2, T1 and C3 modes, and the recording of a scale played on E string contains only the components related to these modes. For this reason, the spectrum of the cross correlation function for the autocorrelation of recordings on G and E strings in the 197 Hz – 650 Hz band has considerable maxima only for the instrument's frequency modes.

The cutoff frequency of the correlation analysis was determined empirically for the instruments in the AMATI database. For these violins, all the modes above 3200 Hz have very low energy values – approximately 90 dB lower than the maximum value in the energy spectrum. This is true for all the violins in the AMATI database.

4.4 Algorithm of Mode Search

As mentioned above, both the autocorrelation and the cross correlation functions can be used to search for modes in the spectrum of the violin sound. The search is implemented in the frequency domain, and the spectrum is calculated with using the DTFT (Discrete Time Fourier Transform). The application of this transformation is determined by the fact that regardless of the number of signal samples in the time domain, it is possible to calculate a spectrum for any frequency. Each recording of a chromatic scale has a different number of samples. Besides, the number of chromatic scale samples for any of the instruments in the database is not a power of 2. Thus, the application of FFT proves very difficult.

Also, due to the fact that the number of samples of each recording is different it was necessary to normalize each spectrum with reference to the maximum energy value found in the band as explained below.

The band where modes were searched for was divided into three sub-bands: the first covers the 197 Hz – 650 Hz frequency range, the second 650 Hz – 810 Hz, and the third 810 Hz – 3200 Hz.

The 197 Hz – 650 Hz band was isolated on the basis of the study results described in Section 4.3. The 650 Hz – 810 Hz frequency band covers the range where the C4 mode should be present. The third sub-band, 810 Hz – 3200 Hz, contains higher violin modes, whose energy values are considerably lower than those of C1, A0, C2, T1, C3 and C4. The recordings of chromatic scales on G and E strings are used to search for modes within the first band, whereas all of the recordings of chromatic scales are used for the other sub-bands. Consequently, three algorithms of mode search were made [21, 22].

As a result two parameters are found for each mode: frequency f_{mod} and the mutual energy factor E_f. The mutual energy factor E_f is the product of the values of normalized energy spectra E_{n_k} for the frequency f; k is the index of normalized energy spectrum.

$$E_f = E_{n_1}(f) \cdot E_{n_2}(f) \cdot \ldots \cdot E_{n_k}(f) \text{ where minimal value of } k \text{ is 2} \qquad (4)$$

The algorithms search for the most important maxima in the product of normalized energy spectra. The first method uses the recordings of chromatic scales on G and E strings in the 197 Hz – 650 Hz band [21]. In the second, all the recordings of chromatic scales are searched in the 650 Hz – 810 Hz frequency range [22]. The algorithm for finding modes in the 810 Hz – 3200 Hz band is the same as the one used for finding modes in the 650 Hz – 810 Hz band.

4.5 Application of BST in the Elimination of Redundant Maxima

Due to the presence of noise in recordings the number of found maxima may exceed the number of modes. This happens most often in the 197 Hz – 650 Hz frequency range, where a scale played on E string is used. For the other frequency range (650 Hz – 3200 Hz), all the modes are marked correctly.

In order to avoid an incorrect recognition of modes it was necessary to determine the maximum band that can be occupied by a single mode in the energy spectrum of the violin sound. Following the analysis of 20 spectra of recordings on E string in the 197 Hz – 650 Hz band, it was found that the maximum relative 10 dB bandwidth occupied by a single mode is approximately that of 100 cents in each spectrum separately.

It was then necessary to find extrema which were less than 50 cents apart, and mark the highest one as the proper mode. This was done with using the BST (Binary Search Tree). The reference value is the absolute value of the ratio of two frequencies expressed in cents. If the value of this measure is lower than 50 cents, the maximum value is saved in the left sub-tree. In any other case the maximum value is saved in the right sub-tree. The maximum value is recorded as an object comprising two variables: mode frequency and mutual energy factor. When the BST is complete, the number of leaves in the right sub-tree equals the number of the modes. The correct frequencies of modes are determined by finding the highest maximum value in the left sub-trees.

4.6   Results and the Interpretation of Parameters

The modes were found for the all instruments in the AMATI database. Recordings from the near field were used, spectra calculations were done with 1 Hz frequency resolution and modes were found using above mentioned method based on the BST. The value of an acceptance threshold variable was 0.1. The acceptance threshold is the variable which decides about an acceptance or a rejection of the found mode. If a quotient of the mutual energy factor of found mode and the maximum value of mutual energy factor for the all found modes is less than the acceptance threshold value, the mode will be rejected. Sample results for two instruments are shown in the table below, where the following symbols are used: the mode frequency f_{mod} and the mutual energy factor E_f (the value of the mutual energy factor for the frequency is indicated on the right).

The C1 mode (198 Hz) is missing from the results shown in Table 2. The same happens for the other instruments in the AMATI database. Also, the A0 mode (about

Table 2 Modes found for two violins in the AMATI database

Instrument no. 23					
f_{mod} [Hz]	E_f	f_{mod} [Hz]	E_f	f_{mod} [Hz]	E_f
265	2.32E-02	279	1.96E-02	420	1.12E-001
489	1.58E-01	558	4.17E-02	758	3.22E-003
901	1.48E-04				
Instrument no. 80					
f_{mod} [Hz]	E_f	f_{mod} [Hz]	E_f	f_{mod} [Hz]	E_f
279	7.47E-02	503	5.92E-02	523	2.80E-01
760	3.95E-03	884	3.90E-04		

270 Hz) was not found for some of the instruments. In both cases the reason is low energy value for C1 and A0 modes. Presumably, a reduced value of the acceptance threshold would have made it possible to find the C1 and A0 modes in each studied case. Nevertheless, if a mode was not found with the value 0.1 of the acceptance threshold, the influence of the undetected mode on the violin tone is negligible.

A mode frequency is interpreted as a standing wave frequency. Additionally, though, these calculations yield another parameter called mutual energy factor, which is a result of calculations involving normalized energy spectra. This parameter can have the values of $\langle 0,1 \rangle$ interval. The higher its value, the higher the energy of a given mode in the spectra of chromatic scales. For example, for many instruments the mutual energy factor reaches its highest values for the T1 mode. This mode also has the highest amplitude in studies involving the impulse response. In the 810 Hz – 3200 Hz band, on the other hand, the values of the mutual energy factor are very low.

In the 197 Hz – 650 Hz band, the mutual energy factor can also be interpreted as a parameter describing the behavior of the modes during the violinist's performance. If, for example, the value of this factor of the mode is close to 1, the mode have the same behavior for different energy levels of excitation.

To sum up, as a result of the implementation of the above-mentioned algorithms, a set of pairs of parameters (frequency, mutual energy factor) is obtained for each violin. These parameters describe the behavior of an instrument being played.

5 Analysis of Jurors' Evaluation of Sound Quality

The jurors' evaluation of sound quality is a subjective parameter in point-based scale. It is possible to make an error analysis of the sound quality evaluation. This section presents results of this analysis for AMATI database.

5.1 Modes and Quality of Sound

The research study was aimed at finding correlations between modes and the appraisal of the violin tone quality. The quality of tone is a subjective parameter, whereas the parameters of the modes of violins are objective. What proves to be a major problem is the subjective description of a sound. The way musicians assess the sound of an instrument depends on the repertoire they play, and may vary accordingly [7]. Presumably, since violins differ from each other in their modes, the modes themselves that may substantially affect on expert's opinion.

Therefore, it would seem that more appropriate to implement the point-based evaluation of sound quality, as used in violin maker competitions, rather than a verbal evaluation, as suggested in Yankovski's work [24]. The assessment in the AMATI database is point-based and was expressed in the form of numeral values during the competition. The jurors may then be regarded as measurement devices, for which an error analysis can be made and based on that a classifier can be build.

The classifier, based on mode parameters (frequency, mutual energy factor), will make it possible to assess the expert evaluation in the instrument sound category.

5.2   Error Analysis of Jurors' Assessment

The AMATI database contains the jurors' complete assessments (in all the categories) of all the instruments that took part in the second stage of the competition. 79 instruments qualified for this stage. Since each instrument constitutes an independent object of measurement, such a number of violins is sufficient for the calculation of the real accuracy value of the jurors' assessments. This kind of analysis was made for the assessments in the sound quality category.

In accordance with the competition regulations, each juror gave marks in the timbre of violin sound in the range of 4 – 20 points, with 1 point accuracy. Such a mark is treated as a numeral value, the ultimate mark being an arithmetic mean of all the jurors' marks. The same statistics as the ones applied in the error analysis [3] should then be applied here.

The following estimators were used in the error analysis:

1. Arithmetic mean μ:

$$\mu = \frac{1}{N} \sum_{i=1}^{N} k_i \tag{5}$$

2. Standard deviation σ:

$$\sigma = \sqrt{\frac{1}{N-1} \sum_{i=1}^{N} (k_i - \mu)^2} \tag{6}$$

In the above formulas, N is the number of random samples, whereas k_i is the i-th value in the random sample. In the case of a small number of measurements the standard deviation σ must be modified by the Student-Fisher factor [3]. The modification involves the multiplication of the value obtain from Formula(6) by the Student-Fisher factor $t_{n,a}$ [3], which value depends on the number of random samples n and the required probability a [3].

The error analysis was made independently for each instrument. The value of the Student-Fisher factor was 1.197, with the assumption that the number of measurements was 4 (the same as the number of jurors), and the probability was 0.6826 (which corresponds to the confidence level standard deviation for normal distribution [3].) Sample results for 15 instruments are shown in Table 3.

On the basis of the calculated statistics, it was found that juror's accuracy was lower than that assumed in the competition regulations. As a result, the average accuracy of the jurors was assessed as the arithmetic mean of all the values of standard deviations. The measurement error of the jurors' evaluation in the sound quality category is ± 1.98 points.

Table 3 Sample error analysis for 15 instruments in the AMATI database in the sound quality category

Instrument	Mark Juror I	Mark Juror II	Mark Juror III	Mark Juror IV	μ	σ	$\sigma \cdot t_{4,0.6826}$
15	10	10	9	10	9.75	0.5	0.6
23	11	11	9	11	10.5	1	1.2
24	10	9	8	10	9.25	0.96	1.15
30	18	17	15	18	17	1.41	1.69
32	15	14	12	15	14	1.41	1.69
43	9	12	4	9	8.5	3.32	3.97
66	12	12	8	12	11	2	2.39
72	16	16	12	16	15	2	2.39
80	14	16	11	14	13.75	2.06	2.47
85	15	14	12	19	15	2.94	3.52
89	17	16	16	19	17	1.41	1.69
100	12	14	9	12	11.75	2.06	2.47
109	11	12	9	11	10.75	1.26	1.51
111	12	11	8	12	10.75	1.9	2.27
118	17	15	17	17	16.5	1	1.2

6   Sound Quality Appraisal

Classifier for sound quality evaluation calculates a value of this subjective parameter basing on objective properties of violin sound and information about hearing process . This section describes assumptions of estimating experts' judgements and results of comparing calculated sound quality evaluation with juror's evaluation in this category.

6.1   Harmony and Four Categories of Intervals

The research study shows that the entire complex spectrum of the violin tone includes mode-related components, and that these components can be isolated from the spectrum. The next task is to investigate the problem of how a listener reacts to modes and which mechanism in the hearing process is responsible for this reaction. The answer to this question can be found in the results of neurobiological studies concerning the hearing process and the distinction of the musical and non-musical sound. The latest studies make use of neuron activity imaging of MRI (Magnetic Resonance Imaging) [16]. On the basis of the results of these studies it is possible to predict a listener's reaction to the presence of modes in the violin sound.

In the numerous studies of musical hearing [16], it has been observed that the main reason for deciding whether a sound possesses a musical quality are the harmonic relations between the elements of spectrum [16]. The ability to effectively distinguish these harmonic relations depends on listener's musical education as well as energy relationships in a spectrum. A similar reaction may then be expected when listening to a violin sound, because the spectrum of this instrument contains mode-related components. As a result, the appraisal of the violin tone quality should depend on the type of harmony and energetic relationships between the modes of an instrument. This hypothesis can be verified by designing an appropriate classifier that will estimate the jurors' appraisals of tone quality, and which will take into account mode parameters (mutual energy factor and frequency), energetic relationships between these modes and the principles of harmony.

Musical harmony is a set of rules with regard to chords and progressions of sounds. The concept of harmony is by all means subjective – there are a number of theories of harmony with distinct definitions of consonance, dissonance, and the rules of how they should be combined [2, 6, 7, 13]. The concept of harmony keeps changing and may depend on the historical period or geographical location. In the western world, however, the principal harmony evolved on the basis of the equal tone temperament system where the semitone, always has the same size (100 cents), is the smallest unit [2, 13].

Consonances and dissonances are axioms in harmony. In classic harmony, the unison interval (0 cents), the major third interval (400 cents), the perfect fifth interval (700 cents) and the octave interval (1200 cents) are all consonances. The minor third (300 cents) is also a consonance, as it is part of a minor chord. The other intervals are dissonances. However, due to the existence of other harmony rules, dissonances can be divided into three categories [13]: strong dissonance, dissonance, and light dissonance. The tritone (600 cents), the augmented octave (1300 cents) and the minor second (100 cents) should be regarded as strong dissonances. The major sixth interval (900 cents), in turn, should be regarded as a light dissonance. The reasons for such a division of intervals are quite complex – the fact that the tritone, for example, belongs to strong dissonances can be traced back to the Middle Ages, when it was referred to as devilish (*diabolus in musica*) and its use was forbidden [13].

The above-mentioned distinction is common in the western musical tradition and is seen as something obvious. Consequently, it can be assumed that musicians trained in this tradition will respond accordingly.

Therefore, a hypothesis was put forward that, when assessing the violin tone quality, the jurors mainly took into account the consonance between the instrument's modes, subconsciously applying the interval classification that follows from the rules of harmony. The division into the four groups of intervals – strong dissonance, dissonance, light dissonance and consonance – was also assumed to be a rule.

6.2   Weighted Average as Classifier of Violin Tone Quality

The classifier which used to evaluate the jurors' appraisals in the tone quality category should take into account both subjective and objective parameters. The objective parameters are the instrument's modes – frequency and the mutual energy factor. The subjective qualities are the harmony relationships between mode frequencies.

The relationships between mode frequencies were arranged in accordance with the rules of the equal temperament system and were then assigned to one of the four categories: strong dissonance, dissonance, light dissonance and consonance. All the intervals expressed in cents were normalized by modulo 1200 cents to allow all the relationships to be octave-related.

The intervals were classified in accordance with the rules described in Table 4.

Table 4 Classification of intervals used in the classifier for the appraisal of violin tone quality

Interval [cent]	Name	Class
$\langle 0,50 \rangle$	Unison	Consonance
$\langle 50,150 \rangle$	Minor second	Strong dissonance
$\langle 150,250 \rangle$	Major second	Dissonance
$\langle 250,350 \rangle$	Minor third	Consonance
$\langle 350,450 \rangle$	Major third	Consonance
$\langle 450,550 \rangle$	Perfect fourth	Dissonance
$\langle 550,650 \rangle$	Tritone	Strong dissonance
$\langle 650,750 \rangle$	Perfect fifth	Consonance
$\langle 750,850 \rangle$	Minor sixth	Dissonance
$\langle 850,950 \rangle$	Major sixth	Light dissonance
$\langle 950,1050 \rangle$	Minor seventh	Dissonance
$\langle 1050,1150 \rangle$	Major seventh	Dissonance
$\langle 1150,1200 \rangle$	Octave	Consonance

Each category was assigned a value defined as the weight of interval evaluation: W_{SD} for strong dissonance, W_D for dissonance, W_{LD} for light dissonance and W_{CN} for consonance.

In order to describe the energetic relationships, an importance factor, with the $\langle 0,1 \rangle$ value range, was formulated. The 0 value means that a given interval is not significant in terms of its energetic qualities, whereas 1 means that the interval must be considered in the analysis. Intermediate values between 0 and 1 are the measure of the importance of a given interval. The importance factor is described with the following formula:

$$\xi_{k,i} = m_k \cdot m_i \qquad (7)$$

$\xi_{k,i}$ in this formula is the importance factor, m_k is the value of the mutual energy factor for the k-th mode, and m_i is the mutual energy factor for the i-th mode.

The following weight factor array for M modes is calculated for each instrument:

$$\Xi = \begin{bmatrix} \xi_{1,1} & \xi_{2,1} & \cdots & \xi_{M,1} \\ \xi_{1,2} & \xi_{2,2} & \cdots & \xi_{M,2} \\ \vdots & \vdots & \ddots & \vdots \\ \xi_{1,M} & \xi_{2,M} & \cdots & \xi_{M,M} \end{bmatrix} \qquad (8)$$

Since $\xi_{k,i} = \xi_{i,k}$ and the comparison of a mode with itself do net add any new information, only the upper triangular matrix derived from the matrix Ξ is used. Each importance factor is also linked to the musical relationship between mode frequencies. This relationship is expressed in cents and belonging to one of the four categories: strong dissonance, dissonance, light dissonance and consonance.

All the above-mentioned parameters enable a numeral description of objective energetic relationships as well as subjective harmonic relationships between violin modes. A weighted average was used to combine these parameters and to design a classifier for the appraisal of the violin tone quality. This classifier is described by the following formula:

$$e_q = \frac{W_{SD} \sum\limits_{k=1}^{M_{SD}} \xi_k^{SD} + W_D \sum\limits_{k=1}^{M_D} \xi_k^{D} + W_{LD} \sum\limits_{k=1}^{M_{LD}} \xi_k^{LD} + W_{CN} \sum\limits_{k=1}^{M_{CN}} \xi_k^{CN}}{\sum\limits_{k=1}^{M_{SD}} \xi_k^{SD} + \sum\limits_{k=1}^{M_D} \xi_k^{D} + \sum\limits_{k=1}^{M_{LD}} \xi_k^{LD} + \sum\limits_{k=1}^{M_{CN}} \xi_k^{CN}} \qquad (9)$$

The following symbols were used in Formula (9): e_q – estimated value of the appraisal of tone quality, SD – strong dissonance, D – dissonance, LD – light dissonance, CN – consonance, M_{SD} – number of strong dissonances, M_D – number of dissonances, M_{LD} – number of light dissonances, M_{CN} – number of consonances, W_{SD} – weight for the strong dissonance category, W_D – weight for the dissonance category, W_{LD} – weight for the light dissonance category, W_{CN} – weight for the consonance category, ξ_k^{SD} – importance factor for the k-th strong dissonance, ξ_k^{D} – importance factor for the k-th dissonance, ξ_k^{LD} – importance factor for the k-th light dissonance, ξ_k^{CN} – importance factor for the k-th consonance.

The weights for four types of intervals – W_{SD}, W_D, W_{LD} and W_{CN} – cannot be calculated directly with Formula (9). In order to calculate W_{SD}, W_D, W_{LD} and W_{CN}, a Monte Carlo method was used (Section 6.3). The mutual energy factors come from the upper triangular part of the matrix Ξ.

6.3   Calculation of Weights with Monte Carlo Method

As mentioned above, the weights for four types of intervals cannot be calculated directly. In order to find them it is necessary to use the jurors' assessments for a group of instruments, establish their mode parameters and calculate their weights. Algorithms from the Monte Carlo family [10] prove to be useful in this case.

Formula (9) contains four unknown parameters: W_{SD}, W_D, W_{LD} and W_{CN}. To find them, we need to use different violins. Each instrument was assessed for its tone quality, with the final mark being the arithmetic mean of the jurors' four individual marks. This average mark was treated as the required value e_q. The error function used to measure the difference between the calculated values and the jurors' marks was expressed as follows:

$$e_{rr} = \frac{1}{N_v} \cdot \sum_{k=1}^{N_v} |e_{qk} - \tilde{e}_{qk}| \qquad (10)$$

Formula (10) contains the following symbols: e_{rr} – error value, N_v– number of instruments used for the calculation of interval weights, e_{qk} – average jurors' mark in the tone-quality category for the k-th instrument, $\tilde{e}_{qk}$ – mark value calculated with the use of the classifier described by Formula (9) for the k-th instrument.

The algorithm applied in the search for the interval weights is as follows:

I. Input data

   1. Number of instruments: **NV**.
   2. **VIOLINS** array of **PARAM** objects. The **VIOLIN** array has **NV** elements. Each element represents one instrument. The **PARAM** object has the following properties:
      - average jurors' mark: **quality_eval**,
      - number of modes of the violin: **MN**,
      - number assigned to instrument **NUMBER**,
      - **VIOLIN_MODE** array of objects.
        These object have two properties: **mod_freq** (mode frequency) and **mutual_energy_factor**. The number of elements in this array is equal to the number of modes of the instrument.
   3. Initial weight values $\mathbf{W_{SD}}$, $\mathbf{W_D}$, $\mathbf{W_{LD}}$, $\mathbf{W_{CN}}$
   4. Maximum step of modification of $\mathbf{W_{SD}}$, $\mathbf{W_D}$, $\mathbf{W_{LD}}$, $\mathbf{W_{CN}}$ **max_step**,
   5. Assumed minimum error value **err_min**.

II. Output data

   1. Weight values: $\mathbf{W_{SD}}$, $\mathbf{W_D}$, $\mathbf{W_{LD}}$, $\mathbf{W_{CN}}$,
   2. Actual error level **err**,
   3. **EVALUATIONS** array. This array has **NV** elements and containins the calculated mark values for the instruments whose parameters are in the **VIOLINS** array.

III. Method

   1. Enter the input data.
   2. Create matrix Ξ for each of the instruments in the **VIOLINS** array.
   3. Calculate the mark value for each instrument with Formula (9).
   4. Calculate the **err** value with Formula (10).
   5. Is the value **err** < **err_min** ?
   6. If so, complete the calculations and close the program.
   7. If **err** $\leq$ **err_min**, then:
   8. Generate random value of
      modification step of $\mathbf{W_{SD}}$ from the $\langle -\mathbf{max\_step}, \mathbf{max\_step} \rangle$ range.
   9. Generate random value of
      modification step of $\mathbf{W_D}$ from the $\langle -\mathbf{max\_step}, \mathbf{max\_step} \rangle$ range.

10. Generate random value of
    modification step of W_{LD} from the $\langle -\text{max\_step}, \text{max\_step} \rangle$ range.
11. Generate random value of
    modification step of W_{CN} from the $\langle -\text{max\_step}, \text{max\_step} \rangle$ range.
12. Save the current weight values W_{SD}, W_D, W_{LD}, W_{CD}, the value of the all calculated marks for the instruments, and the current value of the variable **err**.
13. Modify all the weights by adding the previously randomized diversion steps.
14. Calculate a new **err** value.
15. If the new **err** value is smaller than the previous value of this variable, return to step 5.
16. If the new **err** value is bigger than the previous one, restore the previous weight values and return to step 8.

The algorithm above was used to calculate the interval weights for 15 instruments from the Table 3. The minimum error value obtained (**err**), calculated with Formula (10), was 1.97 (1.96650644697478). The near-field recordings were used for the purpose of the calculations. When searching for the weights, only the modes from the 197 Hz – 810 Hz band were taken into account. The values of the calculated mode weights in this band for four interval categories are shown in Table 5.

Table 5 Weights for four interval categories calculated with a Monte Carlo method

Interval category	Calculated weight	Symbol
Strong dissonance	14.3990523557032	W_{SD}
Dissonance	6.9185314074039	W_D
Light dissonance	5.7049333116643	W_{LD}
Consonance	17.9895114267725	W_{CN}

The minimum error value calculated with Formula (10) is almost identical to the average value of the standard deviation obtained from the error calculus. It means that the sound quality classifier has the same measurement error (section 5.2) as experts.

6.4   Analysis of Results

In order to verify the effectiveness of the algorithm it was necessary to introduce clear criteria that would help to decide if the calculated mark has a similar value to that of the jurors' within a measurement error. Since the standard deviation of the jurors' appraisal of tone quality is different for each instrument, the value of the measurement error was defined as follows:

1. If the standard deviation of the jurors' assessment is greater than the average standard deviation, the error value is the same as the standard deviation of the jurors' assessment.

2. If the standard deviation of the jurors' assessment is less than or the same as the average standard deviation, the error value is the same as the value of the average standard deviation.

If the calculated mark of sound quality is within the range of the measurement error (section 5.2), or within the range between the jurors' lowest and highest mark, it is thought to be correct. The results obtained for the instruments are shown in Table 6. Recordings from the near field were used for the calculation of the modes.

Table 6 Results of the application of classifier (9) for 15 instruments in the AMATI database

Violin	Min mark	Max mark	μ	$\sigma \cdot t_{4, 0.6826}$	Measurement error	Calculated mark	Decision
15	9	10	9.75	0.60	1.98	12.75	NO
23	9	11	10.50	1.20	1.98	12.40	YES
24	8	10	9.25	1.15	1.98	11.23	YES
30	15	18	17.00	1.69	1.98	12.70	NO
32	12	15	14.00	1.69	1.98	13.47	YES
43	4	12	8.50	3.97	3.97	11.67	YES
66	8	12	11.00	2.39	2.39	13.38	YES
72	12	16	15.00	2.39	2.39	17.37	YES
80	11	16	13.75	2.47	2.47	10.01	NO
85	12	19	15.00	3.52	3.52	10.33	NO
89	16	19	17.00	1.69	1.98	16.56	YES
100	9	14	11.75	2.47	2.47	14.00	YES
109	9	12	10.75	1.51	1.98	12.52	YES
111	8	12	10.75	2.27	2.27	10.72	YES
118	15	17	16.50	1.20	1.98	15.51	YES

In Table 6, the minimum mark is the jurors' minimum mark, and the maximum mark is the jurors' maximum mark. The "Decision" column provides information on whether the calculated mark matches that of the jurors' within the range of the measurement error or within the range between the minimum and the maximum mark made by jurors. The measurement error column is the measurement error of jurors.

The calculations of the tone quality appraisal with the help of the classifier described by Formula (9) were made for 53 instruments in the AMATI multimedia database. Modes found within the 197 Hz – 810 Hz frequency range, as well as the W_{SD}, W_D, W_{LD}, W_{CN} weights from Table 5, were used for this purpose. With this assumption, the calculated values proved to be compatible with the jurors' marks in the case of 40 instruments, i.e. 75.5% of all the instruments in the AMATI database. Despite the lack of any assumptions for the Formula (9), all the calculated marks fell within the range specified by the competition regulations. The minimum value of the calculated mark was 6.92, and the maximum value was 17.98. The average value of all the calculated marks is 12.11. The jurors' minimum an maximum values are 4 and 19 respectively. The average of all jurors' marks is 13.35 points. The standard deviation calculated for all the marks given by all the jurors ($\sigma = 2.63$) is very similar to that of the marks calculated by the program ($\sigma = 2.5$).

On the basis of the obtained results it can be asserted that the classifier is capable of estimating a tone quality mark with an accuracy of 75.5%. With this relatively high value, the proposition that a subjective appraisal of the violin tone quality is influenced mainly by energetic and frequency relationships within the modes of an instrument (with the assumption that the frequency relationships comply with the rules of harmony) appears to be a legitimate hypothesis.

7 Conclusions

The methods presented in this paper are designed for automatic classification within multimedia databases containing the recordings of violins and other bowed string instruments. By employing them, it is possible to create a set of objective parameters for each instrument, where sound quality, which itself is a subjective parameter, can be assessed by the automated classifier. Another advantage offered by the classifier is its possible application for the search for similar instruments within the same database. Moreover it was discovered that the modes which affect this parameter at most are to be found in the 197 Hz – 810 Hz band.

The performance of the jurors as well as that of the sound quality classifier are similar. The range of marks calculated with Formula (9) for the violins in the AM-ATI database falls within the range stipulated by the competition regulations despite the fact that the classifier itself does not have any such restrictions. The standard deviation calculated for all the marks given by all the jurors is very similar to that of the marks calculated by the program. For this reason, it is fair to assert that the methods presented in this paper fulfill their main function, i.e. the ability to find objective parameters of violin sound and to estimate the subjective violin tone quality.

References

1. Andou, Y.: Acoustics of musical instruments, Ongaku-no Tomo-sha, Tokyo (2000) (in Japanese)
2. Duffin, R.W.: How Equal Temperament Ruined Harmony: (And Why You Should Care). W.W. Norton, USA (2008)
3. Dunn, P.F.: Measurement and Data Analysis for Engineering and Science. McGraw-Hill, USA (2005)
4. Goto, M.: RWC Music Database, National Institute of Advanced Industrial Science and Technology (AIST), Japan
5. Hafke, H.: Auditory information processing for sound perception and vocalization in case of tritone paradox. In: 55th Open Seminar on Acoustics, Wrocław-Piechowice (2008)
6. Hall, D.E.: Musical Acoustics, An Introduction. Wadsworth Publishing Conpany, USA (1980)
7. Harnoncourt, N.: The Musical Dialogue: Thoughts on Monteverdi, Bach and Mozart. Amadeus Press, USA (2003)

8. Henryk Wieniawski Musical Society of Poznań Official Web Page,
   http://www.wieniawski.pl/en/
9. Hutchins, C.M., Voskuil, D.: Mode tuning for the violin maker. CAS Journal 2(4) (Series II), 5–9 (1993)
10. Kalos, M.H., Whitlock, P.A.: Monte Carlo Methods. Wiley-VCH, Chichester (2008)
11. Kaminiarz, A., Łukasik, E.: MPEG-7 audio spectrum basis as a signature of violin sound. In: EUSIPCO 2007, Poznań, Poland (2007)
12. Kolneder, W.: The Amadeus Book of the Violin: Construction, History, and Music. Amadeus Press, USA (2003)
13. Levitin, D.J.: This is your brain on music – the science of a human obsession. Plume, USA (2007)
14. Łukasik, E.: AMATI: multimedia database of musical sounds. In: Stockholm Music Acoustics Conference, KTH [Kungliga Tekniska Hogskolan], pp. 79–82 (2003)
15. Łukasik, E., Wrzeciono, P.: Digital recording of musical instruments for studies of their timbre, Report no. RB-03/2001 of Institute of Computer Science, Poznan University of Technology, Poland (2001) (in Polish)
16. Peretz, I., Zatorre, R.: The cognitive neuroscience of music. Oxford University Press, USA (2007)
17. Shirota, Y.: Proposal for Analysis of Likes and Dislikes about Musical Instrument Sounds. In: Data Engineering Workshop DEWS, Miyazaki, Japan (2008)
18. Weyna, S.: Energy propagation of acoustic real source. Wydawnictwa Naukowo-Techniczne, Warszawa, Poland (2005) (in Polish)
19. Wieczorkowska, A., Raś, Z.W.: Do We Need Automatic Indexing of Musical Instruments? In: Bolc, L., Michalewicz, Z., Nishida, T. (eds.) IMTCI 2004. LNCS (LNAI), vol. 3490, pp. 239–245. Springer, Heidelberg (2005)
20. Wrzeciono, P.: Statistical and perceptual analysis of violin tone. In: 52nd Open Seminar on Acoustics, Wagrowiec, Poland (2005) (in Polish)
21. Wrzeciono, P.: A New Method of Searching for Violin Modes, Warszawa. In: The IEEE Region 8 Eurocon 2007 Conference, Poland (2007)
22. Wrzeciono, P.: A method of detecting the C4 violin mode in the energy spectra of chromatic scales. Archives of Acoustics 32(4) (Suppl.), 197–201 (2007)
23. Wrzeciono, P.: Statistical evaluation of violinist's performance based on recordings from the AMATI multimedia database. In: 55th Open Seminar on Acoustics, Wrocław-Piechowice, Poland (2008)
24. Yankovskii, B.A.: Method for the objective appraisal of violin tone quality. Soviet Physics-Acoustics 11(269) (1965) (in English)
25. Yoshida, J., Hasegawa, H., Kasuga, M.: Changes in absolute hearing threshold depending on sound pressure level of a previous sound. Acoustical Science and Technology 29(5), 320–325 (2008); The Acoustical Society of Japan

Emotion Based MIDI Files Retrieval System

Jacek Grekow and Zbigniew W. Raś

Abstract. This chapter presents a query answering system *(QAS)* associated with MIDI music database and a query language which atomic expressions represent various types of emotions. System for automatic indexing of music by emotions is one of the main modules of *QAS*. Its construction required building a training database, manual indexation of learning instances, finding a collection of features describing musical segments, and finally building classifiers. A hierarchical model of emotions consisting of two levels, L1 and L2, was used. A collection of harmonic and rhythmic attributes extracted from music files allowed emotion detection in music with an average of 83% accuracy at the level L1. The presented *QAS* is a collection of personalized search engines *(PSE)*, each one based on a personalized system for automatic indexing of music by emotions. In order to use *QAS*, user profile has to be built and compared to representative profiles of *PSE's*. The nearest one is identified and used in answering user query.

1 Introduction

The development of computer technology in recent years has led to a huge expansion of Internet multimedia databases creating the need to develop tools for searching through their content. Systems searching for files according to the title, author, date of creation, etc. that operate on the basis of manu-

Jacek Grekow
Faculty of Computer Science, Bialystok Technical University, Wiejska 45A, Bialystok 15-351, Poland
e-mail: `grekowj@wi.pb.edu.pl`

Zbigniew W. Raś
Computer Science Dept., University of North Carolina, 9201 University City Blvd., Charlotte, NC 28223, USA
e-mail: `ras@uncc.edu`

Z.W. Raś and A.A. Wieczorkowska (Eds.): Adv. in Music Inform. Retrieval, SCI 274, pp. 261–284.
springerlink.com   © Springer-Verlag Berlin Heidelberg 2010

ally filled descriptions of the given multimedia file are no longer sufficient. Searching tools enabling object identification of higher levels of content have become much more in demand. One of the challenges on the way to create such advanced searching tools is an automatic detection of emotions in music pieces. Since emotions change in the course of a music piece and are differently perceived by people, the task of detecting them is quite complex. This chapter presents a system for a personalized automatic indexing of MIDI music files according to emotions and construction of a personalized query answering system.

1.1 Input Data

Many research papers deal with the problem of emotion detection. Some of them rely on audio files [8], [9], [10], [16], [17], [19] and others on MIDI files [1], [11]. In our research, we concentrated on emotion detection in MIDI files containing symbolic representation of music (key, structure, chords, instrument). The means of representation of musical content in MIDI files is much closer to the description which is used by musicians, composers, and musicologists. To describe music, they use key, tempo, scale, notes, etc. This way, we avoid the difficult stage of extraction of separate notes, tracks, instruments from audio files, and we can concentrate on the deciding element which is the musical content.

1.2 Mood Model

There are several models describing emotions contained in music. One of them is the model proposed by Hevner [6]. It is made up of a list of adjectives grouped in 8 main categories. After modification it was used by Li et al. [8] and Wieczorkowska et al. [17]. This model is quite developed and complex, too complicated to be used in our experiment. However, it illustrates the intricacy of describing emotions.

Another model is the two-dimensional Thayer model [14] in which the main elements are Stress and Energy laid out on 2 perpendicular axes. Stress can change from happy to anxious, and Energy varies from calm to energetic. This way, 4 main categories form on the plain: Exuberance, Anxious, Depression and Contentment. This model was used by Liu et al. [10], DiPaola et al. [1], Wang et al. [16], and Yang et al. [19].

The model we chose for our automatic indexing system is based on Thayer's model (Fig. 1). Following its example, we created a hierarchical model of emotions consisting of two levels, L1 and L2.

The first level L1 contains 4 emotions. To ease the indexing of files, group names were replaced with compound adjectives referencing to Arousal and Valence. Our mood model contains the following groups (Table 1):

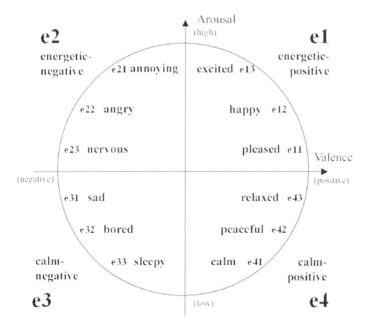

Fig. 1 Arousal-valence emotion plane

Table 1 Description of mood groups in L1, the first level

Abbreviation	Description
e1	energetic-positive
e2	energetic-negative
e3	calm-negative
e4	calm-positive

In the first group (e1), pieces of music can be found which convey positive emotions and have a quite rapid tempo, are happy and arousing (Excited, Happy, Pleased). In the second group (e2), the tempo of the pieces is fast, but the emotions are more negative, expressing Annoying, Angry, Nervous. In the third group (e3) are pieces that have a negative energy and are slow, expressing Sad, Bored, Sleepy. In the last group (e4) are pieces that are calm and positive and express Calm, Peaceful, Relaxed.

The second level is related to the first, and is made up of 12 sub-emotions, 3 emotions for each emotion contained in the first level (Table 2):

Table 2 Description of mood groups in L2, the second level

Abbreviation	Description
e11	pleased
c12	happy
e13	excited
e21	annoying
e22	angry
e23	nervous
e31	sad
e32	bored
e33	sleepy
e41	calm
e42	peaceful
e43	relaxed

1.3 System Construction

Fig. 2 presents the construction of a personalized QAS for music database D where queries relate to emotions. The system deals with enquiries defined as a logical conjunction of emotions in $E = \{E_1, E_2, ...E_n\}$ with corresponding percentage content $C = \{C_1, C_2, ...C_n\}$ in a segment of music piece. User emotional profile $P = \{P_1, P_2, ...P_i\}$ is also taken into consideration. The

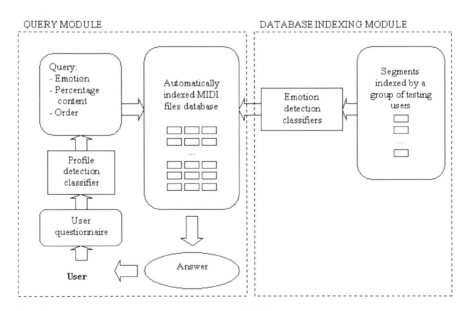

Fig. 2 The construction of personalized QAS for music database

system allows placing enquiries including the order of distribution of emotions dominating in segments $S = \{S_1, S_2, ...S_j\}$. When placing a query, the user may choose the desired level of emotions $L = \{L_1, L_2\}$. One of the possible enquires could be: "using user profile P find all segments in database D which contain emotions E from the level L of the percentage content C" or "using user profile P find all segments in database D which contain emotions E from level L in the order of appearance S". First, the system determines user profile P on the basis of a filled out form. Next, it searches database D for music files matching the query.

2   Mood Detection

2.1   Database

A database with 83 MIDI files of classical music (F. Chopin, R. Schuman, F. Schubert, E. Grieg, F. Mendelssohn-Bartholdy, etc.) was created specifically for the needs of the experiment. Starting from the 5th bar, 16 second segments were isolated from each piece. The shift forward was chosen with the aim of avoiding various, unstable introductions at the beginning of many pieces. Each of these segments was divided into 6 subsegments of 6 seconds each with a mutual overlap (overlapping 2/3). There were 498 resulting 6-second subsegments. Overlapping allows precise tracking of emotions contained within musical segments.

2.2   Indexing

The 498 subsegments were annotated with emotions by a listener-tester, a person with a formal music education/background, who has professional experience in listening to music.

2.3   Feature Extraction

The next stage was to extract features describing the files in music database. Specially written software *AKWET simulator - Features Explorer* jointly with MATLAB was used for that purpose. Finally, every record in the created database was described by 63 features.

2.3.1   Harmony Features

Harmony, along with rhythm, dynamics and melody, is one of the main elements of music upon which emotion in music is dependent. Harmony Features reflect dissonance and consonance of harmony of sounds. They are based on

Table 3 Example of consonance sound frequency ratios

k	Notes	Consonance sound frequency ratios $N_{R1} : N_{R2} : ... : N_{Rk}$
2	C1, G1	2:3
3	C1, E1, G1	4:5:6
4	E1, G1, B♭1, D2	25:30:36:45

previous work by the author [3], [4]. To calculate the harmony parameters, we used the frequency ratio of simultaneously occurring sounds (Table 3).

A given consonance (interval, chord, polyphony) constitutes simultaneously resonant sounds, the frequency ratio of which can be noted as following:

$$N_{R1} : N_{R2} : ... : N_{Rk} \tag{1}$$

where k is the number of sounds comprising the consonance. N_{Ri} is taken from the just intonation tuning system, where the frequencies of the scale notes are related to one another by simple numeric ratios.

From the frequency ratios, we calculated the AkD parameter, which mirrors the degree of dissonance in a single chord. The higher its value, the more dissonant is the consonance, when the AkD value is lower, the consonance is more consonant - more pleasant for the ear.

$$AkD = LCM(N_{R1}, N_{R2}, ..., N_{Rk}) \tag{2}$$

where k is the number of sounds in a given sample. If $k = 1$, then $AkD = 1$. LCM means Least Common Multiple.

From the sequence of consonance samples collected from a musical segment (Fig. 3), we can define $AkD(s)$ as:

$$AkD(s) = (AkD_1, AkD_2, ..., AkD_p) \tag{3}$$

where p is the number of samples collected from a given segment s.

Fig. 3 Process of sample collection from a segment

The moments of sample collection from a segment have been defined accordingly to two criteria. The first is the collection of samples at every eighth, and the second is the collection of samples at every new chord in a segment.

Table 4 Main harmony features

Feature group	Main features
Basic statistical	Average $AkD(s)$
functions	Standard deviation of $AkD(s)$
	Number of samples in $AkD(s)$
Common values	First most frequent value in $AkD(s)$
	Second most frequent value in $AkD(s)$
	Third most frequent value in $AkD(s)$
	Percentage share in the first most frequent value in $AkD(s)$
	Percentage share in the second most frequent value in $AkD(s)$
	Percentage share in the third most frequent value in $AkD(s)$

Harmony features describe what kind of harmony occurs in a given segment, which ones dominate, how many of them occur, etc. (Table 4). Below is a presentation of AkD samples (chords) collected at every eighth in a segment from Etude Op.10 No 5 by F. Chopin (Fig. 4).

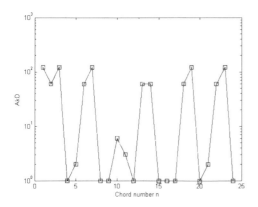

Fig. 4 AkD for fragment of F. Chopin's Etude Op.10 No 5

2.3.2 Rhythmic Features

Rhythmic features represent rhythmic regularity in a given segment of music. These features were obtained from the beat histogram, which was acquired from the calculation of autocorrelation [15].

$$autocorrelation[lag] = \frac{1}{N} \sum_{n=0}^{N-1} x[n]x[n-lag] \qquad (4)$$

where n is the input sample index (in MIDI ticks), N is total number of MIDI ticks in a segment and lag is delay in MIDI ticks $(0 < lag < N)$. The value of $x[n]$ is the velocity of Note On MIDI events. The histogram was transformed so that each bin corresponded to a periodicity unit of beats per minute (Fig. 5). The histogram values were normalized in relation to the highest value of the most frequent beat (beat with the highest bar).

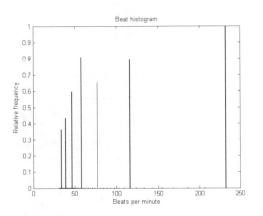

Fig. 5 Beat histogram for fragment of F. Chopin's Etude Op.10 No 5

Rhythmic features describe the strongest pulses in the piece (beats with the highest value in the Beat histogram), relations between them, their quantity, etc. (Table 5). From the example of the image in (Fig. 5), it is apparent that the First Strongest Rhythmic Pulse has a value of 240 BPM, the Second Strongest Rhythmic Pulse - 60 BPM, and the Third Strongest Rhythmic Pulse - 120 BPM.

Table 5 Main rhythmic features

Feature group	Main features
Strongest	First Strongest Rhythmic Pulse (FSRP)
Rhythmic Pulses	Second Strongest Rhythmic Pulse
	Third Strongest Rhythmic Pulse
Pulse Ratios	Ratios of Strongest Pulses
Relatively Strong	Number of Relatively Strong Pulses
Pulses	The number of beats with values greater than 50% of the FSRP value
	The number of beats with values greater than 30% of the FSRP value
	The number of beats with values greater than 10% of the FSRP value
Rhythmic Note	Average Note Duration - Average duration of notes in seconds
values	Note Density - Average number of notes per second

2.3.3 Correlations between Features

Individual features, such as harmony or dynamics, are related to rhythm. They are often correlated. The moment of appearance of a given accent, chord, etc. in the bar is of great significance. The most important and significant parameters were obtained through the correlation of parameters with rhythm.

We created an $AkD(B)$ data table, where B is a beat histogram. It comprises of AkD samples collected from musical segments at moments of the Strongest Pulses (beginnings of bars, repeating accents that dominate in a given fragment).

$$AkD(B) = (AkD_1, AkD_2, ..., AkD_b) \tag{5}$$

where b is the number of collected samples at moments of the Strongest Pulses. All values from the beat histogram which are more than 50% of the First Strongest Rhythmic Pulse in a beat histogram were accepted as the Strongest Pulses. Next, statistical features were calculated, just as with $AkD(s)$ (Table 4).

2.3.4 Dynamic Features

Dynamic features are based on the intensity of sound, the length of sounds, and their development in a segment (Table 6).

Table 6 Main dynamic features

Feature group	Main features
Basic statistical functions	Average of loudness levels of all notes
	Standard deviation of loudness levels of all notes

The last stage consisted of exporting the obtained data to Arff format, allowing for data analysis in the WEKA program.

2.4 Multi-label indexing

Describing emotions contained within a given segment is not always unambiguous. Some segments contain a single emotion, while others can contain several emotions simultaneously. This is why an assignment of more than one emotion per segment was permitted in our system. It allowed testers to assign several emotions to each of the consecutive examples, if needed. Marking an emotion from the lower level, L2, automatically caused the marking of the appropriate emotion from the higher level, L1.

3   Mood Tracking

Emotions in music pieces are not constant. In fragment lasting several seconds
there may be just one emotion or it may change many times. It depends on
the musical content of the music piece. The system described in this chapter
enables tracking emotions present in a given musical fragment during the
playback of the piece as well as the analysis of the diagrams generated by the
program. The following information may be found there:

- the kind of emotions dominating in the piece,
- the kind of emotions in particular fragments,
- the transition from one emotion into another,
- the time when many emotions appear together or when only single emotions are present.

Fig. 6 shows an example of emotion distribution of level L1 and Fig. 7 of
level L2 in Asturias by Isaac Albeniz.

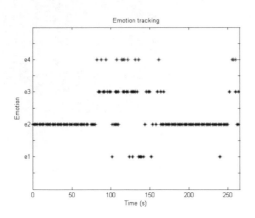

Fig. 6 Emotions of the first level L1 in Asturias by Isaac Albeniz

First level L1 emotions diagram (Fig. 6) leads to the conclusion that Asturias begins with a strong emotion e2 (energetic-negative), next after about
80 seconds there comes a part containing many emotions with dominating
e3 (calm-negative) and e4 (calm-positive). Next around the 170th second
of the piece's duration, just like at the beginning, there comes a fragment
of 80 seconds with one emotion e2 (energetic-negative). In the final part
of the piece (from the 250th second), it is possible to notice the return of
emotions e3 and e4. Considering the emotions of the second level L2 (Fig. 7)
we notice the appearance of emotions e21 (annoying) and e23 (nervous) in the

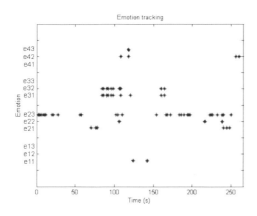

Fig. 7 Emotions of the second level L2 in Asturias by Isaac Albeniz

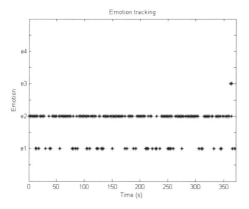

Fig. 8 Emotions of the first level L1 in the Moonlight Sonata part 3 by Ludwig van Beethoven

first fragment (s. 0-80). In the second fragment (s. 80-170) the majority consists of e31 (sad), e32 (bored) with moments of e43 (relaxed), e42 (peaceful), e23 (nervous), e22 (angry) and e11 (pleased). The third fragment is again dominated by e23 (nervous) with e42 (peaceful) in the final part.

Another example (Fig. 8 and Fig. 9) presents the emotions in the Moonlight Sonata part 3 by Ludwig van Beethoven. The e2 emotion (energetic-negative) dominates throughout the entire piece and is interlaced with emotions e21 (annoying), e22 (angry), e23 (nervous). In the course of the piece, emotion e2 is often complemented by emotion e1 (energetic-positive) or more precisely, e12 (happy).

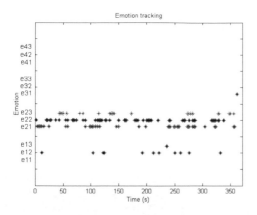

Fig. 9 Emotions of the second level L2 in the Moonlight Sonata part 3 by Ludwig van Beethoven

4 Emotion-Based Music File Searching System

4.1 Database

The process of building automatic indexing system involved using a database consisting of 83 musical files and containing pieces of classical music. This database was also used for creating an emotion based MIDI files searching system. It was created in MySQL technology.

4.2 File Indexation

In order to enable automatic searching of the database according to the emotions it contained, the database needed special adjustments. Indexing large collections of data by one person through listening to and categorizing every piece is practically impossible. The ideal solution would be to automatically index the files according to emotions. The process was divided into several stages:

1. Manual indexation of a learning collection (musical fragments) by testers (45 persons) according to emotions (multi-label indexing),
2. Constructing classifiers for detecting emotions (for every representative user profile),
3. Automatic indexation of entire musical pieces on the basis of decisions made by the classifiers.

4.3 Emotion-Based Musical Piece Searching

The query answering system *(QAS)* not only enables the search according to the composer and title, but also to the percentage content of a particular emotion in the piece as well as the one dominating at the beginning, in the middle, and at the end. Fig. 10 presents emotion based MIDI files searching system. Searching for percentage emotion content in a piece takes place on levels L1 and L2.

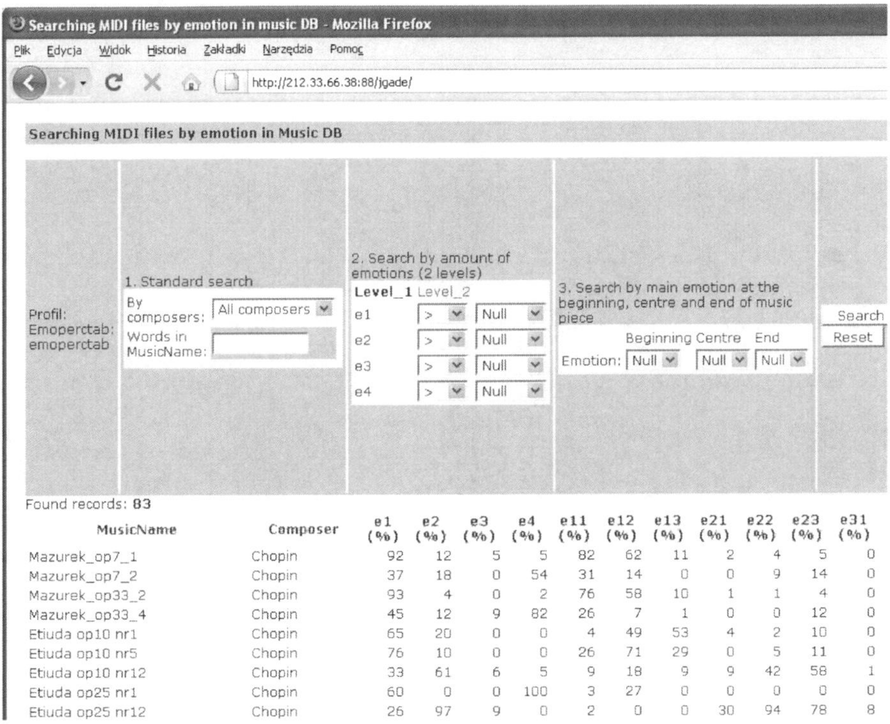

MusicName	Composer	e1 (%)	e2 (%)	e3 (%)	e4 (%)	e11 (%)	e12 (%)	e13 (%)	e21 (%)	e22 (%)	e23 (%)	e31 (%)
Mazurek_op7_1	Chopin	92	12	5	5	82	62	11	2	4	5	0
Mazurek_op7_2	Chopin	37	18	0	54	31	14	0	0	9	14	0
Mazurek_op33_2	Chopin	93	4	0	2	76	58	10	1	1	4	0
Mazurek_op33_4	Chopin	45	12	9	82	26	7	1	0	0	12	0
Etiuda op10 nr1	Chopin	65	20	0	0	4	49	53	4	2	10	0
Etiuda op10 nr5	Chopin	76	10	0	0	26	71	29	0	5	11	0
Etiuda op10 nr12	Chopin	33	61	6	5	9	18	9	9	42	58	1
Etiuda op25 nr1	Chopin	60	0	0	100	3	27	0	0	0	0	0
Etiuda op25 nr12	Chopin	26	97	9	0	2	0	0	30	94	78	8

Fig. 10 A view of the searching system on the first level L1

The *QAS* enables the use of its 3 modules simultaneously: 1. standard search, 2. search according to the amount of emotions, 3. search according to emotion dominating at the beginning, in the middle, and at the end of the piece (Fig. 11).

An example of a query could be the following: "find all F. Chopin's Preludes which are at least 60% exciting". In this case from the collection of 12 preludes present in the database, 2 have been identified by the system - Prelude 5 and 8 (Fig. 12). Another example of a query could be: "find all pieces which begin with the energetic-positive emotion (e1), are calm-positive in the middle (e4) and again energetic-positive at the end (e1)". Fig. 13 presents 5 pieces being a result to the above query. Results of queries depend on the user representative profile that is granted by the system after filling out a questionnaire.

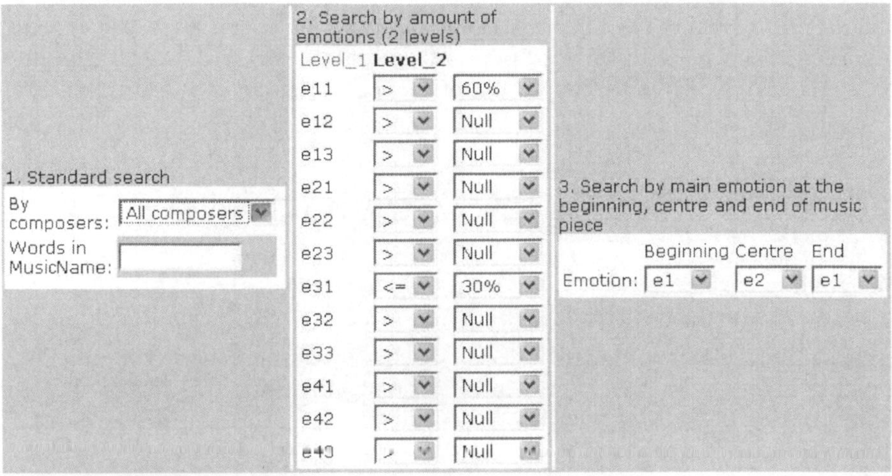

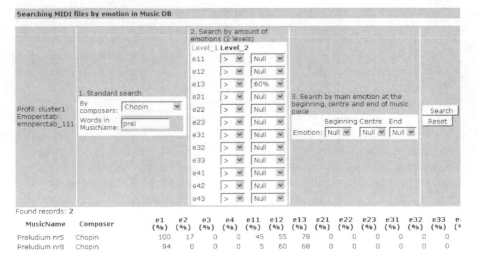

Fig. 11 QAS module searching according to emotion amount on the second level L2

Fig. 12 Search results - example 1

In some situations when we place a complex, "demanding" query, it happens that the system does not find any files. It is due to the fact that the number of pieces in the database is relatively small (83) and it does not guarantee results for all combinations of queries. In order to improve the searching system, it should be enhanced with a generalization module for queries to which a precise result is not found, providing possibly closest answer [2], [13].

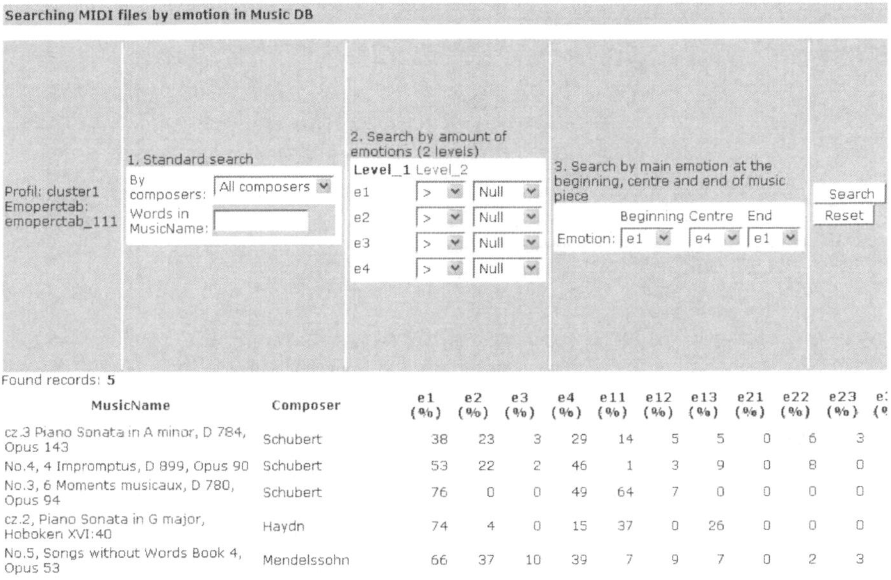

Searching MIDI files by emotion in Music DB

Found records: 5

MusicName	Composer	e1 (%)	e2 (%)	e3 (%)	e4 (%)	e11 (%)	e12 (%)	e13 (%)	e21 (%)	e22 (%)	e23 (%)	e: (%
cz.3 Piano Sonata in A minor, D 784, Opus 143	Schubert	38	23	3	29	14	5	5	0	6	3	
No.4, 4 Impromptus, D 899, Opus 90	Schubert	53	22	2	46	1	3	9	0	8	0	
No.3, 6 Moments musicaux, D 780, Opus 94	Schubert	76	0	0	49	64	7	0	0	0	0	
cz.2, Piano Sonata in G major, Hoboken XVI:40	Haydn	74	4	0	15	37	0	26	0	0	0	
No.5, Songs without Words Book 4, Opus 53	Mendelssohn	66	37	10	39	7	9	7	0	2	3	

Fig. 13 Search results - example 2

4.4 Constructing User Profiles

The emotions contained in musical pieces may be perceived differently by particular listeners. This depends on such factors as musical preferences, education, emotional profile. It may happen that the same piece e.g. Prelude no. 5 by F. Chopin will be qualified by one person as energetic-positive (e1) and as energetic-negative (e2) by another. Yet another listener would choose both emotions simultaneously (e1 and e2). This was confirmed by experiments with tester-listeners conducted for the purpose of working on this chapter. Because people may feel emotions in music differently, it was decided to find and build profiles which would adjust the searching process in the database to the personality of the user.

4.5 Indexation of Musical Fragments by the Testers

The first stage of building the profiles involved indexing a collection of music pieces on the basis of emotions and completing a questionnaire by the listeners. The research was carried out on 45 tester-listeners who listened to 498 musical fragments each. The listener matched one or more emotions from the levels L1 and L2 with every piece.

Table 7 Questions for the tester

Question	Possibile answers
Sex	female, male
Profession	freelancer, specialist, entrepreneur/businessman, farmer, physical worker, services sector worker, housewife, student, pupil, pensioner, unemployed, education, administration, technician
Education	basic, vocational, secondary school, college degree
Age	10-20, 20-30, 30-40, 40-50, 60-70, 70-
How do you feel at the moment?	happy, sad, calm, nervous
Do you like reading books? If yes, give two genres.	classical, professional literature, fantasy, crime fiction, history, hobby, comic book, other
What is your hobby? Give two examples.	music, science, computers and the Internet, fashion and style, tourism, motoring, cinema, drawing, other
What kind of music do you prefer?	classical, pop, rock, metal, jazz, gospel, blues, disco, funk, rap, electro, Latino, world, reggae, soul
What is you favourite instrument? Give two examples.	guitar, violin, cello, bass, drums, flute, oboe, clarinet, saxophone, trumpet, trombone, accordion, piano, harp, keyboard, other
What are your two favourite ice-cream flavours?	vanilla, chocolate, tiramisu, yoghurt, cream, cappuccino, mint, coconut, strawberry, lemon, blueberry, cherry, banana
If you had $3,500.00 what would you most likely spend it on?	a car, tourism, savings, renovation, other
How big is the city where you live?	village, town up to 50 thousand people, town 50 thousand to 100 thousand people, city 100 thousand to 500 thousand people, city over 500 thousand people
Are you happy with you life and what you have so far achieved?	yes, rather yes, rather no, no
Do you consider yourself to be a calm person?	yes, no, I cannot tell
Do you live with your family?	yes, no
Does your job give you pleasure?	yes, rather yes, rather no, no
Do you like playing with and taking care of little children?	yes, rather yes, rather no, no
Do you have or would like to have a little dog at home?	yes, no
How do you see that you have done something well?	because I know, because my boss/friends tell me, because I achieved a goal, I'm never quite sure
What is important to you in carrying out tasks given you by your boss?	doing the task, staying out of trouble
What do you do when you have a new complicated piece of equipment (e.g. electronic device)	you start working it out at once, you read the manual thoroughly, you ask somebody to help you, you leave it for somebody else to work out
When you relate a film you liked to your friend, you talk about it:	in detail, in general (in order not to spoil the pleasure of watching it), you relate a particular scene in detail
You work hard for:	others, yourself, yourself and other, others and yourself
How do you remember your fun vacation?	you see pictures, you hear sounds, you feel the warmth, wind, water, feel everything at once like in a movie

4.6 A Testers Questionnaire

Each tester also filled out a questionnaire used for building user profiles.The set of questions is presented in Table 7. The testers had the option of not responding to particular questions from the questionaire.

5 The Experiment Results

5.1 Mood Detection

The program WEKA was used to carry out the experiments, which allowed for testing data utilizing many methods [18].

Because many musical segments were labeled by many labels simultaneously, multi-label classification in emotion detection was used (multi-label decision attribute was replaced by a set of binary decision attributes representing emotions). The same, we transformed data into several binary types of data and tested one against the rest of the data. For each class, a data set was generated containing a copy of each instance of the original data, but with a modified class value. If the instance had the class associated with the corresponding dataset it was tagged YES, otherwise, it was tagged NO. The classifiers were built for each of these binary data sets. The proposed strategy greatly simplified the process of building classifiers for a decision system with a multi-label decision attribute.

The classification results were calculated using a cross validation evaluation CV-10. We used attribute selection to find the best subset of attributes. The best result was achieved by using Wrapper Subset Evaluator. After testing the data utilizing many methods, one of the best results was achieved with the use of the k-NN classifier (k-nearest neighbor). The use of attribute selection improved the accuracy of classifiers by an average of about 10% (Fig. 14).

A classifier was created for each emotion separately. The results of classification after attribute selection are presented below (Table 8, Table 9). Level one classifiers are more accurate than level two classifiers. This is connected to the fact that the groups of examples containing emotions from the first level are larger. Also, the emotions from level one are much easier to recognize for the listener, since there are only 4. The highest accuracy was attained for emotion classifier e1 - energetic-positive (90%), and the lowest accuracy was attained for emotion classifier e3 - calm-negative (74%).

Table 8 Coverage factor of L1 first level classifiers

Classifier	Emotion	No. of objects	Coverage factor
e1	energetic-positive	151	90%
e2	energetic-negative	172	87%
e3	calm-negative	111	74%
e4	calm-positive	103	82%

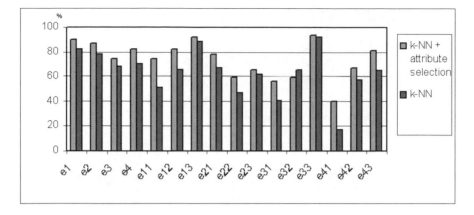

Fig. 14 Comparison of results attained using the k-NN algorithm with and without the use of attribute selection

Table 9 Coverage factor of L2 second level classifiers

Classifier	Emotion	No. of objects	Coverage factor
e11	pleased	66	74%
e12	happy	69	82%
e13	excited	19	92%
e21	annoying	37	78%
e22	angry	52	59%
e23	nervous	82	66%
e31	sad	47	56%
e32	bored	52	59%
e33	sleepy	12	94%
e41	calm	17	40%
e42	peaceful	30	67%
e43	relaxed	56	81%

The accuracy of L2 second level classifiers (Table 9) is somewhat less accurate than L1 first level classifiers, and fluctuates from 40 to 92%. This is connected with the fact that the example groups for specific emotions are smaller as well as that the recognition of these emotions - on this more precise level - is more difficult for the listener. The least accuracy was attained for emotion classifier e41 - calm. Also, e42 is low, which is connected to the fact that the division of emotions into groups e41 - calm and e42 - peaceful is not the most apt. These are rather difficult for the listener to distinguish. In the future, for further research, these two groups should be combined into one. The best results (80-90%) were obtained for emotions e12 (happy), e13 (excited), e33 (sleepy), and e43 (relaxed). These are the most easily recognized emotions by the listeners, and it is rather difficult to confuse them with other emotions.

5.2 Clustering and Searching for User Profiles

5.2.1 Clustering

498 musical fragments were indexed by 45 testers according to emotions. Listeners came from different professional groups (students, musicians, scientists, engineers, teachers, medical workers) and different age groups (from 20 to 50 years old). Each listener assigned one or more emotions from the levels L1 and L2 to every piece. All listeners also filled out the questionnaire which was used to create a table of user profiles. The data was used to create a decision table with rows of tester listeners (45). Each row contained 7969 features received through labeling the pieces by tester-listeners. Features represent the tester's binary (Yes, No) answers to whether or not a given emotion is found in a given musical fragment (16 emotions * 498 fragments = 7969 answers). Searching for representative groups and related classification was conducted with the use of tools in the WEKA package.

In the first stage, the optimal number of groups was established. Cross-validation EM (expectation maximization) [12] and K-means [7] were used for that purpose. When choosing the number of groups, their size was also taken into consideration. Bearing in mind that a relatively small number of different groupings was tested, groups of size smaller than 9 were abandoned. Small groups seemed to be less credible because they are confirmed by a smaller number of examples. In our case, the optional number of groups was set to either 2 or 3. Finally, two grouping sets ZG1 and ZG2 were used (Table 10, Fig. 15, Fig. 16).

Table 10 Grouping sets

Name of grouping set	Grouping algorithm	No. of groups	Number of examples in groups cluster0/cluster1
ZG1	EM	2	36 /9
ZG2	K-means	2	16/29

5.2.2 The Classification of Grouping Sets ZG1 and ZG2

These two grouping sets were linked to the testers' questionnaires and tested with the following classifying algorithms: PART, J48, RandomTree, k-NN, BayesNet [18]. At this point, the sets ZG1 yielded classifiers all having confidence above 70% and on average 15% higher than the sets ZG2 (Fig. 17).

The next step involved selection of attributes. The WrapperSubsetEval evaluator was used (weka.attributeSelection.WrapperSubsetEval) [5] and tested with the algorithms PART, J48, k-NN, NaiveBayes. There were two options for selecting attributes:

- with nominal attributes: 31 nominal attributes found in testers' questionnaires + decision class

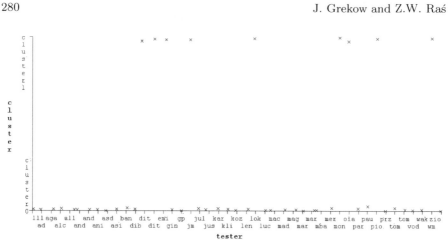

Fig. 15 The visualization of EM algorithm grouping

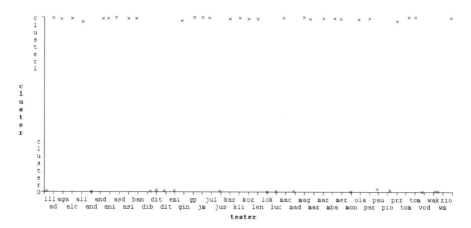

Fig. 16 The visualization of K-means algorithm grouping

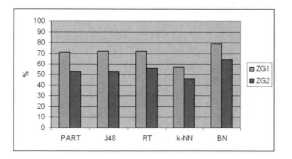

Fig. 17 Comparing the precision of ZG1 and ZG2 classification before selecting
the attributes

- with binary attributes: (all nominal attributes were changed to binary attributes) 170 binary attributes + decision class

In both instances, the decision class is the value of a cluster that had been defined for the tester during clustering the testers' answers pertaining to emotion in musical fragments.

When using the binary attributes, the results were 10% better. In both cases (ZG1 and ZG2) the best collections of attributes were found using WrapperSubsetEval and k-nearest neighbor.

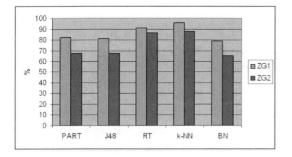

Fig. 18 Comparing the precision of ZG1 and ZG2 classification after selecting attributes

Comparing the results of classification of grouping sets ZG1 and ZG2, it was found that the best results were achieved by the ZG1 set, that is division performed by the EM algorithm (Fig. 18). The best results were achieved with the k-NN classifier (96%) and Random Tree (91%).

The k-NN classifier built from ZG1 set grouping was used in QAS to assign a new user's profile to one of the groups. Classifiers created on the basis of group representatives from grouping set ZG1 were used for indexing files from the database according to emotions.

From the selected attributes, it could be seen what were the most meaningful questions and answers from the questionnaire (Table 11).

Table 11 The most meaningful questions and answers in establishing the user profile for ZG1

Question	Answer
Profession	freelancer, specialist
Age	40-50
What kind of music do sou like?	classical
What is your favourite instrument?	the piano
What are your favourite flavours of ice-cream?	vanilla

Explaining the meaning of questions chosen in the process of attribute selection, one may draw the conclusion that people who do not like classical music, perceive emotions differently than the people who like it (Table 12).

Table 12 Group characteristics for ZG1

Group	Number of people	Description
Cluster0	36	People who: - like music other than classical - do not prefer the piano - like ice-cream other than vanilla
Cluster1	9	People who: - like classical music - like vanilla ice-cream - prefer the piano, represent a freelance profession, specialist

Additionally, the grouping set ZG2, which had a slightly worse compliance with the questionnaires (classifier confidence - 86% for RandomTree and 88% for k-NN), was also tested. In this case the selected attributes showed that the most meaningful questions and answers from the questionnaire are the ones listed in Table 13.

Table 13 The most meaningful questions and answers during establishing the user profile for ZG2

Question	Answer
Profession	Education sector
Age	No decision
What kind of books do like to read?	Criminal-fiction
What is your hobby?	Science
What kind of music do you like?	Pop, Disco
What are your favourite flavours of ice-cream?	Strawberry
Does your work give you pleasure?	Rather no

Table 14 shows the characteristics of groups found through grouping set ZG2.

Because ZG2 grouping gave worse results than ZG1 grouping, it was not used in *QAS*. However it is an interesting and different look at dividing listeners into groups and the meaning of the questions in the questionnaire.

Increasing the number of tester-listeners for indexing musical instances to e.g. 200 people would cause creating a larger collection of data for analysis and could result in:

- a more precise division into groups,
- finding a greater number of groups,

Table 14 The group characteristics for ZG2

Group	Numbe of people	Description
Cluster0	16	People who: - like pop music - do not choose strawberry ice-cream - like Criminal-fiction books - do not choose science as a hobby - do not choose the "rather no" option in the question: Does your work give you pleasure?
Cluster1	29	People who: - do not choose pop and rap music - like strawberry ice-cream - do not choose Criminal-fiction books - do not find pleasure in their work

- finding additional interesting relations between questionnaires and tester groups.

On the other hand, collecting listeners' opinions is a time consuming task for both sides taking part in the experiment, which influenced setting the minimum number of testers who took part in labeling musical pieces to 45.

6 Conclusion

This chapter presented a query answering system connected with MIDI music files database, enabling the search for music files according to emotions. During its construction, a task of automatic file indexation was also attempted. This required building a database, manual indexation of learning instances, finding a collection of features describing musical segments, and constructing classifiers. A hierarchical model of emotions consisting of two levels, L1 and L2, was used. A collection of harmonic and rhythmic attributes extracted from music files made emotion detection with an average of 83% accuracy at level L1 possible. The continuing development of features describing musical segments as well as expanding the database should further improve the precision of classifiers of the lower level L2. In order to base the searching system's results on the user's preferences, a detection of the emotional profile was suggested. The profile alters the system for the user's needs and solves the problem of a subjective emotion perception. The application was enriched with an emotion-tracking user's mood module, which provides information about the distribution of emotions in the course of the musical piece.

One of the directions of continuing works is expanding the searching module through adding a mechanism of generalizing queries to which a precise result is not found and providing possibly closest results.

References

1. DiPaola, S., Arya, A.: Emotional remapping of music to facial animation. In: ACM Siggraph 2006 Video Game Symposium Proceedings (2006)
2. Gaasterland, T.: Cooperative answering through controlled query relaxation. IEEE Expert 12(5), 48–59 (1997)
3. Grekow, J.: Broadening musical perception by AKWETS technique visualization. In: Proceedings of the 9th International Conference on Music Perception and Cognition, ICMPC9 (2006)
4. Grekow, J.: An analysis of the harmonic content - main parameters in the AKWET method. In: Proceedings of II Conference on Technologies of Knowledge Exploration and Representation, TERW (2007)
5. Hall, M.A., Holmes, G.: Benchmarking attribute selection techniques for discrete class data mining. IEEE Transactions on Knowledge and Data Engineering 15(6), 1437 (2003)
6. Hevner, K.: Experimental studies of the elements of expression in music. American Journal of Psychology 48, 246–268 (1936)
7. Larose, D.T.: Discovering Knowledge in Data: An Introduction to Data Mining. Wiley, Chichester (2004)
8. Li, T., Ogihara, M.: Detecting emotion in music. In: Proceedings of the Fifth International Symposium on Music Information Retrieval, pp. 239–240 (2003)
9. Liu, C., Yang, Y., Wu, P., Chen, H.: Detecting and classifying emotion in popular music. In: Proceedings of the 9th Joint Conference on Information Sciences (JCIS)/CVPRIP (2006)
10. Liu, D., Lu, L., Zhang, N.: Automatic mood detection from acoustic music data. In: Proceedings of the Fifth International Symposium on Music Information Retrieval (2003)
11. McKay, C., Fujinaga, I.: Automatic genre classification using large high-level musical feature sets. In: Proceedings of the International Conference on Music Information Retrieval, pp. 525–530 (2004)
12. Nigam, K., McCallum, A.K., Thrun, S., Mitchell, T.M.: Text classification from labeled and unlabeled documents using em. Machine Learning 39(2/3), 103–134 (2000)
13. Ras, Z.W., Dardzinska, A.: Solving failing queries through cooperation and collaboration. Special Issue on Web Resources Access in World Wide Web Journal 9(2), 173–186 (2006)
14. Thayer, R.E.: The biopsychology of mood and arousal. Oxford University Press, Oxford (1989)
15. Tzanetakis, G., Cook, P.: Musical genre classification of audio signals. IEEE Transactions on Speech and Audio Processing 10(5) (2002)
16. Wang, M., Zhang, N., Zhu, H.: User-adaptive music emotion recognition. In: 7th Inter-national Conference on Signal Processing, ICSP (2004)
17. Wieczorkowska, A., Synak, P., Ras, Z.: Multi-label classification of emotions in music. In: Intelligent Information Processing and Web Mining, Advances in Soft Computing, Proceedings of IIS 2006 Symposium, Ustron, Poland, vol. 35, pp. 307–315 (2006)
18. Witten, I.H., Frank, E.: Data Mining: Practical machine learning tools and techniques. Morgan Kaufmann, San Francisco (2005)
19. Yang, Y., Su, Y., Lin, Y., Chen, H.: Music emotion recognition: The role of individuality. In: Proceedings of the international workshop on Human-centered multimedia, HCM 2007 (2007)

On Search for Emotion in Hindusthani Vocal Music

Alicja A. Wieczorkowska, Ashoke Kumar Datta, Ranjan Sengupta,
Nityananda Dey, and Bhaswati Mukherjee

Keywords: Chord Recognition; Machine Learning; Music Analysis.

1 Introduction

Emotions give meaning to our lives. No aspect of our mental life is more important
to the quality and meaning of our existence than emotions. They make life worth living, or sometimes ending. The English word 'emotion' is derived from the French
word *mouvoir* which means 'move'. Great classical philosophers-Plato, Aristotle,
Spinoza, Descartes conceived emotion as responses to certain sorts of events trig-
gering bodily changes and typically motivating characteristic behavior. It is difficult
to find a consensus on the definition of emotion [9]. Most researchers would probably agree that emotions are relatively brief and intense reactions to goal-relevant
changes in the environment that consist of many subcomponents: cognitive appraisal, subjective feeling, physiological arousal, expression, action tendency, and
regulation. It therefore suggests that some part of the brain would be selectively
activated [21]. Origin of emotion may be traced back to 200,000 years ago to semi-
nomadic hunter-gatherer [16]. It is argued that their way of living, which involved
cooperating in such activities as hunting, avoiding predators, finding food, rearing
children, and also competing for resources, could be related to the origin of emotion.
Most emotions are presumably adapted to living this way. Several of the activities

Alicja A. Wieczorkowska
Multimedia Department, Polish-Japanese Institute of Information Technology, Koszykowa
86, 02-008 Warsaw, Poland
e-mail: alicja@pjwstk.edu.pl

Ashoke Kumar Datta · Ranjan Sengupta · Nityananda Dey
Scientific Research Department, ITC Sangeet Research Academy, 1, N S C Bose Road,
Tollygunge, Kolkata 700 040, India

Bhaswati Mukherjee
Center for Development of Advanced Computing, Kolkata, India

Z.W. Raś and A.A. Wieczorkowska (Eds.): Adv. in Music Inform. Retrieval, SCI 274, pp. 285–304.
springerlink.com   © Springer-Verlag Berlin Heidelberg 2010

are associated with basic survival problems that most organisms have in common. These problems, in turn, require specific types of adaptive reactions. A number of authors have suggested that such adaptive reactions are the prototypes of emotions as seen in humans [17], [24].

If various emotions are cognitively differentiable, as is again likely to be publicly agreed upon, there should be differences in sites of brain being excited for different emotions. Bower further suggested [3] that every emotion is associated with autonomic reactions and expressive behaviors. These expressive behaviors or responses to the same stimuli can vary depending on many factors external to the stimuli, like the mood of a person [9], [25], memory association of the person to the applied stimuli [9], [14], etc.

Objects or a sequence of objects elicit feeling through a sequence of psychophysical processes. The sensory organs convert the signals from the objects to neural pulses. These pulse trains are processed in sub-cortical neural structures, which are primarily inherited. These processed signals produce perception in brain. Through the process of learning and experience we cognize the objects or sequence of objects from these perceived signals. Again through the process of learning we learn to associate these with some emotive environment in the past. This ultimately evokes emotion or feeling.

The evaluation of emotional appraisals of stimuli may be done by having the person report the emotions they perceive as reaction to the stimuli. This can be done in several different ways such as verbal descriptions, choosing emotional terms from a list, or rating how well several different emotional terms describe the appraisal [9], [22]. The emotional terms used should be limited in number and as unambiguous as possible. It is also possible to represent these terms in vector forms. The splitting of emotion into dimensions is consistent with Bower's network theory of emotion [8], [18]. However, the number of components and the type of components vary between studies [7], [18], [22], [27].

While sound stimuli may cause general physiological changes ("arousal"), these changes must be interpreted cognitively in order for a specific emotion to emerge. The listener does not come to the listening experience as a blank slate. He or she already has existing musically pertinent knowledge. Even for a musically untrained listener the general exposure to listening to music since childhood is also learning, though not formal.

Thus any emotional behavior, even habitual and seemingly automatic and natural, is actually learned. In case of music this behavior serves as a means of communication, since often emotional behavior is differentiable and intelligible. One major problem that arises in the study of the emotional power of music is that the emotional content of music is very subjective. A piece of music may be undeniably emotionally powerful, and at the same time be experienced in very different ways by each person who hears it. The emotion created by a piece of music may be affected by memories associated with the piece, by the environment it is being played in, by the mood of the person listening and their personality, by the culture they were brought up in; by any number of factors both impossible to control and impossible to quantify.

Under such circumstances, it is extremely difficult to deduce what intrinsic quality of the music, if any, created a specific emotional response in the listener.

It seems that the listeners experience gross emotion through the unfolding of successive events. If the successive events are always predictable, the emotion is boredom, unless an association of past evokes a specific emotion. Again, extensive uncertainties are likely to lead to apprehension and anxiety. These two emotional experiences may therefore be the robust ones. As soon as the unexpected is experienced, the listener attempts to fit it into the general system of beliefs relevant to the theme. This may happen in one of the following three ways: (1) The mind may suspend judgment, expecting that the subsequent event will bring in clarity. (2) If no clarification takes place, irritation will set in. (3) The expected consequent may be seen as a purposeful blunder.

When we expect music to convey emotion, we accept it to be a language, as emotion is built upon the meaning. In order to extract the right emotion, it is essential that the listener is acquainted with the grammar, and thus the extra-musical world of concepts, actions, emotional states, and character. Even if he does not have the appropriate grammar in his mind, he can still extract some emotional meaning out of it, with some of his own stock grammar. In this sense emotion in music may be called referential. A piece of music can be universally pleasant or universally irritating.

On the other hand, a musical stimulus or series of stimuli can be considered to indicate and point to other musical events, which are about to happen, rather than extra-musical concepts and objects. That is, one musical event (be it a tone, a phrase, or a whole section) has meaning because it points to and makes us expect another musical event. Even then the affective experience is still dependent on cognition (involving a process of intellection, conscious or unconscious) that cannot be restricted to the musical concepts alone. The musical expectations and experience grow out of the innate processes of *grouping*, *closure*, and *good continuation* in Gestalt psychology.

Hevner (1936) studied grouping of emotions described by listeners using adjectives through listening experiments [11]. The experiments substantiated a hypothesis that music inherently carries emotional meaning. Hevner discovered the existence of clusters of descriptive adjectives and laid them out in a circle. These are: a) cheerful, gay, happy, b) fanciful, light, c) delicate, graceful, d) dreamy, leisurely, e) longing, pathetic, f) dark, depressing, g) sacred, spiritual, h) dramatic, emphatic, i) agitated, exciting, j) frustrated, k) mysterious, spooky, l) passionate, m) bluesy. Actually, emotion detection in musical information is better considered as a 'Multi-label Classification problem', where the music sounds are classified into multiple classes simultaneously. That means that a single music sound may be characterized by more than one label, e.g. both "dreamy" and "cheerful."

Emotions recognized in music can be represented in a two-dimensional space (Figure 1), with valence (positive vs. negative feelings) and arousal (high-low) as principal axes [20]. These are the dimensions suggested by Russell to describe

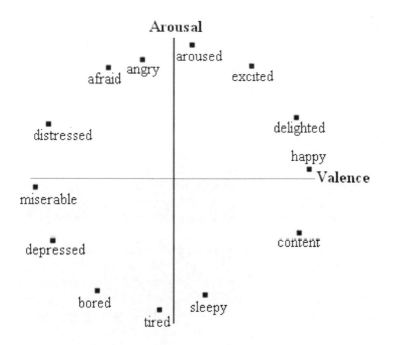

Fig. 1 Possible descriptions of emotion using valence and arousal [12]

emotion [19]. Valence refers to the happiness or sadness of the emotion and arousal is the activeness or passiveness of the emotion [23]. A positive valence corresponds with positive emotions such as joy, happiness, relaxing and a negative valence corresponds with negative emotions such as fear, anger and sadness.

Emotions are denotative signs. When a listener reports an emotion, particularly in case of ragas in Indian Music, he may actually be describing only what he believes the passage is supposed to indicate, not anything he has experienced by himself. Even when a genuine emotional experience is reported, it is liable to become garbled and perverted in the process of verbalization. Some emotional states are much more subtle and varied than are the few crude and standardized words which we use to denote them. In such cases reports may contain a large amount of what psychiatrists call "distortion".

In India, music (*geet*) has been a subject of aesthetic and intellectual discourse since the times of Vedas (*samaveda*). *Rasa* was examined critically as an essential part of the theory of art by Bharata in Natya Sastra, (200 century BC). The *rasa* is considered as a state of enhanced emotional perception produced by the presence of musical energy. It is perceived as a sentiment, which could be described as an aesthetic experience. Although unique, one can distinguish several flavors according to the emotion that colors it. Several emotional flavors are listed, namely erotic love (*sringara*), pathetic (*karuna*), devotional (*bhakti*), comic (*hasya*), horrific (*bhayanaka*), repugnant (*bibhatsa*), heroic (*vira*), fantastic, furious (*roudra*), and peaceful (*shanta*). Italics represent the corresponding emotion given in the Indian treatises. The individual feels immersed in that mood to the exclusion of anything else including himself. It may be noted that during the musical experience, the mind experiences conscious joy even in the representation of painful events because of the integration of perceptual, emotional, and cognitive faculties in a more expanded and enhanced auditory perception, completed by the subtle aesthetic of sensing, feeling, understanding and hearing all at the same time. The Eastern approach to emotional aesthetics and intelligence treats *rasa* as a multi-dimensional principle that explains thoroughly the relation between a sentiment, a mood, the creative process and its transpersonal qualities. This transpersonal domain includes the super-conscious or spiritual state and therefore acts as an interface between individual and collective unconscious states. This transpersonal quality is a germinating power hidden behind aspects of great musical creation that can reveal it, and is able to induce the complete chromatic range of each emotion. *Rasa* conveys the idea of an aesthetic beauty knowable only through the feeling. This aesthetic experience is a transformation of not merely feeling, but equally of cognition, a comprehensive understanding in the mode of ecstasy of the intellect, itself inscrutable and illuminating. In the Vedas the experience of *rasa* is described as a flash of inner consciousness, which appears to whom the knowledge of ideal beauty is innate and intuitive.

Rasa is not the unique property of the art itself. It unites the art with the creator and the observer in the same state of consciousness, and requires the power of imagination and representation and therefore a kind of intellectual sensibility. Indian musicological treaties since Bharata hold that even notes bear the potential of producing emotional effects. Tembe listed eight of them (Table 1). However no rational or scientific scrutiny was provided. It seems that the list was drawn from the proposals presented in Natyashastra.

However, he agrees that only four *rasas*, namely *Karuna*, *Shanta*, *Shringara* and *Vira* may actually be experienced from a single note. He further proposes that when *Shuddha madhyama* dominates a melody, it creates a serene and sublime

Table 1 Emotional attributes of notes according to G.S. Tembe [26]

Notes	Emotional attribute
Shadja	like a *yogi* beyond any attachment
Rishabha (komala)	rather sluggish
Rishabha (shuddha)	reminding of indolence of a person waking up from sleep
Gandhara (komala)	bewildered, helpless and pitiable
Gandhara (shuddha)	fresh and pleasant
Madhyama (shuddha)	grave, noble and powerful
Madhyama (tivra)	sensitive, luxurious
Panchama	brilliant, self composing
Dhaivata (komala)	grief, pathos
Dhaivata (shuddha)	robust, lustful
Nishada (komala)	gentle, happy, affectionate
Nishada (shuddha)	piercing appeal

atmosphere, while a dominant *Panchama* creates an invigorating and erotic feeling. Pandit V.N. Bhatkhande [2] in his work suggested the inadequacy of vadi svara (i.e. the main melodic tone of the raga) in determining the *rasa* of ragas. However, he mentions that *Ragas* employing *Shuddha* (*Rishabha, Dhaivata,* and *Gandhara*) emote *Shringara rasa,* and those employing *Komala* (*Dhaivata* and *Nishad*) emote *Vira rasa.* This view is contradicted by Ratanjankar [26]. According to him, individual notes cannot produce emotion, and they may do so only in a specific context. This implies that expression is born by the melodic content. Konishi et.al [13] reported that listeners can correctly decode emotions like anger, fear, happiness, and sadness from single notes from vibrato effect in Western music.

Having noted all these, it appears that the notion of a single note conveying emotion in general may be somewhat contrived particularly in Indian music. In Indian music, a note does not have a specific frequency. It is related to the scale where the base note *Sa* can be assigned any arbitrary frequency.

Karnani [12] noticed inconsistency between the *rasa* of a raga traditionally prescribed and experienced. He holds that since a raga represents a complex set of feelings, a simple relationship between a raga and *rasa* is unlikely.

Some empirical studies on the relationship of raga and *rasa* in Hindustani music are available. Deva and Virmani [6] reported consistent judgment of Indian listeners on the mood, color, season, and time of day for excerpts from Hindustani ragas.

Gregory and Varney [10] used both Western and Indian listeners to assess the emotional content of Hindustani ragas, Western classical music, and Western new-age music. He used a list of mood terms taken from Hevner for the assessment of the emotional contents. Both the Western and Indian listeners were reported to be sensitive to intended emotions in Western music, but not in the Hindustani ragas. Also the textbook descriptions of ragas did not always reflect the mood intended in a given performance. On the other hand, Balkwill and Thompson [1] reported high ratings in respected categories of correct emotion detection by Western listeners for joy, sadness and anger in Hindustani ragas. They believe that listeners can appreciate affective qualities of unfamiliar music by attending to acoustic cues. Consciously or intuitively, composers and performers draw upon acoustic cues too. When cultural-specific cues are absent, listeners may still attend to acoustic cues such as tempo and loudness. These cues provide listeners with a general understanding of the intended emotion.

In Hindustani music, ragas are said to be associated with different *rasas* (emotions). However, one particular raga is not necessarily associated with one emotion. Moreover, opinion varies; a comprehensive summary is available in *Semiosis in Hindusthani Music* [15]. For the present study we have selected 11 ragas (Table 2) to represent different *rasas*/emotions residing therein. Of the eight emotions listed in the opinion score sheets, only six represent rasas. These are Heroic (Vira), Anger (Raudra), Serenity (Santa), Devotion (Bhakti), Sorrow (Karuna), Romantic (Sringara). Other two emotions namely Joy and Anxiety have been considered additionally.

Table 2 Selected ragas and corresponding *rasas*

Name of the Raga	*Rasas*
Adana	Vira
Bhairav	Raudra, Santa, Bhakti, Karuna
Chayanat	Sringara
Darbari Kannada	Santa
Hindol	Vira, Raudra
Jayjayvanti	Sringara
Jogiya	Karuna, Sringara, Bhakti
Kedar	Santa
Mian-ki-Malhar	Karuna
Mian-ki-Todi	Bhakti, Srigara, Karuna
Shree	Santa

The objective of the present study was to find whether:

1. An oral music segment of short length extracted randomly from a raga elicit any emotion,
2. The elicited emotion from an oral music segment can be specified into prescribed categories,
3. The elicited emotion from different segments from the same raga has some specificity,
4. To what extent the emotional responses from the segments of a raga correspond to those given in Table 2,
5. Whether the elicited response have any cross-cultural similarity,
6. To what extent the melodic sequence (sequence of musical notes) relate with emotional response.

For the purpose of this research, we selected the ragas from ITC Sangeet Research Academy archive, and after signal processing we continued our work in order to find the sequences for listening experiments. These sequences would essentially be the fragments of ragas. In order to find which sequences evoke particular emotions, many possible sequences of various lengths can be taken into account (considering the grammar of the raga). We decided to extract about 30-seconds long sequences from the sound signal, which might evoke emotions, and use them for perceptual tests. The tests were performed by both western and Indian listeners, in order to observe and compare emotions evoked by each sequence. Since each raga has a specific set of notes and sequences used, we could assign short sequences of notes to particular emotions evoked in both Indian and western listeners.

2   Experimental Details

There are different styles for executing a raga in vocal music. The most common is khayal. The performance of khayal has two distinctive parts. The first one is known as alap wherein one tries to establish the image and emotional distinctiveness in slow tempo. The second part is known as bistar where the tempo is faster and faster and the performer tries to expose his skills through use of various embellishments (alankaras) while keeping the mood and emotion of the raga intact.

The present study concentrates on the alap part of professional khayal performances as this really establishes the characteristics of raga. The investigated excerpts represent vocal music (with some accompaniment). Songs by eminent singers in the ragas mentioned in Table 2 were selected from the archives of ITC Sangeet Research Academy. Only the alap portions of the songs were used for the present study. The alap portions had a varying length between 10 to 12 minutes. From the alap part of each song, four segments of about 30 seconds were taken out at different places for the audition test. The selection of the places was random in the sense that no special cognitive procedure was used. The only constraint used was to see that the end notes were not truncated in the middle. The selected segments were not therefore of exactly equal length. The length of the segments varied from 29 seconds to 32 seconds. The collection of these segments was then randomized. The listeners' opinions were collected in the score sheet, presented in Table 3. There were two groups of informants: western listeners (24), and native Indians (12), as we wanted to investigate whether the perceived emotions and the required minimal length of audio segments to listen to were coincident, and also conforming to the theory presented in treatises on ragas.

For western listeners, this music was very different than what they were used to listen, as Indian and western music and melodies are based on different scales. Each raga has specific set of notes used, sequences, prolonged notes, etc. Some western listeners reported difficulties in perceiving emotions when listening to this music.

Initially, the test set consisted of 124 segments, but this test was too long and difficult for all listeners, even educated in music. Therefore, we decided to limit the test to 44 excerpts. These excerpts were chosen through random selection of four (out of around eleven) segments for each raga. The listeners were asked to assign each excerpt to only one emotional category if possible. Two choices were also allowed, but the listener had to order them and mark the first and the second choice. As the score sheets revealed that the respondents did not mark the second choice in a large number of cases, statistical elaboration of results in this research was performed taking into account only the first choices.

Table 3 Opinion Score Sheet

Name of the informant:						Age		Sex: M/F	Knowledge of Music: y/n
No.	Anger	Joy	Sorrow	Heroic	Romantic	Serenity	Devotion	Anxiety	Any other, mention	Nil Emotion
1										
2										
3										
...										
...										
...										
42										
43										
44										

The correlation matrix was obtained using a Pearson product-moment correlation coefficient between different types or category of emotion. The Pearson product-moment correlation coefficient is a common measure of the correlation (linear dependence) between two variables. In our experiment, each category of emotion was compared to every other category, yielding a value in a range [1, -1] and the obtained correlation between the two was plotted in a matrix. The statistics is defined as the sum of the products of the standard scores of the two measures divided by the number of degrees of freedom. If the data comes from a sample, then the Pearson product-moment correlation coefficient r between two series of data X_i, Y_i is given by

$$r = \frac{1}{n-1} \sum_{i=1}^{n} \left(\frac{X_i - \overline{X}}{s_X} \right) \left(\frac{Y_i - \overline{Y}}{s_Y} \right)$$

where

$$\frac{X_i - \overline{X}}{s_X}, \overline{X}, \text{ and } s_X$$

are the standard score, sample mean, and sample standard deviation for the series X_i and similarly for the series Y_i (where n is the number of data in each series).

A summary data sheet is formed from all the opinion score sheets to represent the number of responses in each category of emotion for each sound sequence. Thus we get a series of 44 data (for 44 sequences) for each emotion.

The matrix (Table 4 in Section 3) was derived by calculating r for each category of emotion with every other category used in our experiments, e.g. *Joy* with all the other categories, i.e. *Romantic, Serenity, Devotion, Sorrow, Anxiety, Anger,* and *Heroic*.

Product moment correlation has also been calculated between each pair of emotions as perceived by people of Indian origin against those of non-Indian origin by pooling data of these two groups separately (to form two summary sheets).

The t-test used here is to assess the degree of confidence with which the null hypothesis that a stimulus in the form of audio signal could evoke a designated emotion is false. Here t value is given by

$$t = \frac{x - \mu_0}{s/\sqrt{n}}$$

where s is the standard deviation of the sample and n is the sample size (the number of respondents in this case). The number of degrees of freedom used in this test is $n - 1$. The observed count (x) is obtained here from the score sheet (for a particular stimulus, the number of respondents reporting a particular emotion is the observed count for that emotion for that respondent). μ_0 is the expected value for each category of emotion as per the hypothesis (3.6 for all respondents pooled together). The t-test is sometimes referred to as a sigma test. This test was run on two different values of confidence level, namely 0.05 and 0.01.

Table 4 Pair-wise Pearson's Correlation Coefficients

	Anger	Heroic	Romantic	Joy	Devotion	Serenity	Anxiety	Sorrow
Anger	1	0.211529	-0.17939	-0.20863	0.06554	-0.22597	0.221503	-0.02729
Heroic	0.211529	1	-0.13986	-0.17421	-0.11075	-0.0915	-0.07588	-0.25846
Romantic	-0.17939	-0.13986	1	0.141939	-0.18612	0.066252	-0.21185	-0.39698
Joy	-0.20863	-0.17421	0.141939	1	-0.41156	-0.28621	-0.50181	-0.02176
Devotion	0.06554	-0.11075	-0.18612	-0.41156	1	0.072079	-0.02176	-0.01121
Serenity	-0.22597	-0.0915	0.066252	-0.28621	0.072079	1	-0.10248	-0.15363
Anxiety	0.221503	-0.07588	-0.21185	-0.50181	-0.02176	-0.10248	1	0.363925
Sorrow	-0.02729	-0.25846	-0.39698	-0.02176	-0.01121	-0.15363	0.363925	1

Generation of sequence of notes [5] In order to generate a sequence of notes for each segment, the following procedures were adopted.

1. Pre-processing of the acoustic signal,
2. Pitch detection,
3. Detection of pitch of the tonic for each song,
4. Labeling of each pitch-profile into notes using 12-note western scale intervals using standard grammar for the raga.

The musical notes in Indian system are not frequency-specific; instead, they are interval-specific [4]. Additionally, singers often sing with glissando effect.

Our goal was to extract melodies for each segment, but we decided to keep information about pitch only, and temporal structure of the extracted sequences (i.e. rhythm) was not taken into account in further processing.

There are seven pure notes, namely: Sa (Do), Re (Re), Ga (Mi), ma (Fa), Pa (So/Sol), Dha (La), and Ni (Ti/Si). The five altered notes are 4 flat notes and 1 sharp note: re (Re flat), ga (Ga flat), Ma (ma sharp), dha (Dha flat), and ni (Ni flat). In Section 3, notes in sequences are denoted by the first letter of each of the aforesaid notes, i.e.: S, r, R, g, G, m, M, P, d, D, n, and N.

3 Results

Figure 2 contains eleven plates. Each plate shows the proportion of responses for each of the four segments of a raga. Horizontal axis represents the category of emotion, namely: anger (1), joy (2), sorrow (3), heroic (4), romantic (5), serenity (6), devotion (7), anxiety (8), and nil (9). One can see that most of the ragas exhibit selective emotions. In the following pages we will use different statistical tests to deal with queries posed in the introduction.

Table 4 presents the pair-wise correlation coefficients of eight emotions organized in such a manner that the best correlated pairs come contiguously. A shaded cell represents the highest negatively correlated emotion for the emotion represented by the row. Using this table we tried to organize the emotions in the emotion circle presented in Figure 3.

The valence axis represents the usual arousal dimension. The color axis represents dark (negative) to bright (positive) emotions. It can be seen that each quadrant contains emotions which are positively correlated in Table 4. Moreover, generally each emotion in the circle is positively correlated with the neighboring emotion. The exceptions occur only when two neighboring emotions are in different semicircles in the color axis. Again, in general, the emotions which are oppositely placed in the emotion circle are also negatively correlated. In fact, most of them show highest negative correlations.

Table 5 has been obtained by calculating a t-test between the counts of markings by all the listeners for each category of emotion, for all the samples in the score sheets, pooled together, for listeners of Indian origin and those of non-Indian origin.

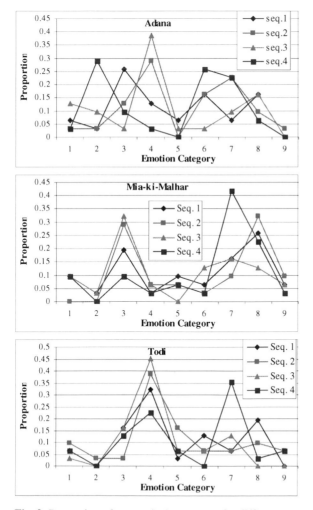

Fig. 2 Proportion of categorical responses for different segments arranged raga-wise

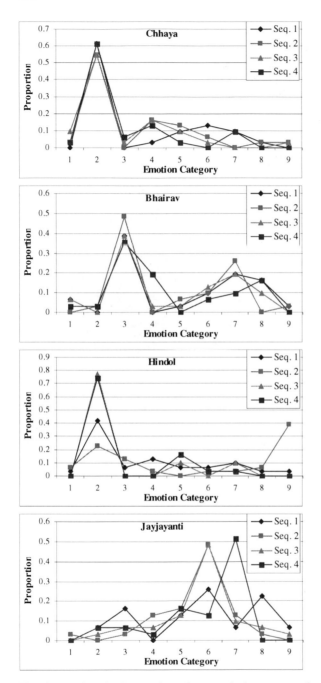

Fig. 2 (continued): Proportion of categorical responses for different segments arranged raga-wise

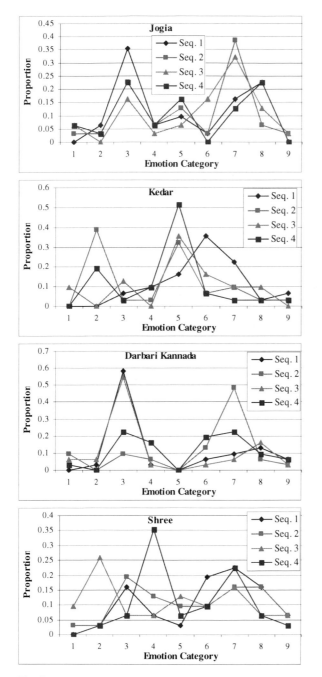

Fig. 2 (continued): Proportion of categorical responses for different segments arranged raga-wise

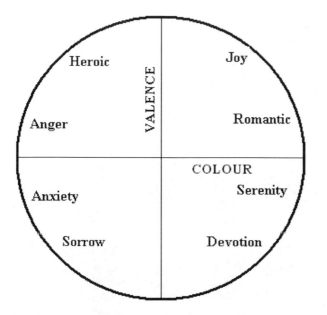

Fig. 3 Circle representing emotions corresponding to rasas reflected in Indian Ragas

Table 5 Results of T-test showing cross-cultural similarity

Category of Emotion	T-test of Indian and Non-Indian origin perception	Level of Signifi-cance 0.05	Level of Signifi-cance 0.01	Remarks about the significance
1. Anger	0.05284118			
2. Joy	0.09924301			
3.Sorrow	0.1685246			
4.Heroic	0.86375077			
5.Romantic	0.983095	2.31	3.36	Not Significant
6 Serenity	0.4438216			
7.Devotion	0.0629127			
8.Anxiety	0.00103316			
9.Any other	0.1253069			
10.Nil	0.94685096			

The results obtained reveal that none of the category of emotion shows significant difference in perception between the listeners of Indian origin and Non-Indian origin. Therefore, these results indicate that culture does not play any significant differentiating role in the perception of emotion in music. E.g. segments, which were perceived by Indians to evoke 'devotion', also evoked the same emotion (devotion) in listeners of non-Indian origin. This result is further confirmed by its high Product-Moment correlation value of **0**.9954 (for devotion).

Table 6 Results of T-test showing emotional preference of each segment

Raga	Seq. No	Anger	Heroic	Romantic	Joy	Devotion	Serenity	Anxiety	Sorrow
	1							**15.6608**	
	44		6.988	3.404441		**10.57169**		1.61263	
	30					**14.11648**			3.01363
Adana	39		6.131	**7.702846**					
	21		4.207					6.421611	**10.85031**
	14							9.274855	**10.84686**
Mia-ki-Malhar	13		1.577	1.576749				**13.84036**	1.576749
	37		**13.64**						5.37472
	31			1.576749		**12.08841**		3.328693	5.080637
	18					**14.51458**	3.098617		
Mia-ki-Todi	12					**14.44331**		2.517641	
	42		**12.76**			6.300541		1.453971	
	33								
	29					2.044736			
	43					2.053277			
Chayanot	36								
	20		4.334					**13.30181**	2.839713
	3		5.663					**13.75244**	
	5		4.442					**13.63331**	
Bhairav	7					4.729469		**12.88373**	3.098617
	40					**1.390392**			
	8	**13.3**						1.345127	
	26								
Hindol	9						**1.478265**		
	35			**9.720197**			1.785342	3.769056	7.736483
	32			**14.46216**			2.309084		
	23			**15.29446**					
Jaijayanti	15		**14.84**				2.185332		
	19		2.935					**12.20455**	6.025033
	28		13.3				1.345127	5.828885	
	2		**12.81**	3.528429				3.528429	1.671361
Jogia	27		1.926				4.066746	**8.347531**	8.347531
	10		5.947	**12.04588**			2.897111		
	17						**8.975132**		
	4			3.223991			**13.40502**	1.527154	
Kedar	38						**14.7674**		
	25							**15.34946**	
	34		**15.12**						
	22							**15.21361**	2.079558
Darbari	16		**7.283**	5.415706		3.548221		**7.283191**	
	6		**8.957**	6.65997				4.363429	4.363429
	41		5.709			2.704365		**8.714066**	5.709216
	24		**4.845**				2.295176		
Shree	11		6.485			**13.13662**			

Table 6 presents the results of T-test, showing significant preference of emotions for all respondents for each music segment. The segments are grouped with respect to the ragas. Blank cells represent non-significant values. The most significant expressed emotion is given in bold values. While a segment may exhibit more than one significant emotion, a close examination of the results reveals that in most cases values for the most significant emotion are ways higher than the others. This signifies

Table 7 Comparison between source-directed emotion and observed emotion

Raga	Anger	Heroic	Romantic	Joy	Devotion	Serenity	Anxiety	Sorrow
Adana		S	O		O		O	
Mia-ki-Malhar		O					O	S, O
Mia-ki-Todi		O	S		S, O			S
Chayanot			S		O			
Bhairav	S				S, O	S	O	S
Hindol	S, O	S			O	O		
Jaijayanti		O	S, O					
Jogia		O	S		S		O	S, O
Kedar			O			S, O		
Darbari		O				S	O	
Shree		O			O	S	O	

that, in general, a particular segment can be thought of as provoking one emotion only. Highest significant emotion revealed is anxiety, next comes devotion. It may be noted that the emotion of anxiety does not find a place in Indian *rasas*.

Table 7 shows the comparison between the emotions expected to be expressed by each raga (S) and the observed emotion (O). The two agree only in seven cases as against non-agreement in 31 cases.

In the present study, sequences consisting of three to seven notes have been considered. The total number of such distinct sequences in all 44 segments is 5943. Of these, a large number of the sequences did not occur more than once in a segment. These were excluded from consideration. The remaining 556 sequences were examined with respect to their abundance in all four segments of a particular raga. As a result of this examination, it was found that usually the most abundant sequences consisted of repetitions of only two notes. These were considered as not significant for the present purpose. After all this sieving out, there were 57 sequences in which at least three are different notes. Of these, only those sequences which elicit responses at least for 6 music segments are presented in Table 8. The numbers shown in bold or italic fonts represent cases where the four segments of one raga elicit at least a total of 6 responses. The italic ones represent the highest number of responses for a sequence in a raga.

Table 9 shows the results yielded from matching Table 8 and Table 6, presenting whether a particular sequence may be associated with some particular emotion or emotions. As we can see, short sequences of 3-4 notes, specific for ragas, can evoke particular emotions.

4   Conclusions

In our experiments, we extracted meaningful emotional sequences of sounds from ragas, and tested what emotions were evoked by these sequences. The listening tests were performed on two groups of listeners: on Hindustani listeners, and on

Table 8 Number of occurrences of sequences in segments under raga. Numbers 1, 2, 3 and 4 represent each of the four different signals from the same raga (they are not the original sequence numbers which range from 1 to 44 - in total, there are 44 sequences from 11 ragas)

Note Sequence	Adana				Mia-ki-Malhar				Mia-ki-Todi				Chayanot				Bhairav				Hindol				Jaijayanti				Jogia				Kedar				Darbari				Shree			
	1	2	3	4	1	2	3	4	1	2	3	4	1	2	3	4	1	2	3	4	1	2	3	4	1	2	3	4	1	2	3	4	1	2	3	4	1	2	3	4	1	2	3	4
mSN	0	0	0	0	0	0	0	0	0	0	0	0	0	0	0	0	0	0	0	0	1	2	1	0	0	0	0	0	0	0	0	0	2	2	1	0	1	0	0	0
RgM	1	0	0	0	0	0	0	1	0	0	0	0	0	0	0	0	0	0	0	0	2	1	1	0	2	1	2	1	0	1	0	0	2	2	1	1	0	0	0	0
SgR	0	0	0	0	0	0	0	0	0	0	0	0	0	0	0	0	0	0	0	0	0	0	0	0	0	0	0	0	0	0	0	0	2	1	1	0	0	0	0	0
GMmP	0	0	0	0	0	0	0	1	0	0	0	0	0	1	1	0	0	0	0	0	0	0	0	0	0	0	0	0	1	1	2	1	0	0	0	1	0	0	0	0
Pmd	2	0	0	0	0	0	0	0	0	0	0	0	0	0	1	0	0	0	0	0	0	0	0	0	0	0	0	0	0	0	2	1	0	0	0	0	0	0	0	0
SND	0	0	0	0	0	0	1	0	0	0	0	0	0	0	1	0	0	0	0	0	3	1	0	0	0	0	0	0	1	0	0	0	0	0	0	0	0	3	2	1
DPmP	0	0	0	0	0	0	0	0	0	0	0	0	0	0	0	0	0	0	0	1	0	1	0	1	0	0	0	0	1	0	2	3	0	0	0	0	0	0	0	0
rGrS	0	0	0	0	0	0	0	0	0	0	0	0	0	0	0	0	0	0	0	0	0	0	0	0	0	0	0	0	0	0	0	0	0	0	0	0	0	0	0	0
GMR	0	0	0	0	0	0	0	0	0	0	0	0	4	2	1	0	0	0	0	0	0	0	0	0	0	0	0	0	0	0	1	0	0	0	0	0	0	0	0	0
MGR	0	0	0	0	0	0	0	0	0	0	0	0	1	2	4	0	0	0	0	0	0	0	0	0	0	0	0	0	0	0	1	0	0	0	0	0	0	0	0	0
NDP	0	0	0	0	0	0	0	0	0	0	0	0	0	0	0	0	0	1	1	2	0	0	0	0	0	0	0	0	0	0	0	0	0	0	0	0	0	0	0	2
NDm	0	0	0	0	0	0	0	0	0	0	0	0	0	0	0	0	0	1	0	0	1	1	3	2	0	0	0	0	2	1	2	1	0	0	1	1	3	1	3	0
mPmG	0	0	0	0	0	0	0	0	3	1	3	3	0	0	0	0	0	0	0	0	0	0	0	0	0	0	0	0	0	0	0	0	0	0	0	0	3	0	3	4
mGr	0	0	0	0	0	0	0	0	0	1	1	0	0	0	0	0	1	1	0	0	1	1	0	0	0	0	0	0	0	0	0	0	0	0	0	0	0	0	0	0
dPMG	2	0	0	0	0	0	0	0	0	0	0	2	0	0	0	0	1	0	0	2	0	0	0	0	0	0	0	0	0	0	0	0	0	0	0	0	0	0	0	0
grS	0	0	0	0	0	0	0	0	3	1	3	0	0	0	0	0	0	0	0	0	0	0	0	0	0	0	0	0	0	0	0	1	0	0	0	0	0	0	0	0
PmG	0	0	0	0	0	0	0	0	0	0	0	0	3	3	0	1	0	1	0	0	0	0	0	0	0	0	0	0	0	1	0	0	0	0	0	0	0	0	3	0
RGM	0	0	0	0	0	0	0	0	0	0	0	0	3	3	0	1	0	1	0	0	0	0	0	0	2	2	2	2	0	1	2	1	0	1	2	3	0	3	0	0
gRS	0	1	0	2	0	0	0	0	0	0	0	0	0	0	0	0	1	0	0	0	0	0	0	0	2	2	2	2	0	0	0	0	0	0	0	0	0	0	0	0
GMP	0	0	0	0	0	0	0	0	0	0	0	0	2	2	0	0	2	2	0	2	0	0	0	0	0	0	0	0	0	0	0	1	0	0	0	0	0	0	0	0
DPm	2	0	0	0	0	0	0	0	0	0	0	0	1	2	1	0	0	0	0	0	0	0	0	0	4	1	3	1	1	2	2	3	0	0	0	0	0	3	3	0
MdP	2	0	0	0	0	0	0	0	0	0	0	0	0	0	0	0	2	0	0	2	0	0	0	0	1	1	0	0	0	2	3	2	0	0	1	0	0	3	3	0
MPdP	1	0	0	0	0	0	0	0	0	0	0	0	0	0	0	1	2	0	0	1	0	0	0	0	0	0	0	0	0	0	2	2	0	0	0	0	0	0	0	0
GrS	0	0	0	0	0	0	0	0	0	0	0	0	0	0	0	0	1	3	0	1	0	0	0	0	0	0	0	0	1	0	1	0	0	0	1	0	0	0	0	0
PMGM	0	1	1	0	0	0	0	0	0	0	0	0	0	0	0	0	1	3	0	2	0	0	0	0	0	1	1	0	0	2	2	2	0	0	0	0	3	3	2	0
MPd	1	1	2	0	0	0	0	0	0	0	0	0	0	0	0	0	2	0	0	0	0	0	0	0	1	1	1	1	0	0	1	0	0	1	0	0	0	0	0	0
NSR	1	2	2	2	0	0	0	0	0	0	0	0	3	3	1	0	1	3	0	3	0	0	0	0	2	2	3	3	0	0	1	0	0	0	1	0	0	0	0	0
PMG	0	0	0	0	0	0	0	0	0	0	0	0	3	1	0	0	1	3	0	3	0	0	0	0	3	2	5	3	0	0	1	0	0	2	0	0	3	3	0	0
dPM	1	0	1	0	0	0	0	0	0	0	0	0	3	0	0	0	2	2	0	2	0	0	0	0	0	4	3	0	0	0	1	0	0	2	0	0	0	0	0	0

Table 9 Relationship between note sequences and emotions

Note Sequence	Major emotional Response
mSN	Anger, Devotion, Anxiety
RgM	Heroic, Anxiety
SgR	Heroic, Romantic
GMmP	Romantic, Serenity, Anxiety
Pmd	Heroic, Anxiety, Sorrow
SND	Anger, Devotion
DPmP	Romantic, Serenity
rGrS	Heroic, Devotion, Anxiety
GMR	Devotion
MGR	Devotion
NDP	Heroic, Anxiety, Sorrow
NDm	Anger
mPmG	Heroic, Devotion, Anxiety
mGr	Heroic, Devotion, Anxiety
dPMG	Heroic, Anxiety, Sorrow
grS	Heroic, Devotion
PmG	Heroic, Devotion, Anxiety
RGM	Devotion
gRS	Romantic, Serenity
GMP	Anxiety
DPm	Romantic, Serenity
MdP	Heroic, Anxiety, Sorrow
MPdP	Heroic, Anxiety, Sorrow
GrS	Heroic, Devotion, Anxiety
PMGM	Anxiety
MPd	Heroic, Anxiety, Sorrow
NSR	Devotion
PMG	Devotion
dPM	Anxiety

Western listeners not familiar with Hindustani music. For both groups, we investigated what emotions were evoked, for the audio segments used in listening test, and for sequences of notes of minimal length, specific for each raga.

The results of the experiments described in this paper can be summarized as follows.

1. An oral music segment of length 3 seconds (a few notes) elicit specific emotion,
2. The elicited emotion can be assigned into prescribed categories,
3. The elicited emotion from different segments from the same raga has some specificity, i.e. the segments of a raga have shown a specific emotion; it might be that four segments from the same raga show different emotions,
4. The emotional response from the segments of a raga does not generally correspond to those prescribed in Indian treaties,

5. The cross-cultural similarity of the elicited response is significant,
6. The melodic sequence (sequence of musical notes) vaguely relate with emotional response.

Since the number of listening experiments we performed is not really large, these outcomes can be considered as a coarse estimate. Still, cross-cultural understanding of emotions in music seems to be quite clearly visible, because the same excerpts evoked similar emotions in both western and Indian listeners. Also, evoked emotions may differ from those described in treatises. However, one piece of music usually consists of many phrases and motives, evoking sometimes a variety of emotions. We find it interesting to observe that short excerpts (a few seconds, a few notes) are sufficient to evoke emotions which can be shared by people of different culture and place of living.

Acknowledgements

This work is based in part upon work supported by the National Science Foundation under Grant Number IIS-0414815. The Indian authors gratefully acknowledge the financial support received from ITC Ltd and also to the Department of Scientific and Industrial Research (DSIR), Ministry of Science and Technology, Government of India, for their recognition to ITC Sangeet Research Academy under SIRO scheme (11/41/88-TU-V dated 23.4.2008). Any opinions, findings, and conclusions or recommendations expressed in this material are those of the authors and do not necessarily reflect the views of the National Science Foundation, ITC Ltd. and DSIR.

This work was also supported by the Research Center of PJIIT, supported by the Polish National Committee for Scientific Research (KBN).

References

1. Balkwill, L.L., Thompson, W.F.: A crosscultural investigation of the perception of emotion in music: Psychophysical and cultural cues. Music Perception 17, 43–64 (1999)
2. Bhatkhande, V.N.: Hindustani Sangeet Paddhati, vol. IV. Sangit Karyalay, Hathras (1970)
3. Bower, G.H.: Mood and memory. American Psychologist 36, 129–148 (1981)
4. Datta, A.K., Sengupta, R., Dey, N., Nag, D., Mukherjee, A.: Objective Analysis of the Interval Boundaries and Swara-Shruti relations in Hindustani vocal music from actual performances. Ninaad (J. ITC Sangeet Research Academy) 20 (2006)
5. Datta, A.K., Sengupta, R., Dey, N., Nag, D., Mukherjee, A.: Study of Melodic Sequences in Hindustani Ragas: A cue for Emotion? In: Proc. Frontiers of Research on Speech and Music (FRSM 2007), All India Institute of Speech and Hearing, Mysore, Karnataka, India, January 8-9 (2007)
6. Deva, B.C., Virmani, K.G.: A Study in the psychological response to Ragas. Research Report II of Sangeet Natak Akademy. Indian Musicological Society, New Delhi (1975)
7. Ekman, P., Friesen, W.V.: What emotion categories or dimensions can observers judge from facial behaviour? In: Ekman, P. (ed.) Emotions in the Human Face. Cambridge University Press, London (1982)

8. Fischer, K.W., Shaver, P.R., Carnochan, P.: How emotions develop and how they organise development. Cognition and Emotion 4, 81–127 (1990)
9. Gabrielsson, A.: Perceived emotion and felt emotion: Same or different? Musicae Scientiae, Special Issue 2001-2002 (2002) ISSN 1029-8649
10. Gregory, A.H., Varney, N.: Cross-Cultural Comparisons in the Affective Response to Music. Psychology of Music 24(1), 47–52 (1996)
11. Hevner, K.: Experimental studies of the elements of expression in music. American Journal of Psychology 48, 246–268 (1936)
12. Karnani, C.: Listening in Hindustani Music. Orient Longman (1976)
13. Konishi, T., Imaizumi, S., Niimi, S.: Vibrato and emotion in singing voice. In: Woods, C., Luck, G., Brochard, R., Seddon, F., Sloboda, J.A. (eds.) Proceedings of the Sixth International Conference for Music Perception and Cognition (CD-ROM). Keele University, Keele, England (2000)
14. Lavy, M.M.: Emotion and the Experience of Listening to Music: A Framework for Empirical Research. PhD thesis, Jesus College, Cambridge (2001)
15. Martinez, J.L.: Semiosis in Hindustani Music. Motilal Banarasidas Publishers Pvt. Ltd., Delhi (2001)
16. Oatley, K., Jenkins, J.M.: Understanding emotions. Blackwell, Oxford (1996)
17. Plutchik, R.: The psychology and biology of emotion. Harper-Collins, New York (1994)
18. Russell, J.A., Mehrabian, A.: Evidence for a three-factor theory of emotions. Journal of Research in Personality 11, 273–294 (1977)
19. Russell, J.A.: A circumplex model of affect. Journal of Personality and Social Psychology 39, 1161–1178 (1980)
20. Russell, J.A.: Measures of emotion. In: Plutchik, R., Kellerman, H. (eds.) Emotion: Theory Research and Experience, vol. 4, pp. 81–111. Academic Press, New York (1989)
21. Scherer, K.R.: Psychological models of emotion. In: Borod, J. (ed.) The neuropsychology of emotion, pp. 137–162. Oxford University Press, New York (2000)
22. Schubert, E.: Measurement and Time Series Analysis of Emotion in Music. PhD thesis, University of New South Wales (1999)
23. Schubert, E.: Measuring emotion continuously: Validity and reliability of the two dimensional emotion-space. Australian Journal of Psychology 51(3), 154–165 (1999)
24. Scott, J.P.: The function of emotions in behavioral systems: A systems theory analysis. In: Plutchik, R., Kellerman, H. (eds.) Emotion: Theory, research, and experience. Theories of emotion, vol. 1, pp. 35–56. Academic Press, New York (1980)
25. Sloboda, J.A.: Empirical studies of emotional response to music. In: Jones, M., Holleran, S. (eds.) Cognitive Bases of Musical Communication. American Psychological Association, Washington (1992)
26. Tembe, G.S.: Aspects of music. Ministry of Broadcasting and Information, New Delhi (1957)
27. Wedin, L.: Multidimensional study of perceptual-emotional qualities in music. Scandinavian Journal of Psychology 13, 241–257 (1972)

Part IV
Music Similarity

Audio Cover Song Identification and Similarity: Background, Approaches, Evaluation, and Beyond

Joan Serrà, Emilia Gómez, and Perfecto Herrera

1 Introduction

A cover version[1] is an alternative rendition of a previously recorded song. Given that a cover may differ from the original song in timbre, tempo, structure, key, arrangement, or language of the vocals, automatically identifying cover songs in a given music collection is a rather difficult task. The music information retrieval (MIR) community has paid much attention to this task in recent years and many approaches have been proposed. This chapter comprehensively summarizes the work done in cover song identification while encompassing the background related to this area of research. The most promising strategies are reviewed and qualitatively compared under a common framework, and their evaluation methodologies are critically assessed. A discussion on the remaining open issues and future lines of research closes the chapter.

1.1 Motivation

Cover song identification has been a very active area of study within the last few years in the MIR community, and its relevance can be seen from multiple points of view. From the perspective of audio content processing, cover song identification yields important information on how musical similarity can be measured and modeled. Music similarity is an ambiguous term and, apart from musical facets themselves, may also depend on different cultural (or contextual) and personal (or subjective) aspects [24]. The purpose of many studies is to define and evaluate the concept of music similarity, but there are many factors involved in this problem, and

Joan Serrà · Emilia Gómez · Perfecto Herrera
Music Technology Group, Department of Information and Communication Technologies, Universitat Pompeu Fabra. Tànger 122-140, office 55.318, 08018 Barcelona Spain
e-mail: {joan.serraj,emilia.gomez,perfecto.herrera}@upf.edu

[1] We use the term cover or version interchangeably.

Z.W. Raś and A.A. Wieczorkowska (Eds.): Adv. in Music Inform. Retrieval, SCI 274, pp. 307–332.
springerlink.com   © Springer-Verlag Berlin Heidelberg 2010

some of them (maybe the most relevant ones) are difficult to measure [6]. Still, the relationship between cover songs is context-independent and can be qualitatively defined and objectively measured, as a "canonical" version exists and any other rendition of it can be compared to that.

The problem of identifying covers is also challenging from the point of view of music cognition, but apparently it has not attracted much attention by itself. When humans are detecting a cover, they have to derive some invariant representation of the whole song or maybe of some of its critical sections. We do not know precisely what is the essential information that has to be encoded in order for the problem to be solved by human listeners. Nevertheless, it seems relevant the knowledge gained about the sensitivity or insensitivity to certain melodic transformations, for example [17, 87]. In addition, when the cover is highly similar in terms of timbre, it seems that this cue can do the job to help us to identify the song even using very short snippets of it [71]. An additional issue that is called for by cover identification is that of the memory representation of the songs in humans. It could be either the case that the canonical song acts as a prototype for any possible version, and that the similarity of the covers is computed in their encoding step, or either that all the songs are stored in memory (as exemplary-based models would hypothesize) and their similarity is computed at the retrieval phase. For example, Levitin [46] presents evidence in favor of absolute and detailed coding of song specific information (at least for the original songs). On the other hand, Deliege [14] has discussed the possibility of encoding processes that abstract and group by similarity certain musical cues.

From a commercial perspective, it is clear that detecting cover songs has a direct implication to musical rights' management and licenses. Furthermore, quantifying music similarity is key to searching, retrieving, and organizing music collections. Nowadays, online digital music collections are in the order of ten [59] to a few hundred million tracks[2] and they are continuously increasing. Therefore, one can hypothesize that the ability to manage this huge amount of digital information in an efficient and reliable way will make the difference in tomorrow's music-related industry [10, 85]. Personal music collections, which by now can easily exceed the practical limits on the time to listen to them, might benefit as well from efficient and reliable search and retrieval engines.

From a user's perspective, finding all versions of a particular song can be valuable and fun. One can state an increasing interest for cover songs just by looking at the emergence of related websites, databases, and podcasts in the internet such as Second Hand Songs[3], Coverinfo[4], Coverville[5], Midomi[6], Fancovers[7], or YouTube[8].

---

[2] See for example http://www.easymp3downloader.com/,
http://blog.wired.com/music/2007/04/lastfm_subscrip.html, or
http://www.qsrmagazine.com/articles/news/story.phtml?id=5852.

[3] http://www.secondhandsongs.com

[4] http://www.coverinfo.de

[5] http://www.coverville.com

[6] http://www.midomi.com

[7] http://www.fancovers.com

[8] http://www.youtube.com

Frequently, these sites also allow users to share/present their own (sometimes home-made) cover songs, exchange opinions, discover new music, make friends, learn about music by comparing versions, etc. Thus, cover songs are becoming part of a worldwide social phenomena.

1.2 Types of Covers

Cover songs were originally part of a strategy to make profit from 'hits' that had achieved significant commercial success by releasing them in other commercial or geographical areas without remunerating the original artist or label. Little promotion and highly localized record distribution in the middle of the 20th century favored that. Nowadays, the term has nearly lost these purely economical connotations. Musicians can play covers as a homage or a tribute to the original performer, composer or band. Sometimes, new versions are rendered for translating songs to other languages, for adapting them to a particular country/region tastes, for contemporizing old songs, for introducing new artists, or just for the simple pleasure of playing a familiar song. In addition, cover songs represent the opportunity (for beginners and consolidated artists) to perform a radically different interpretation of a musical piece. Therefore, today, and perhaps not being the proper way to name it, a cover song can mean any new version, performance, rendition, or recording of a previously recorded track [42].

Many distinctions between covers can be made (see [27, 79, 89] for some MIR-based attempts). These usually aim at identifying different situations where a song was performed in the context of mainstream popular music. Considering the huge amount of tags and labels related to covers (some of them being just buzzwords for commercial purposes), and according to our current understanding of the term cover version, we advocate for a distinction based on musical features instead of using commercial, subjective, or situational tags. But, just in order to provide an overview, some exemplary labels associated with versions are listed below [42].

- Remaster: Creating a new master for an album or song generally implies some sort of sound enhancement (compression, equalization, different endings, fade-outs, etc.) to a previous, existing product.
- Instrumental: Sometimes, versions without any sung lyrics are released. These might include karaoke versions to sing or play with, cover songs for different record-buying public segments (e.g. classical versions of pop songs, children versions, etc.), or rare instrumental takes of a song in CD-box editions specially made for collectors.
- Live performance: A recorded track from live performances. This can correspond to a live recording of the original artist who previously released the song in a studio album, or to other performers.
- Acoustic: The piece is recorded with a different set of acoustical instruments in a more intimate situation.
- Demo: It is a way for musicians to approximate their ideas on tape or disc, and to provide an example of those ideas to record labels, producers, or other artists.

Musicians often use demos as quick sketches to share with band mates or arrangers. In other cases, a songwriter might make a demo in order to be send to artists in hopes of having the song professionally recorded, or a music publisher may need a simplified recording for publishing or copyright purposes.

- Duet: A successful piece can be often re-recorded or performed by extending the number of lead performers outside the original members of the band.
- Medley: Mostly in live recordings, and in the hope of catching listeners' attention, a band covers a set of songs without stopping between them and linking several themes.
- Remix: This word may be very ambiguous. From a 'traditionalist' perspective, a remix implies an alternate master of a song, adding or subtracting elements, or simply changing the equalization, dynamics, pitch, tempo, playing time, or almost any other aspect of the various musical components. But some remixes involve substantial changes to the arrangement of a recorded work and barely resemble the original one. Finally, a remix may also refer to a re-interpretation of a given work such as a hybridizing process simultaneously combining fragments of two or more works.
- Quotation: The incorporation of a relatively brief segment of existing music in another work, in a manner akin to quotation in speech or literature. Quotation usually means melodic quotation, although the whole musical texture may be incorporated. The borrowed material is presented exactly or nearly so, but is not part of the main substance of the work.

1.3   Involved Musical Facets

With nowadays' concept of cover song, one might consider the musical dimensions in which such a piece may vary from the original one. In classical music, different performances of the same piece may show subtle variations and differences, including different dynamics, tempo, timbre, articulation, etc. On the other hand, in popular music, the main purpose of recording a different version can be to explore a radically different interpretation of the original one. Therefore, important changes and different musical facets might be involved. It is in this scenario where cover song identification becomes a very challenging task. Some of the main characteristics that might change in a cover song are listed below:

- Timbre: Many variations changing the general color or texture of sounds might be included into this category. Two predominant groups are:
  - Production techniques: Different sound recording and processing techniques (e.g. equalization, microphones, dynamic compression, etc.) introduce texture variations in the final audio rendition.
  - Instrumentation: The fact that the new performers can be using different instruments, configurations, or recording procedures, can confer different timbres to the cover version.

- Tempo: Even in a live performance of a given song from its original artist, tempo might change, as it is not so common to control tempo in a concert. In fact, this might become detrimental for expressiveness and contextual feedback. Even in classical music, small tempo fluctuations are introduced for different renditions of the same piece. In general, tempo changes abound (sometimes on purpose) with different performers.
- Timing: In addition to tempo, the rhythmical structure of the piece might change depending on the performer's intention or feeling. Not only by means of changes in the drum section, but also including more subtle expressive deviations by means of swing, syncopation, pauses, etc.
- Structure: It is quite common to change the structure of the song. This modification can be as simple as skipping a short 'intro', repeating the chorus, introducing an instrumental section, or shortening one. But it can also imply a radical change in the musical section ordering.
- Key: The piece can be transposed to a different key or tonality. This is usually done to adapt the pitch range to a different singer or instrument, for 'aesthetic' reasons, or to induce some mood changes on the listener.
- Harmonization: While maintaining the key, the chord progression might change (adding or deleting chords, substituting them by relatives, modifying the chord types, adding tensions, etc.). This is very common in introduction and bridge passages. Also, in instrument solo parts, the lead instrument voice is practically always different from the original one.
- Lyrics and language: One purpose of performing a cover song is for translating it to other languages. This is commonly done by high-selling artists to be better known in large speaker communities.
- Noise: In this category we consider other audio manifestations that might be present in a song recording. Examples include audience manifestations such as claps, shouts, or whistles, audio compression and encoding artifacts, speech, etc.

Notice that, in some cases, the characteristics of the song might change, except, perhaps, a lick or a phrase that is on the background, and that it is the only thing that reminds of the original song (e.g. remixes or quotations). In these cases, it becomes a challenge to recognize the original song, even if the song is familiar to the listener. Music characteristics that may change within different types of covers are shown in table 1.

1.4 Scientific Background

In the literature, one can find plenty of approaches addressing song similarity and retrieval, both in the symbolic and the audio domains[9]. Within these, research

---

[9] As symbolic domain we refer to the approach to music content processing that uses, as starting raw data, symbolic representations of musical content (e.g. MIDI or **kern files, data extracted from printed scores). Contrastingly, the audio domain processes the raw audio signal (e.g. WAV or MP3 files, real-time recorded data).

Table 1 Musical changes that can be observed in different cover song categories. Stars indicate that the change is possible, but not necessary.

	Timbre	Tempo	Timing	Structure	Key	Harm.	Lyrics	Noise
Remaster	★							
Instrumental	★						★	★
Live	★	★	★					★
Acoustic	★	★	★		★	★		★
Demo	★	★	★	★	★	★	★	★
Medley	★	★	★	★	★			★
Remix	★	★	★	★	★	★	★	★
Quotation	★			★				★

done in areas such as query-by-humming systems, content-based music retrieval, genre classification, or audio fingerprinting, is relevant for addressing cover song similarity.

Many ideas for cover song identification systems come from the symbolic domain [45, 56, 67, 81], and query-by-humming systems [12] are paradigmatic examples. In query-by-humming systems, the user sings or hums a melody and the system searches for matches in a musical database. This query-by-example situation is parallel to retrieving cover songs from a database. In fact, many of the note encoding or alignment techniques employed in query-by-humming systems could be useful in future approaches for cover song identification. However, the kind of musical information that query-by-humming systems manage is symbolic (usually MIDI files), and the query, as well as the music material, must be transcribed into the symbolic domain. Unfortunately, transcription systems of this kind do not yet achieve a significantly high accuracy on real-world audio music signals. Current state-of-the-art algorithms yield overall accuracies around 75%[10], even for melody estimation[11], indicating that there is still much room for improvement in these areas. Consequently, we argue that research in the symbolic domain cannot be directly applied to audio domain cover song similarity systems without incurring several estimation errors in the first processing stages of these. These errors, in turn, may have dramatic consequences in final system's accuracy.

Content-based music retrieval is organized around use cases which define a type of query, the sense of match, and the form of the output [10, 18]. The sense of match implies different degrees of specificity: it can be exact, retrieving music with specific content, or approximate, retrieving near neighbors in a musical space where proximity encodes different senses of musical similarity [10]. One prototypical use case is genre classification [70]. In this case, one generally tries to group songs according to a commercially or culturally established label, the genre, where

---

[10] http://www.music-ir.org/mirex/2008/index.php/
Multiple_Fundamental_Frequency_Estimation_&_Tracking_Results
[11] http://www.music-ir.org/mirex/2008/index.php/
Audio_Melody_Extraction_Results

certain characteristics might be more or less the same but many others might radically change (category-based song grouping). Therefore, genre classification is considered to have a low match specificity [10]. On the other hand, audio fingerprinting [7] is an example of a task with a highly specific match. This essentially consists in identifying a particular performance of a concrete song (exact duplicate detection). In contrast to many prototypical use cases, cover song identification is representative of an intermediate specificity region [10]. It goes beyond audio fingerprinting in the sense that it tries to approximate duplicate detection while allowing many musical facets to change. In addition, it is more specific than genre classification in the sense that it goes beyond timbral similarity to include the important idea that musical works retain their identity notwithstanding variations in many musical dimensions [19].

It must be noted that many studies approach the aforementioned intermediate match specificity. This is the case, for instance, of many audio fingerprinting algorithms using tonality-based descriptors instead of the more routinely employed timbral ones (e.g. [8, 51, 66, 84]). These approaches can also be named with terms such as audio identification, audio matching, or simply, polyphonic audio retrieval. The adoption of tonal features adds some degrees of invariance (timbre, noise) to audio fingerprinting algorithms which are, by nature, invariant with respect to song structure changes. In spite of that, many of them might still have a low recall for cover versions. This could be due to an excessively coarse feature quantization [66], and to the lack of other desirable degrees of invariance to known musical changes like tempo variations or key transpositions [76].

Like recent audio identification algorithms, many other systems derived from the genre classification task or from traditional music similarity approaches may also fall into the aforementioned intermediate specificity region. These, in general, differ from traditional systems of their kind in the sense that they also incorporate tonal information (e.g. [48, 61, 82, 90]). However, these systems might also fail in achieving invariance to key or tempo modifications. In general, they do not consider full sequences of musical events, but just statistical summarizations of them, which might blur/distort valuable information for assessing the similarity between cover songs.

Because of the large volume of existing work it is impossible to cover every top in this area. We focus on algorithms designed for cover song identification, that, in addition, include several modules explicitly designed to achieve invariance to characteristic musical changes among versions[12].

2 Approaches

The standard approach to measuring similarity between cover songs is essentially to exploit music facets shared between them. Since several important characteristics

---

[12] Even considering this criteria, it is difficult to present the complete list of methods and alternatives. We apologize for possible omissions/errors and, in any case, we assert that these have not been intentional.

are subject to variation (timbre, key, harmonization, tempo, timing, structure, and so forth, Section 1.3), cover song identification systems must be robust against these variations.

Extracted descriptors are often in charge of overcoming the majority of musical changes among covers, but special emphasis is put on achieving tempo, key, or structure invariance, as these are very frequent changes that are not usually managed by extracted descriptors themselves. Therefore, one can group the elements of existing cover song identification systems into four basic functional blocks: feature extraction, key invariance, tempo invariance, and structure invariance. An extra block can be considered at the end of the chain for the final similarity measure used (figure 1 illustrates these blocks). A summary table for several state-of-the-art approaches, and the different strategies they follow in each functional block, is provided at the end of the present section (table 2).

2.1   Feature Extraction

In general, we can assume that different versions of the same piece mostly preserve the main melodic line and/or the harmonic progression, regardless of its main key. For this reason, tonal or harmonic content is a mid-level characteristic that should be considered to robustly identify covers.

The term tonality is commonly used to denote a system of relationships between a series of pitches, which can form melodies and harmonies, having a tonic (or central pitch class) as its most important (or stable) element [42]. In its broadest possible sense, this term refers to the arrangements of pitch phenomena. Tonality is ubiquitous in Western music, and most listeners, either musically trained or not, can identify the most stable pitch while listening to tonal music [11]. Furthermore, this process is continuous and remains active throughout the sequential listening experience [72].

A tonal sequence can be understood, in a broad sense, as a sequentially-played series of different note combinations. These notes can be unique for each time slot (a melody) or can be played jointly with others (chord or harmonic progressions). From a MIR point of view, clear evidence about the importance of tonal sequences for music similarity and retrieval exists [9, 22, 34]. In fact, almost all cover song identification algorithms exploit tonal sequence representations extracted from the raw audio signals: they either estimate the main melody, the chord sequence, or the harmonic progression. Only early systems, which, e.g., work with the audio signal's energy or with spectral-based timbral features, are an exception [25, 89].

Melody is a salient musical descriptor of a piece of music [73] and, therefore, several cover song identification systems use melody representations as a main descriptor [49, 50, 68, 78, 79]. As a first processing step, these systems need to extract the predominant melody from the raw audio signal [62]. Melody extraction is strongly related to pitch tracking, which itself has a long and continuing history [13]. However, in the context of complex mixtures, the pitch tracking problem becomes further complicated because, although multiple pitches may be present at the same time, at

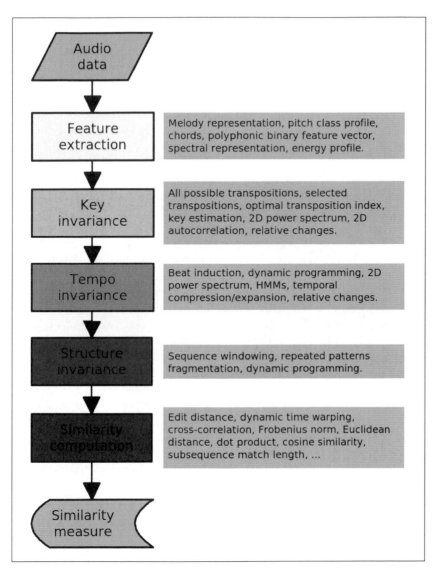

Fig. 1 Generic block diagram for cover song identification systems

most just one of them will be the melody. This and many other facets [62] make melody extraction from real-world audio signals a difficult task (see Section 1.4). To refine the obtained representation, cover detection systems usually need to combine a melody extractor with a voice/non-voice detector and other post-processing modules in order to achieve a more reliable representation [68, 78, 79]. Another possibility is to generate a so-called "mid-level" representation for these melodies [49, 50], where the emphasis is not only put on melody extraction, but also on the feasibility to describe audio in a way that facilitates retrieval.

Alternatively, cover song similarity can be assessed by harmonic, rather than melodic, sequences using so-called chroma features or pitch class profiles (PCP) [26, 27, 63, 82]. These mid-level features might provide a more complete, reliable, and straightforward representation than, e.g., melody estimation, as they do not need to tackle the pitch selection and tracking issues outlined above. PCP features are derived from the energy found within a given frequency range (usually from 50 to 5000 Hz) in short-time spectral representations (typically 100 msec) of audio signals extracted on a frame-by-frame basis. This energy is usually collapsed into a 12-bin octave-independent histogram representing the relative intensity of each of the 12 semitones of an equal-tempered chromatic scale. Reliable PCP features should, ideally, (a) represent the pitch class distribution of both monophonic and polyphonic signals, (b) consider the presence of harmonic frequencies, (c) be robust to noise and non-tonal sounds, (d) be independent of timbre and played instrument, (e) be independent of loudness and dynamics, and (f) be independent of tuning, so that the reference frequency can be different from the standard A 440 Hz [27]. This degree of invariance with respect to several musical characteristics make PCP features very attractive for cover song identification systems. Hence, the majority of systems use a PCP-based feature as primary source of information [20, 21, 23, 28, 29, 36, 37, 38, 39, 40, 41, 55, 53, 76, 74].

An interesting variation of using PCP features for characterizing cover song similarity is proposed in [9]. In this work, PCP sequences are collapsed into string sequences using vector quantization, i.e. summarizing several features vectors by a close representative, done via the K-means algorithm [88] (8, 16, 32, or 64 symbols). In [55], vector quantization is performed by computing binary PCP feature vector components in such a way that, with 12 dimensional feature vectors, a codebook of $2^{12} = 4096$ symbols is generated (so-called polyphonic binary feature vectors). Sometimes, the lack of interpretability of the produced sequences and symbols makes the addition of some musical knowledge to these systems rather difficult. This issue is further studied in [41] where, instead of quantizing in a totally unsupervised way, a codebook of PCP features based on musical knowledge (with a size of 793 symbols) is generated. In general, vector quantization, indexing, and hashing techniques, result in highly efficient algorithms for music retrieval [8, 41, 55, 66], even though their accuracy has never been formally assessed for the specific cover song identification task. It would be very interesting to see how these systems perform on a benchmark cover song training set (e.g. MIREX [18]) in comparison to specifically designed approaches. More concretely, it is still an issue if PCP quantization strongly degrades cover song retrieval. Some preliminary results suggest that this is the case [66].

Instead of quantizing PCP features, one can use chord or key template sequences for computing cover song similarity [2, 4, 35, 43]. Estimating chord sequences from audio data has been a very active research area in recent years [5, 44, 60, 77]. The common process for chord estimation consists of two steps: pre-processing the audio into a feature vector representation (usually a PCP feature), and approximating the most likely chord sequence from these vectors (usually done via template-matching or expectation-maximization trained Hidden Markov Models

[64]). Usually, 24 chords are used (12 major and 12 minor), although some studies incorporate more complex chord types, such as 7th, 9th, augmented, and diminished chords [26, 32]. This way, the obtained strings have a straightforward musical interpretation. However, chord-based tonal sequence representations might be too coarse for the task at hand if one considers previously mentioned PCP codebook sizes, and might be also error-prone.

2.2 Key Invariance

As mentioned in section 1.3, cover songs may be transposed to different keys. Transposed versions are equivalent to most listeners, as pitches are perceived relative to each other rather than in absolute categories [16]. In spite of being a common change between versions, some systems do not explicitly consider transpositions. This is the case for systems that do not specifically focus on cover songs, or that do not use a tonal representation [25, 35, 53, 89].

Transposition is reflected as a ring-shift with respect to the "pitch axis" of the feature representation. Several strategies can be followed to tackle transposition, and their suitability may depend on the chosen feature representation. In general, transposition invariance can be achieved by relative feature encoding, by key estimation, by shift-invariant transformations, or by applying different transpositions.

The most straightforward way to achieve key invariance is to test all possible feature transpositions [21, 23, 36, 38, 39, 41, 50, 55]. In the case of an octave-independent representation, this implies the computation of a similarity measure for all possible circular (or ring) shifts in the pitch axis for each song. This strategy usually guarantees a maximal retrieval accuracy [75] but, on the other hand, either the time or the size (or both) of the database to search in increases.

Recently, some speeding-up approaches for this process have been presented [75, 76]. Given a tonal representation for two songs, these algorithms basically compute the most probable relative transpositions given an overall representation of the tonal content of each song (the so-called optimal transposition index) [20, 74, 76]. This process is very fast since this overall representation can be, e.g., a simple averaging of the PCP features over the whole sequence, and can be calculated off-line. Finally, only the K most probable shifts are chosen. Further evaluation suggests that, for 12 bin PCP-based representations, a near-optimal accuracy can be reached with just two shifts [75], thus reducing six times the computational load. Some systems do not follow these strategy and predefine a certain number of transpositions to compute. These can be chosen arbitrarily [78, 79], or based on some musical and empirical knowledge [4]. Decisions of this kind are very specific for each system.

An alternative approach is to off-line estimate the main key of the song and then apply transposition accordingly [28, 29, 49]. In this case, errors propagate faster and can dramatically worsen retrieval accuracy [75, 76] (e.g. if the key for the original song is not correctly estimated, no covers will be retrieved as they might have been estimated in the correct one). However, it must be noted that a similar procedure to choosing the K most probable transpositions could be employed.

If a symbolic representation such as chords is used, one can further modify it in order to just describe relative chord changes. This way, a key-independent feature sequence is obtained [2, 43, 68]. This idea, which is grounded in existing research on symbolic music processing [12, 45, 56, 67, 81], has been recently extended to PCP sequences [40, 38] by using the concept of optimal (or minimizing) transposition indices [52, 76].

A very interesting approach to achieve transposition invariance is to use a 2D power spectrum [50] or a 2D autocorrelation function [37]. Autocorrelation is a well-known operator for converting signals into a delay or shift-invariant representation [58]. Therefore, the power spectrum (or power spectral density), which is formally defined as the Fourier transform of the autocorrelation, is also shift-invariant. Other 2D transforms (e.g. from image processing) could be also used, specially shift-invariant operators derived from higher-order spectra [33].

2.3 Tempo Invariance

Different renditions of the same piece may vary in the speed they have been played, and any descriptor sequence extracted in a frame-by-frame basis from these performances will reflect this variation. For instance, in case of doubling the tempo, frames $i, i+1, i+2, i+3$ might correspond to frames $j, j, j+1, j+1$, respectively. Consequently, extracted sequences cannot be directly compared. Some cover song identification systems fail to include a specific module to tackle tempo fluctuations [2, 38, 39, 90, 91]. The majority of these systems generally focus on retrieval efficiency and treat descriptor sequences as statistical random variables. Thus, they throw away much of the sequential information that a given representation can provide (e.g. a representation consisting of a 4 symbol pattern like ABABCD, would yield the same values as AABBCD, ABCABD, etc., which is indeed a misleading oversimplification of the original data).

In case of having a symbolic descriptor sequence (e.g. the melody), one can encode it by considering the ratio of durations between two consecutive notes [68]. This strategy is employed in query-by-humming systems [12] and, combined with relative pitch encoding (section 2.3), leads to a representation that is key and tempo independent. However, for the reasons outlined in section 2.1, extracting a symbolic descriptor sequence is not straightforward and may lead to important estimation errors. Therefore, one needs to look at alternative tempo-invariance strategies.

One way of achieving tempo invariance is to estimate the tempo and then aggregate the information contained within comparable units of time. In this manner, the usual strategy is to estimate the beat [30] and then aggregate the descriptor information corresponding to the same beat. This can be done independently of the descriptor used. Some cover song identification systems based on a PCP [23, 55] or a melodic [49, 50] representation use this strategy, and extensions with chords or other types of information could be easily devised. If the beat does not provide enough temporal resolution, a finer representation (e.g. half-beat, quarter-beat, etc.) might be employed [21]. However, some studies suggest that systems using beat-averaging strategies can be outperformed by others, specially the ones employing dynamic programming [4, 76].

An alternative to beat induction is doing temporal compression/expansion [41, 53]. This straightforward strategy consists in re-sampling the signal into several musically plausible compressed/expanded versions and then comparing all of them in order to empirically discover the correct re-sampling.

Another interesting way to achieve tempo independence is again the 2D power spectrum or the 2D autocorrelation function [36, 37, 50]. These functions are usually designed for achieving both tempo as well as key independence, but 1D versions can also be designed (section 2.2).

If one wants to perform direct frame to frame comparison, a sequence alignment/similarity algorithm must be used to determine frame to frame correspondence between two song's representations. Several alignment algorithms for MIR have been proposed (e.g. [1, 15, 52]) which, sometimes, derive from general string and sequence alignment/similarity algorithms [31, 65, 69]. In cover song identification, dynamic programming [31] is a routinely employed technique for aligning two representations and automatically discovering their local correspondences [4, 20, 25, 28, 29, 35, 43, 49, 55, 74, 76, 78, 79, 89]. Overall, one iteratively constructs a cumulative distance matrix by considering the optimal alignment paths that can be derived by following some neighboring constraints (or patterns) [54, 65]. These neighboring constraints determine the allowed local temporal deviations and they have been evidenced to be an important parameter in the final system's accuracy [54, 76]. One might hypothesize that this importance relies on the ability to track local timing variations between small parts of the performance (section 1.3). For cover song identification, dynamic programming algorithms have been found to outperform beat induction strategies [4, 76]. The most typical algorithms for dynamic programming alignment/similarity are dynamic time warping algorithms [65, 69] and edit distance variants [31]. Their main drawback is that they are computationally expensive (i.e., quadratic in the length of the song representations), but several fast implementations may be derived [31, 56, 83].

2.4 Structure Invariance

The difficulties that a different song structure may pose in the computation of a cover song similarity measure are very often neglected. However, this has been demonstrated to be a key factor [76] and actually, recent cover song identification systems thoughtfully consider this aspect, especially many of the best-performing ones[13].

A classic approach to structure invariance consists in summarizing a song into its most repeated or representative parts [29, 49]. In this case, song structure analysis [57] is performed in order to segment sections from the song's representation used. Usually, the most repetitive patterns are chosen and the remaining patterns are disregarded. This strategy might be prone to errors since structure segmentation algorithms still have much room for improvement [57]. Furthermore, sometimes the

---

[13] For accuracies please see section 3 and references therein.

most identifiable or salient segment for a song is not the most repeated one, but the introduction, the bridge, and so forth.

Some dynamic programming algorithms deal with song structure changes. These are basically the so-called local alignment algorithms [31], and have been successfully applied to the task of cover song identification [20, 74, 76, 89]. These systems solely consider the best subsequence alignment found between two song's representation for similarity assessment, what has been evidenced to yield very satisfactory results [76].

However, the most common strategy for achieving structure invariance consists in windowing the descriptors representation (sequence windowing) [41, 50, 53, 55]. This windowing can be performed with a small hop size in order to faithfully represent any possible offset in the representations. This hop size has not been found to be a critical parameter for accuracy, as near-optimal values are found for a considerable hop size range [50]. Sequence windowing is also used by many audio fingerprinting algorithms using tonality-based descriptors [8, 51, 66], and it is usually computationally less expensive than dynamic programming techniques for achieving structural invariance.

2.5 Similarity Computation

The final objective of a cover song identification system is, given a query, to retrieve a list of cover songs from a music collection. This list is usually ranked according to some similarity measure so that first songs are the most similar to the query. Therefore, cover song identification systems output a similarity (or dissimilarity[14]) measure between pairs of songs. This similarity measure operates on the obtained representation after feature extraction, key invariance, tempo invariance, and structure invariance modules.

Common dynamic programming techniques used for achieving tempo invariance (section 2.3) already provide a similarity measure as an output [31, 65, 69]. Accordingly, the majority of the cover song identification systems following a dynamic programming approach use the similarity measure these approaches provide. This is the case for systems using edit distances [4, 68] or dynamic time warping algorithms [25, 28, 29, 35, 43, 78, 79]. These similarity measures usually contain an implicit normalization depending on the representation's lengths, which can generate some conflicts with versions of very different durations. In the case of local alignment dynamic programming techniques (section 2.4), the similarity measure usually corresponds to the length of the found subsequence match [20, 55, 74, 76, 89].

Conventional similarity measures like cross-correlation [21, 23, 49], the Frobenius norm [36], the Euclidean distance [37, 50], or the dot product [38, 39, 41, 53] are also used. They are sometimes normalized depending on compared representation's lengths. In the case of adopting a sequence windowing strategy for dealing

---

[14] For the sake of generality, we use the term similarity to refer to both the similarity and the dissimilarity. In general, a distance measure can also be considered a dissimilarity measure, which, in turn, can be converted to a similarity measure.

Table 2 Cover song identification methods and their ways to overcome departures from the "canonical" song. A blank space denotes no specific treatment for them. Abbreviations for extracted features are PBFV for polyphonic binary feature vector, and PCP for pitch class profile. Abbreviation for key invariance is OTI for optimal transposition index. Abbreviations for tempo invariance are DP for dynamic programming, and HMM for Hidden Markov Models. Abbreviations for similarity computation are DTW for dynamic time warping, MLSS for most likely sequence of states, and NCD for normalized compression distance.

Reference(s)	Extracted feature	Key invariance	Tempo invariance	Structure invariance	Similarity computation
Ahonen & Lemstrom [2]	Chords	Relative changes			NCD
Bello [4]	Chords	K transpositions	DP		Edit distance
Egorov & Linetsky [20]	PCP	OTI	DP	DP	Match length
Ellis et al. [21, 23]	PCP	All transpositions	Beat		Cross-correlation
Foote [25]	Energy + Spectral		DP		DTW
Gómez & Herrera [28]	PCP	Key estimation	DP		DTW
Gómez et al. [29]	PCP	Key estimation	DP	Repeated patterns	DTW
Izmirli [35]	Key templates		DP		DTW
Jensen et al. [36]	PCP	All transpositions	Fourier transform		Frobenius norm
Jensen et al. [37]	PCP	2D autocorrelation	2D autocorrelation		Euclidean distance
Kim et al.[38, 39]	PCP + Delta PCP	All transpositions			Dot product
Kim & Perelstein [40]	PCP	Relative changes	HMM		MLSS
Kurth & Muller [41]	PCP	All transpositions	Temporal comp./exp.	Sequence windowing	Dot product
Lee [43]	Chords	Key estimation	DP		DTW
Marolt [49]	Melodic	Key estimation	DP	Repeated patterns	Cross-correlation
Marolt [50]	Melodic	2D spectrum	Beat + 2D spectrum	Sequence windowing	Euclidean distance
Müller et al. [53]	PCP	All transpositions	Temporal comp./exp.	Sequence windowing	Dot product
Nagano et al. [55]	PBFV		Beat + DP	Seq. windowing + DP	Match length
Sailer & Dressler [68]	Melodic		Relative changes		Edit distance
Serrà et al. [74, 76]	PCP	OTIs	DP	DP	Match length
Tsai et al. [78, 79]	Melodic	K transpositions	DP		DTW
Yang [89]	Spectral		DP	Linearity filtering	Match length

with structure changes (section 2.4), these similarity measures are usually combined with multiple post-processing steps such as threshold definition [41, 53, 50], TF-IDF[15] weights [50], or mismatch ratios [41]. Less conventional similarity measures include the normalized compression distance [2], and the Hidden Markov Model-based most likely sequence of states [40]. In table 2 we show a summary of all the outlined approaches and their strategies for overcoming musical changes among cover versions and for similarity computation.

3   Evaluation

The evaluation of cover song identification and similarity systems is a complex task, and it is difficult to find in the literature a common methodology for that. The only existing attempt to compare version identification systems is found in the Music Information Retrieval Evaluation eXchange (MIREX[16]) initiative [18, 19]. Nevertheless, the MIREX framework only provides an overall accuracy of each system. A valuable improvement would be to implement independent evaluations for the different processes involved (feature extraction, similarity computation, etc.), in order to analyze their contributions to the global system behavior.

The evaluation of cover song identification systems is usually set up as a typical information retrieval "query and answer" task [3], where one submits a query song and the system returns a ranked set (or list) of answers retrieved from a given collection [19]. Then, the main purpose of the evaluation process is to assess how precise the retrieved set is. We discuss two important issues regarding the evaluation of cover song retrieval systems: the evaluation measures and the music material used.

3.1   Evaluation Measures

A referential evaluation measure might be the mean of average precision (MAP). This measure is routinely employed in various information retrieval disciplines [3, 47, 86] and some works on cover song identification have recently started reporting results based on it [2, 20, 74]. In addition, it has been also used to evaluate the cover song identification task in the MIREX [19].

Although MIREX defines some evaluation measures, in the literature there is no agreement on which one to use. Therefore, in addition to MAP, several other measures have been proposed. These include the R-Precision (R-Prec, [4, 35]), variants of Precision or Recall at different rank levels (P@X, R@X, [21, 23, 25, 36, 37, 38, 39, 41, 78, 79, 89]), the average of Precision and Recall (Avg PR, [55]), and the F-measure (Fmeas, [28, 29, 76]).

---

[15] The TF-IDF weight (term frequency-inverse document frequency) is a weight often used in information retrieval and text mining. For more details we refer to [3].

[16] http://www.music-ir.org/mirexwiki/index.php/Main_Page

3.2   Music Material: Genre, Variability, and Size Issues

One relevant issue when dealing with evaluation is the considered music material. Both the complexity of the problem and the selected approach largely depend on the studied music collection and the types of versions we want to identify, which might range from remastered tracks, to radically different songs (section 1.2). In this sense, it is very difficult to compare two systems evaluated in different conditions and designed to solve different problems. Some works solely analyze classical music [35, 38, 39, 41, 53], and it is the case that all of them obtain very high accuracies. However, classical music versions might not present strong timbral, structural, or tempo variations. Therefore, one might hypothesize that, when only classical music is considered, the complexity of the cover song identification task decreases. Other works use a more variated style distribution in their music collections [2, 4, 21, 23, 36, 37, 49, 50] but many times it is still unclear which types of versions are used. These are usually mixed and may include remastered tracks (which might be easier to detect), medleys (where invariance towards song structure changes may be a central aspect), demos (with substantial variations with respect to the finally released song), remixes, or quotations (which might constitute the most challenging scenario due to their potentially low duration and distorted harmonicity). The MIREX music collection is meant to include a wide variety of genres (e.g. classical, jazz, gospel, rock, folk-rock, etc.), and a sufficient variety of styles and orchestrations [19]. However, the types of covers that are present in the MIREX collection are unknown[17]. In our view, a big variety in genres and types of covers is the only way to ensure the general applicability of the method being developed.

Apart from the qualitative aspects of the considered music material, one should also care with the quantitative aspects of it. The total amount of songs and the distribution of these can strongly influence final accuracy values. To study this influence, one can decompose a music collection into cover sets (i.e. each original song is assigned to a separate cover set). Then, their cardinality (number of covers per set, i.e., the number of covers for each original song) becomes an important parameter. A simple test was performed with the system described in [74] in order to assess the influence of these two parameters (number of cover sets, and their cardinality) on the final system's accuracy. Based on a collection of 2135 cover songs, 30 random selections of songs were carried out according to the aforementioned parameters. Then, average MAP for all runs was computed and plotted (figure 2). We can see that considering less than 50 cover sets or even just a cardinality of 2 yields unrealistically high results, while higher values for these two parameters at the same time all fall in a stable accuracy region[18]. This effect can also be seen if we plot the standard deviations of the evaluation measure across all runs (figure 3). Finally, it can be observed that using less than 50 cover sets introduces a high variability in the

---

[17] As the underlying datasets are not disclosed, information of this kind is unavailable.

[18] It is not the aim of the experiment to provide explicit accuracy values. Instead, we aim at illustrating the effects that different configurations of the music collection might have for final system's accuracy.

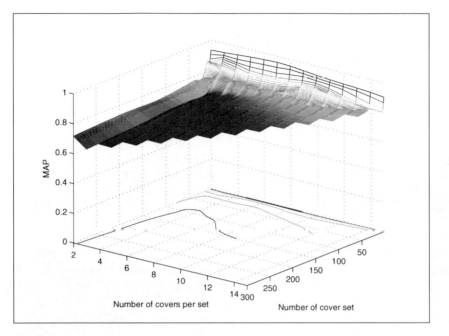

Fig. 2 Accuracy of a cover song identification system depending on the number of cover sets, and the number of covers per set

evaluated accuracy, which might then depend on the chosen subset. This variability becomes lower as the number of cover sets and their cardinality increase.

With the small experiment above, we can see that an insufficient size of the music collection could potentially lead to abnormally high accuracies, as well as to parameter over fitting (in the case of requiring a training procedure for some of them). Unfortunately, many reported studies use less than 50 cover sets [25, 28, 29, 35, 55, 78, 79]. Therefore, one cannot be confident about the reported accuracies. This could even happen with the so-called *covers80* cover song dataset[19] (a freely available dataset that many researchers use to test system's accuracy and to tune their parameters [2, 21, 23, 36, 37]), which is composed of 80 cover sets with a cardinality of 2.

In case when the evaluation dataset is not large enough, one may try to compensate the artifacts this might produce by adding so-called 'noise' or 'control' songs [4, 20, 49, 50]. The inclusion of these songs in the retrieval collection might provide an extra dose of difficulty to the task, as the probability of getting relevant items within the first ranked elements becomes then very low [19]. This approach is also followed within the MIREX framework. Therefore, test data is composed of thirty cover sets, each one consisting of eleven different versions. Accordingly, the total cover song collection contains 330 songs. In order to make the detection task more difficult, 670 individual songs, i.e., cover sets of cardinality 1, are added [19].

[19] http://labrosa.ee.columbia.edu/projects/coversongs/covers80

Table 3 Cover song identification methods and their evaluation strategies. Accuracies (including MIREX) correspond to best result achieved. Blank space, '~', and '^' denote unknown, approximate, and average values, respectively. Legend for genres is (B) blues, (C) classical, (CO) country, (E) electronic, (J) jazz, (HH) hip-hop, (M) metal, (P) pop, (R) rock, and (W) world. Legend for types of covers is (A) acoustic, (DE) demo, (DU) duet, (I) instrumental, (L) live, (M) medley, (RR) remaster, (RX) remix, and (Q) quotation.

Reference(s)	Cover sets	Covers per set	Total	Musical styles	Types of covers	Evaluation measure	Accuracy	MIREX MAP
Ahonen & Lemström [2]	80	2	160	B, CO, M, P, R	A, DU, I, L	MAP	0.18	
Bello [4]	36	4.4~	3208	P, R,	L,	R-Prec	0.25	0.27
Egorov & Linetsky [20]	30	11	1000	C, CO, E, HH, MT, R	A, DU, I, L, RR,	MAP	0.72	0.55
Ellis et al. [21, 23]	80	2	160	B, CO, M, P, R	A, DU, I, L	P@1	0.68	0.33
Foote [25]	28		82	C, P	A, I, L,	P@3	0.80	
Gómez & Herrera [28]	30	3.1~	90			Fmeas	0.39	
Gómez et al. [29]	30	3.1~	90			Fmeas	0.41	
Izmirli [35]	12		125	C		R-Prec	0.93	
Jensen et al. [36]	80	2	160	B, CO, M, P, R	A, DU, I, L	P@1	0.38	0.24
Jensen et al. [37]	80	2	160	B, CO, M, P, R	A, DU, I, L	P@1	0.48	0.23
Kim et al.[38, 39]	1000^	2~	2000	C		P@1	0.79	
Kim & Perelstein [40]								0.06
Kurth & Muller [41]								
Lee [43]			1167	C		R@1	0.97	0.13
Marolt [49]	8	4.5~	1820	P, R		P@5	0.22	
Marolt [50]	34	4.3~	2424	P, R		MAP	0.40	
Müller et al. [53]			1167	C		P@15	0.93	
Nagano et al. [55]	8	27~	216		L	Avg PR	0.89	
Sailer & Dressler [68]								0.07*
Serrà et al. [74, 76]	525	4.1~	2135	B, C, CO, E, J, M, P, R, W	A, DE, DU, I, L, M, RX, Q	MAP	0.66	0.66
Tsai et al. [78, 79]	47	2	794			P@1	0.77	
Yang [89]			120	C, P, R		P@2	0.99	

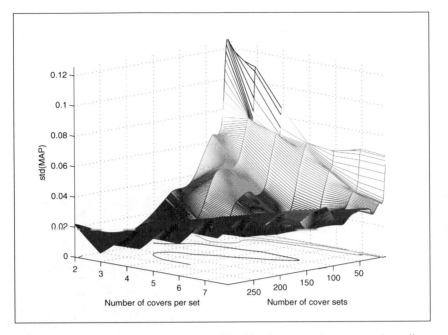

Fig. 3 Standard deviation of a cover song identification system's accuracy depending on the number of cover sets, and the number of covers per set

As a corollary, we could hypothesize that the bigger and more varied the music collection is, the more similar the out-of-sample results (and therefore better scalability) we shall obtain. In addition, one should stress that the usage of an homogeneous and small music collection, apart from leading to abnormal accuracies, could also lead to incorrect parameter estimates. In table 3 we show a summary of the evaluation strategies and accuracies reported by the cover song identification systems outlined in section 2.

4 Concluding Remarks

We have summarized here the work done for addressing the problem of automatic cover song identification. Even though different approaches have been tried, it seems quite clear that a well-crafted system has to be able to exploit tonal, temporal, and structural invariant representations of music. We have also learnt that there are methodological issues to be considered when building music collections used as ground-truth for developing and evaluating cover identification systems.

Once we have concluded this exhaustive overview, some conceptual open issues can be remarked. Even though the main invariances to be computed correspond to tonal and rhythm information, we still ignore the role (if any) of timbre, vocal features, or melodic similarity. Timbre similarity, including vocal similarity for sung music, could have some impact for identifying those covers intended to be close matches to a given query. In other situations this type of similarity would be misleading, though. Finding an automated way for deciding on that is still an open research issue.

In order to determine a similarity measure between cover songs, the usual approach pays attention solely to the musical facets that are shared among them (section 2). This makes sense if we suppose that these changes do not affect the similarity between covers. For instance, if two songs are covers and have the same timbre characteristics, and a third song is also a cover but does not exhibit the same timbre, they will score the same similarity. This commonality-based sense of similarity contrasts with the feature contrast similarity model [80], wherein similarity is determined by both common and distinctive features of the objects being compared. Future works approaching cover song similarity in a stricter sense might benefit from considering also differences between them, so that, in the previous example, the third cover is less similar than the two first ones.

Determining cover song similarity in a stricter sense would have some practical consequences and would be a useful feature for music retrieval systems. Therefore, depending on the goals of the listeners, different degrees of similarity could be required. Here we have a new scenario where the ill-defined but typical music similarity problem needs to be addressed. Research reported in this chapter could provide reasonable similarity metrics for this, but preservation of timbral and structural features would be required in addition, in order to score high in similarity with respect to the query song.

Another avenue for research is that of detecting musical quotations. In classical music, there is a long tradition of composers citing phrases or motives from other composers (e.g. Alban Berg quoting Bach's chorale *Es ist genug* in his *Violin Concerto*, or Richard Strauss quoting Beethoven's *Eroica symphony* in his 'Metamorphosen for 23 solo strings'). In popular music there are also plenty of quotations (e.g. The Beatles' ending section of *All you need is love* quotes the French anthem *La Marseillaise* and Glen Miller's *In the mood*, or Madonna's *Hung up* quoting Abba's *Gimme, Gimme, Gimme*), and even modern electronic genres massively borrow loops and excerpts from any existing recording. As the quoted sections are usually of short duration, special adaptations of the reviewed algorithms would be required to detect them. In addition to facilitating law enforcement procedures, linking this way diverse musical works opens new interesting ways for navigating across huge music collections.

Beyond many conceptual open issues, there are still some technical aspects that deserve effort to improve the efficiency of a system. First, perfecting a music processing system requires careful examination and analysis of errors. When errors are patterned they can reveal specific deficiencies or shortcomings in the algorithm. We are still lacking of that kind of in-depth analysis. Second, rigorously evaluating a cover song similarity metric would require the ground truth songs to be categorized according to the musical facets involved (section 1.3) and, maybe, according to the cover song category they belong to (section 1.2). Third, achieving a robust, scalable, and efficient method is still an issue. It is outstanding that systems achieving the highest accuracies are quite computationally expensive, and that fast retrieval systems fail in recognizing many of the cover songs a music collection might contain (section 2.1). We hypothesize that there exists a trade-off between system's accuracy and efficiency. However, we believe that these and many other technical as well as conceptual issues might be overcome in next years.

Acknowledgments

The authors want to thank their colleagues and staff at the Music Technology Group (UPF) for their support and work. This research has been partially funded by the EU-IP project PHAROS[20] (IST-2006-045035) and the e-Content Plus project VARIAZIONI[21] (ECP-2005-CULT-038264).

References

1. Adams, N.H., Bartsch, N.A., Shifrin, J.B., Wakefield, G.H.: Time series alignment for music information retrieval. In: Int. Symp. on Music Information Retrieval (ISMIR), pp. 303–310 (2004)
2. Ahonen, T.E., Lemstrom, K.: Identifying cover songs using normalized compression distance. In: Int. Workshop on Machine Learning and Music, MML (July 2008)
3. Baeza-Yates, R., Ribeiro-Neto, B.: Modern Information Retrieval. ACM Press Books, New York (1999)
4. Bello, J.P.: Audio-based cover song retrieval using approximate chord sequences: testing shifts, gaps, swaps and beats. In: Int. Symp. on Music Information Retrieval (ISMIR), September 2007, pp. 239–244 (2007)
5. Bello, J.P., Pickens, J.: A robust mid-level representation for harmonic content in music signals. In: Int. Symp. on Music Information Retrieval (ISMIR), pp. 304–311 (2005)
6. Berenzweig, A., Logan, B., Ellis, D.P.W., Whitman, B.: A large scale evaluation of acoustic and subjective music similarity measures. In: Int. Symp. on Music Information Retrieval, ISMIR (2003)
7. Cano, P., Batlle, E., Kalker, T., Haitsma, J.: A review of audio fingerprinting. Journal of VLSI Signal Processing 41, 271–284 (2005)
8. Casey, M., Rhodes, C., Slaney, M.: Analysis of minimum distances in high-dimensional musical spaces. IEEE Trans. on Audio, Speech, and Language Processing 16(5), 1015–1028 (2008)
9. Casey, M., Slaney, M.: The importance of sequences in musical similarity. In: IEEE Int. Conf. on Acoustics, Speech, and Signal Processing (ICASSP), May 2006, vol. 5, p. V (2006)
10. Casey, M., Veltkamp, R.C., Goto, M., Leman, M., Rhodes, C., Slaney, M.: Content-based music information retrieval: current directions and future challenges. Proceedings of the IEEE 96(4), 668–696 (2008)
11. Dalla Bella, S., Peretz, I., Aronoff, N.: Time course of melody recognition: a gating paradigm study. Perception and Psychophysics 7(65), 1019–1028 (2003)

---

[20] http://www.pharos-audiovisual-search.eu

[21] http://www.variazioniproject.org

12. Dannenberg, R.B., Birmingham, W.P., Pardo, B., Hu, N., Meek, C., Tzanetakis, G.: A comparative evaluation of search techniques for query-by-humming using the musart testbed. Journal of the American Society for Information Science and Technology 58(5), 687–701 (2007)
13. de Cheveigné, A.: Pitch perception models. In: Plack, C.J., Oxenham, A., Fay, R.R., Popper, A.N. (eds.) Pitch – Neural coding and perception, pp. 169–233. Springer, New York (2005)
14. Deliège, I.: Cue abstraction as a component of categorisation processes in music listening. Psychology of Music 24(2), 131–156 (1996)
15. Dixon, S., Widmer, G.: Match: A music alignment tool chest. In: Int. Symp. on Music Information Retrieval (ISMIR), pp. 492–497 (2005)
16. Dowling, W.J.: Scale and contour: two components of a theory of memory for melodies. Psychological Review 85(4), 341–354 (1978)
17. Dowling, W.J., Harwood, J.L.: Music cognition. Academic Press, London (1985)
18. Downie, J.S.: The music information retrieval evaluation exchange (2005–2007): a window into music information retrieval research. Acoustical Science and Technology 29(4), 247–255 (2008)
19. Downie, J.S., Bay, M., Ehmann, A.F., Jones, M.C.: Audio cover song identification: MIREX 2006-2007 results and analyses. In: Int. Symp. on Music Information Retrieval (ISMIR), September 2008, pp. 468–473 (2008)
20. Egorov, A., Linetsky, G.: Cover song identification with IF-F0 pitch class profiles. In: MIREX extended abstract (September 2008)
21. Ellis, D.P.W., Cotton, C.: The 2007 labrosa cover song detection system. In: MIREX extended abstract (September 2007)
22. Ellis, D.P.W., Cotton, C., Mandel, M.: Cross-correlation of beat-synchronous representations for music similarity. In: IEEE Int. Conf. on Acoustics, Speech, and Signal Processing (ICASSP), April 2008, pp. 57–60 (2008)
23. Ellis, D.P.W., Poliner, G.E.: Identifying cover songs with chroma features and dynamic programming beat tracking. In: IEEE Int. Conf. on Acoustics, Speech, and Signal Processing (ICASSP), April 2007, vol. 4, pp. 1429–1432 (2007)
24. Ellis, D.P.W., Whitman, B., Berenzweig, A., Lawrence, S.: The quest for ground truth in musical artist similarity. In: Int. Symp. on Music Information Retrieval (ISMIR), October 2002, pp. 518–529 (2002)
25. Foote, J.: Arthur: Retrieving orchestral music by long-term structure. In: Int. Symp. on Music Information Retrieval (ISMIR) (October 2000)
26. Fujishima, T.: Realtime chord recognition of musical sound: a system using common lisp music. In: Int. Computer Music Conference (ICMC), pp. 464–467 (1999)
27. Gómez, E.: Tonal description of music audio signals. PhD thesis, Universitat Pompeu Fabra, Barcelona, Spain (2006), http://mtg.upf.edu/node/472
28. Gómez, E., Herrera, P.: The song remains the same: identifying versions of the same song using tonal descriptors. In: Int. Symp. on Music Information Retrieval (ISMIR), October 2006, pp. 180–185 (2006)
29. Gómez, E., Ong, B.S., Herrera, P.: Automatic tonal analysis from music summaries for version identification. In: Conv. of the Audio Engineering Society (AES) (October 2006); CD-ROM, paper no. 6902
30. Gouyon, F., Klapuri, A., Dixon, S., Alonso, M., Tzanetakis, G., Uhle, C., Cano, P.: An experimental comparison of audio tempo induction algorithms. IEEE Trans. on Speech and Audio Processing 14(5), 1832–1844 (2006)
31. Gusfield, D.: Algorithms on strings, trees and sequences: computer sciences and computational biology. Cambridge University Press, Cambridge (1997)

32. Harte, C.A., Sandler, M.B.: Automatic chord identification using a quantized chroma-gram. In: Conv. of the Audio Engineering Society (AES), pp. 28–31 (2005)
33. Heikkila, J.: A new class of shift-invariant operators. IEEE Signal Processing Magazine 11(6), 545–548 (2004)
34. Hu, N., Dannenberg, R.B., Tzanetakis, G.: Polyphonic audio matching and alignment for music retrieval. In: IEEE Workshop on Apps. of Signal Processing to Audio and Acoustics (WASPAA), pp. 185–188 (2003)
35. Izmirli, Ö.: Tonal similarity from audio using a template based attractor model. In: Int. Symp. on Music Information Retrieval (ISMIR), pp. 540–545 (2005)
36. Jensen, J.H., Christensen, M.G., Ellis, D.P.W., Jensen, S.H.: A tempo-insensitive distance measure for cover song identification based on chroma features. In: IEEE Int. Conf. on Acoustics, Speech, and Signal Processing (ICASSP), April 2008, pp. 2209–2212 (2008)
37. Jensen, J.H., Christensen, M.G., Jensen, S.H.: A chroma-based tempo-insensitive distance measure for cover song identification using the 2d autocorrelation. In: MIREX extended abstract (September 2008)
38. Kim, S., Narayanan, S.: Dynamic chroma feature vectors with applications to cover song identification. In: IEEE Workshop on Multimedia Signal Processing (MMSP), October 2008, pp. 984–987 (2008)
39. Kim, S., Unal, E., Narayanan, S.: Fingerprint extraction for classical music cover song identification. In: IEEE Int. Conf. on Multimedia and Expo (ICME), June 2008, pp. 1261–1264 (2008)
40. Kim, Y.E., Perelstein, D.: MIREX 2007: audio cover song detection using chroma features and hidden markov model. In: MIREX extended abstract (September 2007)
41. Kurth, F., Müller, M.: Efficient index-based audio matching. IEEE Trans. on Audio, Speech, and Language Processing 16(2), 382–395 (2008)
42. Larkin, C. (ed.): The Encyclopedia of Popular Music, 3rd edn. (November 1998)
43. Lee, K.: Identifying cover songs from audio using harmonic representation. In: MIREX extended abstract (September 2006)
44. Lee, K.: A system for acoustic chord transcription and key extraction from audio using hidden Markov models trained on synthesized audio. PhD thesis, Stanford University, USA (2008)
45. Lemstrom, K.: String matching techinques for music retrieval. PhD thesis, University of Helsinki, Finland (2000)
46. Levitin, D.: This is your brain on music: the science of a human obsession. Penguin (2007)
47. Manning, C.D., Prabhakar, R., Schutze, H.: An introduction to Information Retrieval. Cambridge University Press, Cambridge (2008),
    http://www.informationretrieval.org
48. Mardirossian, A., Chew, E.: Music summarization via key distributions: analyses of similarity assessment across variations. In: Int. Symp. on Music Information Retrieval, ISMIR (2006)
49. Marolt, M.: A mid-level melody-based representation for calculating audio similarity. In: Int. Symp. on Music Information Retrieval (ISMIR), October 2006, pp. 280–285 (2006)
50. Marolt, M.: A mid-level representation for melody-based retrieval in audio collections. IEEE Trans. on Multimedia 10(8), 1617–1625 (2008)
51. Miotto, R., Orio, N.: A music identification system based on chroma indexing and statistical modeling. In: Int. Symp. on Music Information Retrieval (ISMIR), September 2008, pp. 301–306 (2008)
52. Müller, M.: Information Retrieval for Music and Motion. Springer, Heidelberg (2007)

53. Müller, M., Kurth, F., Clausen, M.: Audio matching via chroma-based statistical features. In: Int. Symp. on Music Information Retrieval (ISMIR), pp. 288–295 (2005)
54. Myers, C.: A comparative study of several dynamic time warping algorithms for speech recognition. Master's thesis, Massachussets Institute of Technology, USA (1980)
55. Nagano, H., Kashino, K., Murase, H.: Fast music retrieval using polyphonic binary feature vectors. In: IEEE Int. Conf. on Multimedia and Expo (ICME), vol. 1, pp. 101–104 (2002)
56. Navarro, G., Mäkinen, V., Ukkonen, E.: Algorithms for transposition invariant string matching. Journal of Algorithms (56) (2005)
57. Ong, B.S.: Structural analysis and segmentation of music signals. PhD thesis, Universitat Pompeu Fabra, Barcelona, Spain (2007), http://mtg.upf.edu/node/508
58. Oppenheim, A.V., Schafer, R.W., Buck, J.B.: Discrete-Time Signal Processing, 2nd edn. Prentice Hall, Englewood Cliffs (1999)
59. Pachet, F.: Knowledge management and musical metadata. Idea Group (2005)
60. Papadopoulos, H., Peeters, G.: Large-scale study of chord estimation algorithms based on chroma representation and hmm. In: Int. Conf. on Content-Based Multimedia Information, pp. 53–60 (2007)
61. Pickens, J.: Harmonic modeling for polyphonic music retrieval. PhD thesis, University of Massachussetts Amherst, USA (2004)
62. Poliner, G.E., Ellis, D.P.W., Ehmann, A., Gómez, E., Streich, S., Ong, B.S.: Melody transcription from music audio: approaches and evaluation. IEEE Trans. on Audio, Speech, and Language Processing 15, 1247–1256 (2007)
63. Purwins, H.: Proles of pitch classes. Circularity of relative pitch and key: experiments, models, computational music analysis, and perspectives. PhD thesis, Berlin University of Technology, Germany (2005)
64. Rabiner, L.R.: A tutorial on hidden markov models and selected applications in speech recognition. Proc. of the IEEE (1989)
65. Rabiner, L.R., Juang, B.H.: Fundamentals of speech recognition. Prentice Hall, Englewood Cliffs (1993)
66. Riley, M., Heinen, E., Ghosh, J.: A text retrieval approach to content-based audio retrieval. In: Int. Symp. on Music Information Retrieval (ISMIR), September 2008, pp. 295–300 (2008)
67. Robine, M., Hanna, P., Ferraro, P., Allali, J.: Adaptation of string matching algorithms for identification of near-duplicate music documents. In: ACM SIGIR Workshop on Plagiarism Analysis, Authorship Identification, and Near-Duplicate Detection (PAN), pp. 37–43 (2007)
68. Sailer, C., Dressler, K.: Finding cover songs by melodic similarity. In: MIREX extended abstract (September 2006)
69. Sankoff, D., Kruskal, J.: Time warps, string edits, and macromolecules. Addison-Wesley, Reading (1983)
70. Scaringella, N., Zoia, G., Mlynek, D.: Automatic genre classification of music content: a survey. IEEE Signal Processing Magazine 23(2), 133–141 (2006)
71. Schellenberg, E.G., Iverson, P., McKinnon, M.C.: Name that tune: identifying familiar recordings from brief excerpts. Psychonomic Bulletin and Review 6(4), 641–646 (1999)
72. Schulkind, M.D., Posner, R.J., Rubin, D.C.: Musical features that facilitate melody identification: how do you know it's your song when they finally play it? Music Perception 21(2), 217–249 (2003)
73. Selfridge-Field, E.: Conceptual and representational issues in melodic comparison. MIT Press, Cambridge (1998)

74. Serrà, J., Serra, X., Andrzejak, R.G.: Cross recurrence quantification for cover song iden-tification. New Journal of Physics 11, art. 093017 (September 2009)
75. Serrà, J., Gómez, E., Herrera, P.: Transposing chroma representations to a common key. In: IEEE CS Conference on The Use of Symbols to Represent Music and Multimedia Objects, October 2008, pp. 45–48 (2008)
76. Serrà, J., Gómez, E., Herrera, P., Serra, X.: Chroma binary similarity and local align-ment applied to cover song identification. IEEE Trans. on Audio, Speech, and Language Processing 16(6), 1138–1152 (2008)
77. Sheh, A., Ellis, D.P.W.: Chord segmentation and recognition using em-trained hid-den markov models. Int. Symp. on Music Information Retrieval (ISMIR), pp. 183–189 (2003)
78. Tsai, W.H., Yu, H.M., Wang, H.M.: A query-by-example technique for retrieving cover versions of popular songs with similar melodies. In: Int. Symp. on Music Information Retrieval (ISMIR), pp. 183–190 (2005)
79. Tsai, W.H., Yu, H.M., Wang, H.M.: Using the similarity of main melodies to identify cover versions of popular songs for music document retrieval. Journal of Information Science and Engineering 24(6), 1669–1687 (2008)
80. Tversky, A.: Features of similarity. Psychological Review 84, 327–352 (1977)
81. Typke, R.: Music retrieval based on melodic similarity. PhD thesis, Utrecht University, Netherlands (2007)
82. Tzanetakis, G.: Pitch histograms in audio and symbolic music information retrieval. In: Int. Symp. on Music Information Retrieval (ISMIR), pp. 31–38 (2002)
83. Ukkonen, E., Lemstrom, K., Mäkinen, V.: Sweepline the music! Comp. Sci. in Perspec-tive, 330–342 (2003)
84. Unal, E., Chew, E.: Statistical modeling and retrieval of polyphonic music. In: IEEE Workshop on Multimedia Signal Processing (MMSP), pp. 405-409 (2007)
85. Vignoli, F., Paws, S.: A music retrieval system based on user-driven similarity and its evaluation. In: Int. Symp. on Music Information Retrieval (ISMIR), pp. 272–279 (2005)
86. Voorhees, E.M., Harman, D.K.: Trec: Experiment and evaluation in information retrieval (2005)
87. White, B.W.: Recognition of distorted melodies. American Journal of Psychology 73, 100–107 (1960)
88. Xu, R., Wunsch, D.C.: Clustering. IEEE Press, Los Alamitos (2009)
89. Yang, C.: Music database retrieval based on spectral similarity. Technical report (2001)
90. Yu, Y., Downie, J.S., Chen, L., Oria, V., Joe, K.: Searching musical audio datasets by a batch of multi-variant tracks. In: ACM Multimedia, October 2008, pp. 121–127 (2008)
91. Yu, Y., Downie, J.S., Mörchen, F., Chen, L., Joe, K., Oria, V.: Cosin: content-based re-trieval system for cover songs. In: ACM Multimedia, October 2008, pp. 987–988 (2008)

Multimodal Aspects of Music Retrieval: Audio, Song Lyrics – and Beyond?

Rudolf Mayer and Andreas Rauber

Abstract. Music retrieval is predominantly seen as a problem to be tackled in the acoustic domain. With the exception of symbolic music retrieval and score-based systems, which form rather separate sub-disciplines on their own, most approaches to retrieve recordings of music by content rely on different features extracted from the audio signal. Music is subsequently retrieved by similarity matching, or classified into genre, instrumentation, artist or other categories. Yet, music is an inherently multimodal type of data. Apart from purely instrumental pieces, the lyrics associated with the music are as essential to the reception and the message of a song as is the audio. Album covers are carefully designed by artists to convey a message that is consistent with the message sent by the music on the album as well as by the image of a band in general. Music videos, fan sites and other sources of information add to that in a usually coherent manner. This paper takes a look at recent developments in multimodal analysis of music. It discusses different types of information sources available, stressing the multimodal character of music. It then reviews some features that may be extracted from those sources, focussing particularly on audio and lyrics as sources of information. Experimental results on different collections and categorisation tasks will round off the chapter. It shows the merits and open issues to be addressed to fully benefit from the rich and complex information space that music creates.

1 Introduction

Multimedia data by definition incorporates multiple types of content. However, often a strong focus is put on one view only, disregarding many other opportunities and exploitable modalities. In the same way as video, for instance, incorporates

Rudolf Mayer
Institute of Software Technology and Interactive Systems, Vienna University of Technology
e-mail: mayer@ifs.tuwien.ac.at

Andreas Rauber
Institute of Software Technology and Interactive Systems, Vienna University of Technology
e-mail: rauber@ifs.tuwien.ac.at

Z.W. Raś and A.A. Wieczorkowska (Eds.): Adv. in Music Inform. Retrieval, SCI 274, pp. 333–363.
springerlink.com   © Springer-Verlag Berlin Heidelberg 2010

visual, auditory, and text info (in the case of subtitles or extra information about the current programme via TV text and other channels), music data itself is not limited solely to its sound. Yet, a strong focus is put on audio based feature sets throughout the music information retrieval community, as music perception itself is based on sonic characteristics to a large extent. For many people, acoustic content is the main property of a song and makes it possible to differentiate between acoustic styles. For many examples or even genres this is true, for instance 'Hip-Hop' or 'Techno' music being dominated by a strong bass. Specific instruments very often define different types of music – once a track contains trumpet sounds it will most likely be assigned to genres like 'Jazz', traditional Austrian/German 'Blasmusik', 'Classical', or 'Christmas'.

However, a great deal of information is to be found in extra information in the form of text documents, be it about artists, albums, or song lyrics. Many musical genres are rather defined by the topics they deal with than a typical sound. 'Christmas' songs, for instance, are spread over a whole range of musical genres. Many traditional 'Christmas' songs were interpreted by modern artists and are heavily influenced by their style; 'Punk Rock' variations are recorded as well as 'Hip-Hop' or 'Rap' versions. What all of these share, though, is a common set of topics to be sung about. Another example is 'Christian Rock', which has a sound indistinguishable from other Rock music, but has highly religious topics (the same holds true for 'Christian Hip-Hop'). These simple examples show that there is a whole level of semantics inherent in song lyrics, that can not be detected by audio based techniques alone.

We assume that a song's text content can help in better understanding its meaning. In addition to the mere textual content, song lyrics exhibit a certain structure, as they are organised in blocks of choruses and verses. Many songs are organised in rhymes, patterns which are reflected in a song's lyrics and easier to detect from text than audio. Whether or not rhyming structures occur at all, and the level of complexity of the patterns used, may be highly characteristic for certain genres. In some cases, for example when thinking about very 'ear-catching' songs, maybe even the simplicity of rhyme structures are the common denominator.

For similar reasons, musical similarity can also be defined on textual analysis of certain parts-of-speech (POS) characteristics. Quiet or slow songs could, for instance, be discovered by rather descriptive language which is dominated by nouns and adjectives, whereas we assume a high number of verbs to express the nature of lively songs. In this paper, we further show the influence of so called text statistic features on song similarity. We employ a range of simple statistics such as the average word or line lengths as descriptors. Analogously to the common beats-per-minute (BPM) descriptor in audio analysis, we introduce the words-per-minute (WPM) measure to identify similar songs. The rationale behind WPM is that it can capture the 'density' of a song and its rhythmic sound in terms of similarity in audio and lyrics characteristics.

We therefore stress the importance of taking into account several of the afore-mentioned properties of music by means of a combinational approach. We want to point out that there is much to be gained from such a combinational approach as single genres may be best described in different feature sets. Musical genre classification therefore is heavily influenced by these modalities and can yield better overall results. We show the applicability of our approach with a detailed analysis of both the distribution of text and audio features, and genre classification on two test collections. One of our test collections consists of manually selected and cleansed songs subsampled from a real-world collection. We further use a larger collection which again is subsampled, but not manually cleansed, to show the stability of our approach.

This remainder of this paper is structured as follows. We start by giving an overview on related work in Section 2. We then give a detailed description of our approach and the feature sets we use for analysing song lyrics and audio tracks alike in Section 3. In Section 4 we apply our techniques to several audio corpora. We provide a summary of previous as well as novel results for the musical genre classification task, and a wide range of experimental settings. Finally, we analyse our results, conclude, and give a short outlook on future research in Section 5.

2   Related Work

Music information retrieval is a discipline of information retrieval, concerned with adequately accessing (digital) audio. Its major research topics include, but are not limited to, musical genre classification (and classification into other types of categories, such as mood or situations), similarity retrieval, or music analysis and knowledge representation. Comprehensive overviews of music information retrieval research are given in [8, 27].

The still dominant method of processing audio files in music information retrieval is by analysis of the audio signal, which is computed from plain wave files or via a preceding decoding step from other wide-spread audio formats such as MP3 or the (lossless) FLAC format. A wealth of different descriptive features for the abstract representation of audio content have been presented. Early overviews on content-based music information retrieval and experiments are given in [10] and [36, 38], focussing mainly on automatic genre classification of music.

Mel-Frequency Cepstral Coefficients (MFCC) [31] are a perceptually motivated set of features developed in context of speech recognition. The Mel scale, which is a perceptual scale found empirically through human listening tests, and models perceived pitch distances, is applied to the logarithmic spectrum before applying a discrete cosine transform (or an inverse Fourier transform) to obtain the MFCCs. An investigation about their adoption in the MIR domain was presented in [19]. Content-based audio retrieval based on K-Means clustering of MFCC features is performed in [21]. A comparison of MFCC and MPEG-7 features on sports audio classification is presented in [39].

Daubechies Wavelet Coefficient Histograms as a feature set suitable for music genre classification are proposed in [16]. The feature set characterises amplitude variations in the audio signal.

Chroma features [11] extract the harmonic content (e.g, keys, chords) of music by computing the spectral energy present at frequencies that correspond to each of the 12 notes in a standard chromatic scale.

The MARSYAS system [36], besides new graphical user interfaces for browsing and interacting with audio signals, introduces a number of new algorithms for audio description: a general multifeature audio texture segmentation methodology, feature extraction from MP3 compressed data, beat detection based on the discrete Wavelet transform and musical genre classification combining timbral, rhythmic and harmonic features.

The Moving Picture Experts Group (MPEG) released the MPEG-7 standard, which defines the Multimedia Content Description Interface, and is a standard for description and search of audio and visual content. Part 4 of said standard describes 17 low-level audio temporal and spectral descriptors, divided into seven classes, including silence. Some of the features are based on basic wave-form or spectral information, while others use harmonic or timbral information. In [1] these features are used for audio fingerprinting, i.e. using signatures based on various properties of audio signal for the robust identification of audio material. A classification approach with MPEG-7 features is done in [6].

Rhythm Patterns [33, 29] are a set of audio features which model modulation amplitudes on critical frequency bands. To this end, they consider and employ a set of psycho-acoustic models. Two other feature sets have been derived from and are based on different parts of the computation of the Rhythm patterns, namely the Rhythm Histograms and Statistical Spectrum Descriptors [17] feature sets.

In this paper, the MFCC, Marsyas, Chroma, Rhythm Patterns, Rhythm Histograms and Statistics Spectrum Descriptors are combined with and compared to our set of lyrics features. Therefore, these audio feature sets will be described in more detail in Section 3.1.

Several research teams have further begun working on adding textual information to the retrieval process, predominantly in the form of song lyrics and an abstract vector representation of the term information contained in text documents. A semantic and structural analysis of song lyrics is conducted in [22]. It focuses on aspects such as structure detection, e.g. verses and chorus, classification into thematic categories such as 'love', 'violent', 'christian', and similarity search. The correlation between artist similarity and song lyrics is studied in [20]. It is pointed out that acoustic similarity is superior to textual similarity, yet a combination of both approaches might lead to better results. A promising approach targeted at large-scale recommendation engines is presented in [14]. Lyrics are gathered from multiple sources on the Web, and are subsequently aligned to each other for matching sequences, to filter out errors like typing errors, or retrieved parts not actually belonging to the lyrics of the song, such as commercials.

Also, the analysis of karaoke music is an interesting new research area. A multimodal lyrics extraction technique for tracking and extracting karaoke text from video frames is presented in [41]. Some effort has also been spent on the automatic synchronisation of lyrics and audio tracks at a syllabic level [12]. A multi-modal approach to query music, text, and images with a special focus on album covers is presented in [4]. Other cultural data is included in the retrieval process e.g. in the form of textual artist or album reviews [3]. Cultural data is also used to provide a hierarchical organisation of music collections on the artist level in [28]. The system describes artists by terms gathered from web search engine results.

Another area were lyrics have also been employed is the field of emotion detection and classification, for example [40], which aims at disambiguating music emotion with lyrics and social context features. More recent work combined both audio and lyrics-based feature for mood classification [15].

In [13], additional information like web data and album covers are used for labelling, showing the feasibility of exploiting a range of modalities in music information retrieval. A three-dimensional musical landscape via a Self-Organising Maps (SOMs) is created and applied to small private music collections. Users can then navigate through the map by using a video game pad. An application of visualisation techniques for lyrics and audio content based on employing two separate SOMs is given in [26]. It demonstrates the potential of lyrics analysis for clustering collections of digital audio. The similarity of songs is visualised according to both modalities, and a quality measures with respect to the differences in distributions across the two maps is computed, in order to identify interesting genres and artists.

Experiments on the concatenation of audio and bag-of-words features were reported in [25]. The results showed potential for dimensionality reduction when using different types of features.

First results for genre classification using the rhyme and style features used later in this paper are reported in [24]; these results particularly showed that simple lyrics features may well be worthwile. This approach has further been extended on two bigger test collections, and to combining and comparing the lyrics features with audio features in [23].

3 Employed Feature Sets

Figure 1 shows an overview of the processing architecture. We start from plain audio files. The preprocessing/enrichment step involves decoding of audio files to plain wave format as well as lyrics fetching. We then apply the audio and lyrics-based feature extraction described in the following subsections. Finally, the results of both feature extraction processes are used for musical genre classification.

Input Data   Preprocessing/Enrichment   Feature Extraction   Classification

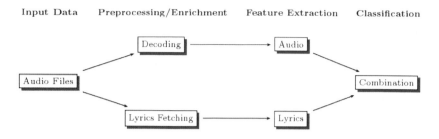

Fig. 1 Processing architecture for combined audio and lyrics analysis stretching from a set of plain audio files to combined genre classification

3.1 Audio Features

In our study, we employ several different sets of features extracted from the audio content of the songs, to compare them to and combine them with our newly designed set of features based on the song lyrics. To give comprehensive evidence that our feature set can improve classification results of audio-only feature sets, we extended the experiments presented in [23], which made use of the Rhythm Patterns, Statistical Spectrum Descriptors, and Rhythm Histograms audio feature sets. To those, we add analysis of the combination of the lyrics-based features with other popular and widely used feature sets, namely the Mel Frequency Cepstral Coefficients (MFCCs), MARSYAS and Chroma features. All these feature sets will be described below.

3.1.1 MFCC Features

Mel Frequency Cepstral Coefficients (MFCCs) originated in research for speech processing [31], and soon gained popularity in the field of music information retrieval [19]. A cepstrum is defined as the Discrete Cosine Transform (DCT) or inverse Fourier transform of the logarithm of the spectrum. If the Mel scale is applied to the logarithmic spectrum before applying the DCT (or inverse Fourier transform), the result is called Mel Frequency Cepstral Coefficients. The Mel scale is a perceptual scale that models perceived pitch distances, and was found empirically through human listening tests. With increasing frequency, the intervals in Hz producing equal increments in perceived pitch are getting larger and larger. Thus, the Mel scale is approximately a logarithmic scale; it corresponds more closely to the human auditory system than the linearly spaced frequency bands of a spectrum. A related scale is e.g. the Bark Scale, used in the Rhythm Patterns features (c.f. Section 3.1.4). From the MFCCs, commonly only the first few (for instance 5 to 20) Coefficients are used as features. In this work, we use the MFCC features extracted by the MARSYAS system, which provides four statistical values (means and variances over a texture window of one second) for the first 13 coefficients, thus resulting in 52 dimensions.

3.1.2 MARSYAS Features

The MARSYAS system [36] is a software framework for audio analysis and provides a number of feature extractors, all of which compute statistics over a texture window of approximately one second.

The *Short-Time Fourier Transform (STFT) Spectrum based Features* provide standard temporal and spectral low-level features, such as Spectral Centroid, Spectral Rolloff, Spectral Flux, Root Mean Square (RMS) energy and Zero Crossings.

A set of *MPEG Compression based features* is extracted directly from MPEG compressed audio data (e.g. from mp3 files) [37]. This approach utilises the fact that MPEG compression already performs a lot of analysis in the encoding stage, including a time-frequency analysis. The spectrum is divided into 32 sub-bands of equal size, via an analysis filterbank, wherefrom features such as the centroid, rolloff, spectral flux and RMS are directly computed from. Note that these features are not equal to the MPEG-7 standard features.

The *Wavelet Transform* is an alternative to the Fourier Transform, overcoming the trade-off between time and frequency resolution. It provides low frequency resolution and high time resolution for high frequency ranges, while in low frequency ranges, it provides high frequency and lower time resolution. This is a closer representation of the human perception of a sound. A set of features is extracted by computing the mean absolute values and standard deviation of the coefficients in each frequency band, and ratios of the mean absolute values between adjacent bands. The features represent 'sound texture' and provide information about the frequency distribution of the signal and its evolution over time.

For the *Beat Histogram* computation, a Discrete Wavelet Transform, which decomposes the signal into octave frequency bands, is applied before a time-domain amplitude envelope extraction and periodicity detection. The time domain amplitude envelope are extracted separately for each band. The sum of the normalised envelopes is then processed through an autocorrelation function to detect the dominant periodicities of the signal. The amplitude values of the dominant peaks are then accumulated over the whole song into the Beat Histogram, which not only captures the dominant beat in a sound, but more detailed information about the rhythmic content of a piece of music. The relative amplitude (of the sum of amplitudes) of the first and second peak, the ratio of the amplitude of the second to the first peak, the period of the first and second beat (in beats per minute), and the overall sum of the histogram, as indication of beat strength, are computed as features.

The *Pitch Histogram* feature computation decomposes the signal into two frequency bands (below and above 1000 Hz). For each band, amplitude envelopes are extracted, which are then summed up and an autocorrelation function is used to detect the main pitches. The three dominant peaks are accumulated into a histogram, where each bin corresponds to a musical note. The histogram thus contains information about the pitch range of a piece of music. A folded version of the histogram, obtained by mapping the notes of all octaves onto a single octave, contains information about the pitch classes or the harmonic content. The amplitude of the maximum peak of the folded histogram (i.e. magnitude of the most dominant pitch class), the period of the maximum peak of the unfolded (i.e. octave range of the dominant pitch) and folded histogram (i.e. main pitch class), the pitch interval between the two most prominent peaks of the folded histogram (i.e. main tonal interval relation) and the overall sum of the histogram are computed as features.

3.1.3 Chroma Features

Chroma features aim at representing the harmonic content (e.g, keys, chords) of a short-time window of audio. The Chroma vector is a perceptually motivated feature vector[11]. It uses the concept of *chroma* in the cyclic helix representation of musical pitch perception [35]. Chroma therein refers to the position of a pitch within an octave. The chroma vector thus represents magnitudes in twelve pitch classes in a standard chromatic scale (e.g., black and white keys within one octave on a piano). The feature vector is extracted from the magnitude spectrum by using a short-time Fourier transform (STFT). We specifically employ the feature extractor implemented in the MARSYAS system, which computes four statistical values (means and variances over a texture window of one second), for each of the 12 chromatic notes, thus finally resulting in a 48-dimensional feature vector.

3.1.4 Rhythm Patterns

Rhythm Patterns (RP), also called Fluctuation patterns, are a feature set for handling audio data based on analysis of the spectral audio data and psycho-acoustic transformations [32, 17]. The feature set has been employed e.g. in the SOM-enhanced jukebox (SOMeJB) [29] digital music library system. Rhythm patterns are basically a matrix representation of fluctuations on several critical bands. An overview of the computational steps is given in Figure 2, which also depicts the process for obtaining the Statistical Spectrum Descriptions and Rhythm Histograms, which are derived from the Rhythm Patterns features, and skip or modify some of the processing steps; further, they exhibit a different feature dimensionality, and represent different aspects of the audio signal.

If needed, a set of preprocessing steps is applied before the actual feature computation: multiple channels are averaged to one, and the audio is segmented into parts of six seconds. Often, it can be of advantage to leave out possible lead-in and fade-out segments, which might greatly differ from the rest of the song. Depending on the processing capability available, also further segments maybe be skipped, e.g. only processing every third segment.

The feature extraction process for a Rhythm Pattern is then composed of two stages, indicated as steps $S1–S6$ and $R1–R3$ in Figure 2. First, the spectrogram of the audio is computed for each segment, utilising the short time Fast Fourier Transform (STFT), and applying a Hanning window (cf. S1). Next we employ the Bark scale, a perceptual scale that groups frequencies to *critical bands* according to perceptive pitch regions. Applying the scale to the spectrograms results in an aggregation to 24 frequency bands (S2). A Spectral Masking spreading function is applied to the signal, which models the occlusion of one sound by another sound (S3). Then, the Bark scale spectrogram is transformed into the decibel scale (S4), and further psycho-acoustic transformations are applied: computation of the Phon scale (S5) incorporates equal loudness curves, which account for the different perception of loudness at different frequencies. Subsequently, the values are transformed into the unit Sone (S6), which relates to the Phon scale in the way that a doubling on the Sone scale sounds to the human ear like a doubling of the loudness. This results in a psycho-acoustically modified Sonogram representation that reflects human loudness sensation.

In the second stage, the varying energy on a critical band in the Bark-scale Sonogram is regarded as a modulation of the amplitude over time. A discrete Fourier transform is applied to this Sonogram, resulting in a time-invariant spectrum of loudness amplitude modulation per modulation frequency for each individual critical band (R1). After additional weighting (R2) and smoothing steps using a gradient filter and Gaussian smoothing (R3), a Rhythm Pattern finally exhibits the magnitude of modulation for 60 frequencies on 24 bands, and has thus 1440 dimensions.

In order to summarise the characteristics of an entire piece of music, the median of the Rhythm Patterns of the six-second segments is computed.

3.1.5   Statistical Spectrum Descriptors

Statistical Spectrum Descriptors (SSD) features are derived based on the first stage of the Rhythm Patterns computation, i.e. on the Bark-scale representation of the frequency spectrum (cf. steps S1–S6 in Figure 2). In order to describe fluctuations within the critical bands, from this representation of perceived loudness, seven statistical measures are subsequently computed for each segment per critical band: the mean, median, variance, skewness, kurtosis, min- and max-values, resulting in a Statistical Spectrum Descriptor for a segment. The SSD feature vector for a piece of audio is then again calculated as the median of the descriptors of its segments.

In contrast to the Rhythm Patterns feature set, the dimensionality of the feature space is much lower: SSDs have $24 \times 7 = 168$ instead of 1440 dimensions, and this at matching performance regarding genre classification accuracies [17], on specific data sets even outperforming the Rhythm Patterns [18].

3.1.6   Rhythm Histogram Features

The Rhythm Histogram (RH) features are capturing rhythmical characteristics in a piece of music. Contrary to the Rhythm Patterns and the Statistical Spectrum Descriptor, information is not stored per critical band. Instead, early in the second stage of the RP calculation process (after step R1 in Figure 2), the magnitudes of each modulation frequency of all 24 critical bands are summed up, forming a histogram of 60 bins of 'rhythmic energy' per modulation frequency between 0.168 and 10 Hz. For a given piece of music, the Rhythm Histogram feature set is again calculated by taking the median of the histograms of every single segment processed. Rhythm Histogram features represent similar information as the Beat Histogram of MARSYAS, but have a different extraction approach.

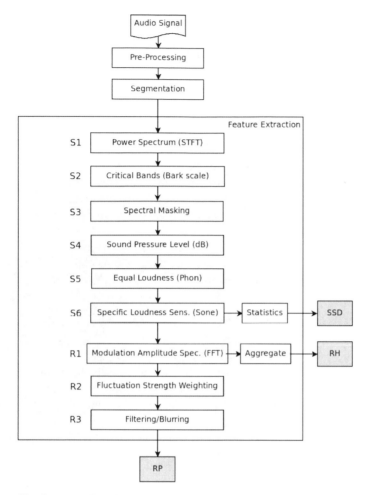

Fig. 2 Steps of the feature extraction process for Rhythm Patterns (RP), Statistical Spectrum Descriptors (SSD), and Rhythm Histograms (RH)

We further utilise the beats per minute (BPM) feature, computed from the modulation frequency of the peak of a Rhythm Histogram, to give a comparison to the lyrics-based words per minute (WPM) feature (cf. Section 3.2.2).

3.2 Lyrics Features

In this section we describe the four types of lyrics features we use in the experiments throughout the remainder of the paper: a) bag-of-words features computed from tokens (terms) occurring in documents, b) rhyme features taking into account the rhyming structure of lyrics, c) features considering the distribution of certain parts-of-speech, and d) text statistics features covering average numbers of words and particular characters. The latter three feature sets are referred to as rhyme and style features.

3.2.1 Bag-of-Words

Classical bag-of-words indexing at first tokenises all text documents in a collection, most commonly resulting in a set of words representing each document. Let the number of documents in a collection be denoted by N, each single document by d, and a term or token by t. Accordingly, the *term frequency* $tf(t,d)$ is the number of occurrences of term t in document d and the *document frequency* $df(t)$ the number of documents term t appears in. From this, an *inverse document frequency* idf can be computed.

The process of assigning weights to terms according to their importance or significance for the classification is called 'term-weighing'. The basic assumptions are that terms which occur very often in a document are more important for classification, whereas terms that occur in a high fraction of all documents are less important. The weighing we rely on is the most common model of *term frequency* times *inverse document frequency* [34], computed as:

$$tf \times idf(t,d) = tf(t,d) \cdot ln(N/df(t)) \qquad (1)$$

This results in vectors of weight values for each document d in the collection, i.e. each song lyrics document. This representation also introduces a concept of similarity, as lyrics that contain a similar vocabulary are likely to be semantically related. We do not perform term stemming in this setup, as earlier experiments showed only negligible differences for stemmed and non-stemmed features [24]; the rationale behind using non-stemmed terms is the occurrence of slang language in some genres, which we aim to preserve.

Selecting all terms present in a document collection will in most cases yield a vocabulary too large to be adequately processed by machine learning algorithms. Further, some terms might rather add noise than helping to distinguish documents from different genres. Thus, *feature (or term) selection* is an important pre-processing step. In this work, we employ a *frequency thresholding* technique: we omit terms that occur too frequent, and thus are likely stop-words, and terms that occur in too few documents, and therefore likely have less discriminative power.

3.2.2 Text Statistic Features

Text documents can also be described by simple statistical measures based on term (word) or character frequencies. Measures such as the average length of words or the ratio of unique words in the vocabulary capture aspects of the complexity of the texts, and are expected to vary over different genres. Further, the usage of punctuation marks such as exclamation or question marks may be specific for some genres. We further expect some genres to make increased use of apostrophes when

Table 1 Overview of text statistic features

Feature Name	Description
exclamation mark, colon, quote, comma, question mark, dot, hyphen, semicolon	simple counts of occurrences
d0 - d9	occurrences of digits
WordsPerLine	words / number of lines
UniqueWordsPerLine	unique words / number of lines
UniqueWordsRatio	unique words / words
CharsPerWord	number of chars / number of words
WordsPerMinute	the number of words / length of the song

omitting the correct spelling of word endings. The list of extracted features is given in Table 1.

All features that simply count character occurrences are normalised by the number of words of the song text to accommodate for different lyrics lengths. 'WordsPerLine' and 'UniqueWordsPerLine' describe the words per line and the unique number of words per line. The 'UniqueWordsRatio' is the ratio of the number of unique words and the total number of words. 'CharsPerWord' denotes the simple average number of characters per word. The last feature, 'WordsPerMinute' (WPM), is computed analogously to the well-known beats-per-minute (BPM) value[1]. Even though the computation is similar, the two features may still take very different values in various genres – as such, both 'Hip-Hop' and e.g. 'Techno' music may have similar BPM, but the latter generally way less song text, and thus much lower WPM values.

3.2.3 Part-of-Speech Features

Part-of-speech tagging is a lexical categorisation or grammatical tagging of words according to their definition and the textual context they appear in. Different part-of-speech categories are for example nouns, verbs, articles or adjectives. We presume that different genres will differ also in the category of words they are using, and therefore we additionally extract several part of speech descriptors from the lyrics. To this end, we employ the 'LingPipe' suite of libraries[2]. We in particular count the numbers of: *nouns, verbs, pronouns, relational pronouns* (such as 'that' or 'which'), *prepositions, adverbs, articles, modals,* and *adjectives.* To account for different document lengths, all of these values are normalised by the number of words of the respective lyrics document.

---

[1] Actually we use the ratio of the number of words and the song length in seconds to keep feature values in the same range. Hence, the correct name would be 'WordsPerSecond', or WPS.

[2] http://alias-i.com/lingpipe/

Table 2 Rhyme features for lyrics analysis

Feature Name	Description
Rhymes-AA	A sequence of two (or more) rhyming lines ('Couplet')
Rhymes-AABB	A block of two rhyming sequences of two lines ('Clerihew')
Rhymes-ABAB	A block of alternating rhymes
Rhymes-ABBA	A sequence of rhymes with a nested sequence ('Enclosing rhyme')
RhymePercent	The percentage of blocks that rhyme
UniqueRhymeWords	The fraction of unique terms used to build the rhymes

3.2.4   Rhyme Features

Rhyme denotes the the consonance or similar sound of two or more syllables or whole words. This linguistic style is most commonly used in poetry and songs. The rationale behind the development of rhyme features is that different genres of music should exhibit different styles of lyrics. We assume the rhyming characteristics of a song to be given by the degree and form of the rhymes used. 'Hip-Hop' or 'Rap' music, for instance, makes heavy use of rhymes, which (along with a dominant bass) leads to their characteristic sound. To automatically identify such patterns we introduce several descriptors from the song lyrics to represent different types of rhymes.

For the analysis of rhyme structures we do not rely on lexical word endings, but rather apply a more correct approach based on phonemes – the sounds, or groups thereof, in a language. Hence, we first need to transcribe the lyrics to a phonetic representation. The words 'sky' and 'lie', for instance, both end with the same phoneme /ai/. Phonetic transcription is language dependent, thus the language of song lyrics first needs to be identified, using e.g. the text categoriser 'TextCat' [5] to determine the correct transcriptor to apply. However, for our test collections presented in this paper we considered only songs in English language, and we therefore exclusively use English phonemes. For the transcription step, we utilise the 'Analysing Sound Patterns' software package [3]. This package includes a phoneme transcriptor, which is derived from early work on text-to-speech translation [9], which introduced a set of 329 letter-to-sound rules that translate from English text to the international phonetic alphabet (IPA).

After transcribing the lyrics into this phoneme representation, we distinguish two basic patterns of subsequent lines in a song text: *AA* and *AB*. The former represents two rhyming lines, while the latter denotes non-rhyming. Based on these basic patterns, we extract the features described in Table 2.

As the simplest structure, a 'Couplet' *AA* describes the rhyming of two or more subsequent pairs of lines. It usually occurs in the form of a 'Clerihew', i.e. several blocks of Couplets such as *AABBCC*. Another common pattern is the alternating rhyme, in the form of *ABAB*. An *enclosing rhyme*, defined as *ABBA*, denotes the rhyming of the first and fourth, as well as the second and third out of four lines. Based on these structure, we further measure 'RhymePercent', the percentage of

---

[3] http://www2.eng.cam.ac.uk/ tpl/asp/

lines with rhyming patterns versus the total number of lines in a song. Besides, we define the unique rhyme words as the fraction of unique terms used to build rhymes 'UniqueRhymeWords', which describes whether rhymes are frequently formed using the same word pairs, or a wide variety of words is used for the rhymes.

For our initial studies, we do not take into account rhyming schemes based on assonance, semirhymes, or alliterations. We also do not yet incorporate more elaborate rhyme patterns, especially not the less obvious ones, such as the 'Ottava Rhyme' of the form *ABABABCC*, and others. Also, we assign to all the rhyme forms the same weights, i.e. we for example do not give more importance to complex rhyme schemes. Experimental results lead to the conclusion that some of these patterns may well be worth studying. An experimental study on the frequency of occurrences might be a good starting point first, as modern popular music does not seem to contain many of these patterns.

4   Experiments

In this section we first introduce the test collections we use, followed by an illustration of some selected characteristics of our new features on these collections. We further present the results of our experiments, where we compare the performance of audio features and text features using various classifiers.

4.1   Test Collections

Music information retrieval research in general suffers from a lack of standardised benchmark collections, which is mainly attributable to copyright issues. Nonetheless, some collections have been used frequently in the literature, such as the collections provided for the ISMIR 2004 'rhythm' and 'genre' contest tasks, or the collection presented in [36]. However, for the first two collections, hardly any lyrics are available, as they are either instrumental songs, or their lyrics were not published electronically. For the latter, no meta-data is available revealing the song titles, making the automatic fetching of lyrics impossible. The collection used in [14] turned out to be infeasible for our experiments. It consists of only about 260 pieces, and was not initially used for genre classification: it was compiled from only about 20 different artists, and it was not well distributed over several genres (we specifically wanted to circumvent unintentionally classifying artists rather than genres).

To elude these limitations, we opted to compile our own test collections; more specifically, we first constructed two test collections different in size, first presented in [23]. For the first of these databases, we selected a total number of 600 songs (*collection_600*) as a random sample from a private collection. We aimed at having a high number of different artists, represented by songs from different albums, in order to prevent biased results by too many songs from the same artist and album. This collection thus comprises songs from 159 different artists, stemming from 241 different albums. The ten genres listed in the left-hand side of Table 3 are repre-

Table 3 Composition of the two small (*collection_600* and *collection_660*) and two large (*collection_3000* and *collection_3120*) test collections

Genre	collection_600			collection_3000		
	Artists	Albums	Songs	Artists	Albums	Songs
Country	6	13	60	9	23	227
Folk	5	7	60	11	16	179
Grunge	8	14	60	9	17	181
Hip-Hop	15	18	60	21	34	380
Metal	22	37	60	25	46	371
Pop	24	37	60	26	53	371
Punk Rock	32	38	60	30	68	374
R&B	14	19	60	18	31	373
Reggae	12	24	60	16	36	181
Slow Rock	21	35	60	23	47	372
Total	159	241	600	188	370	3009
	collection_660			collection_3120		
Children's music	7	5	60	7	5	109
Total	166	246	660	195	375	3118

sented by 60 songs each. Note that the number of different artists and albums is not equally spread, which is closer to a real-world scenario, though.

We then automatically fetched lyrics for this collection from the Internet using the lyrics scripts provided for the Amarok Music Player[4]. These scripts are simple wrappers for popular lyrics portals on the Web. To obtain all lyrics we used one script after another until all lyrics were available, regardless of the quality of the texts with respect to content or structure. Thus, the collection is named *collection_600_uncleansed*.

In order to evaluate the impact of proper lyrics preprocessing, we then manually cleansed the automatically collected lyrics. This is a tedious task, which first involves checking whether the fetched lyrics were matching the song at all. Then, we corrected the lyrics both in terms of structure and content, i.e. all lyrics were manually corrected in order to remove additional markup like '[2x]', '[intro]' or '[chorus]', and to include the unabridged lyrics for all songs. We payed special attention to completeness in terms of the resultant text documents being as adequate and proper transcriptions of the songs' lyrics as possible. This collection, which differs from *collection_600_uncleansed* only in the song lyrics quality, is thus called *collection_600_cleansed*. Effects of manually cleansing lyrics as opposed to automatic crawling from the Web on the performance of the lyrics features, as well as the impact of stemming, were studied in [23] and [24]. As their impact has been found to be rather small, and not consistently improving or degrading the classification results, detailed studies on this issue are thus omitted here, and in the following experiments we only employ the cleansed version of the collection.

---

[4] http://amarok.kde.org

To evaluate our findings from the smaller test collection on a larger one, we constructed a more diversified database. This collection includes all the songs of the smaller collection, and consists of 3.010 songs, which can be seen as prototypical for a private collection. The numbers of songs per genre range from 179 in 'Folk' to 381 in 'Hip-Hop'. Detailed figures about the composition of this collection can be taken from the right-hand side in Table 3. To be able to better relate and match the results obtained for the smaller collection, we only selected songs belonging to the same ten genres as in the *collection_600*.

In a novel set of experiments, we added one more genre to these existing collections, namely children's music, consisting of nursery rhymes and similar songs. The pieces of music in this genre in general have very distinctive acoustical properties, with a strong focus on vocals, and little instrumentation, which is often limited to the same instruments, such as guitars. Therefore, they already achieve high classification accuracies with audio-only features, and are thus an interesting challenge to test whether the lyrics features are able to improve performance also on genres that have distinctive acoustical properties. We therefore extended our smaller test database by 60 more songs, thus creating the new database *collection_660*, and added a total of 109 songs to the larger collection, thus resulting in *collection_3120*, both of which are illustrated also in Table 3.

4.2   Analysis of Selected Features

To demonstrate the ability of the newly proposed lyrics-based features to discriminate between different genres, we illustrate the distribution of the numerical values for these new features across the different genres. We focus on the most interesting features from each bag-of-words, rhyme, part-of-speech, and text statistic features, for the *collection_600_cleansed*.

First, plots for selected features from the bag-of-words set, all of which were among the highest ranked by the Information Gain feature selection method[5], are presented in Figure 3. Of those high ranked terms, we selected some that have interesting characteristics regarding different classes. It can be generally said that notably 'Hip-Hop' seems to have a lot of commonly used terms, especially from swear and cursing language (subsequently obscured), or slang terms. This can be seen in Figure 3(a) and 3(b), showing the terms 'n*gga' and 'f*ck'. While 'n*gga' is used almost solely in 'Hip-Hop' (in many types – singular and plural forms, with ending 's' and 'z'), 'f*ck' is also used in 'Metal' and to some lesser extent in 'Punk-Rock'. On the contrary, 'R&B' and 'Pop' do not use the term at all, and other genres just very rarely employ it. Regarding the dominant topics, 'Hip-Hop' also frequently has *violence* and *crime* as content of their songs, which is exemplified in the terms 'gun' and 'police' in Figures 3(c) and 3(d), respectively. Both terms are also used in 'Grunge' and 'Reggae'.

---

[5] Information Gain is a popular feature selection criterion, measuring the information obtained by a single term for category classification [30].

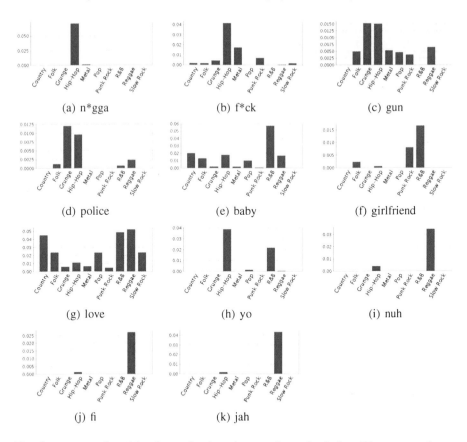

Fig. 3 Average $tf \times idf$ values of selected terms from the lyrics. Obscene words are obscured.

By contrast, 'R&B' has several songs concerning *relationships*, which is illustrated in Figures 3(e) and 3(f). Several genres deal with *love*, but to a very varying extent. In 'Country', 'R&B', and 'Reggae', this is a dominant topic, while it hardly occurs in 'Grunge', 'Hip-Hop', 'Metal' and 'Punk-Rock'.

Another interesting aspect is the use of slang and colloquial terms, or more generally using a transcription of the phonetic sound of some words. This is especially used in the genres 'Hip-Hop' and 'Reggae', but also in 'R&B'. Figure 3(h), for instance, shows that both 'Hip-Hop' and 'R&B' make use of the word 'yo', while 'Reggae' often uses a kind of phonetic transcription, as e.g. the word 'nuh' for 'not' or 'no', or many other examples, such as 'mi' (me), 'dem' (them), etc. 'Reggae' further employs a lot of particular terms, such es 'jah', which stands for 'god' in the Rastafari movement, or the Jamaican dialect word 'fi', which is used instead of 'for'.

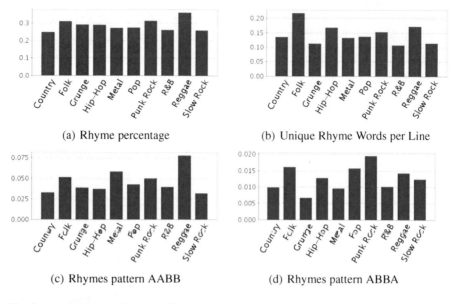

(a) Rhyme percentage

(b) Unique Rhyme Words per Line

(c) Rhymes pattern AABB

(d) Rhymes pattern ABBA

Fig. 4 Average values for selected rhyme features

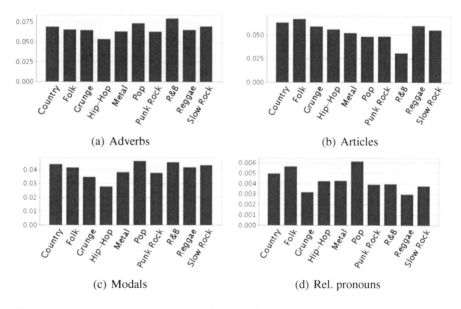

(a) Adverbs

(b) Articles

(c) Modals

(d) Rel. pronouns

Fig. 5 Average values for selected part-of-speech features

Summarising, a seemingly high amount of terms that are specific for 'Hip-Hop' and 'Reggae' can be observed, which should render those two genres well distinguishable from the others regarding bag-of-words features.

Figure 4 depicts selected rhyme features. 'Reggae' has the highest value of percentage of rhyming lines, while the other genres have rather equal usage of rhymes. 'Folk' may seem as using the most creative language for building those rhymes, which is manifested in the clearly higher number of unique words forming the rhymes, rather than repeatidly using the same words. 'Grunge' and 'R&B' seem to have distinctively lower values than the other genres. The distribution across the actual rhyme patterns used is also quite different over the genres, where 'Reggae' lyrics use a lot of *AABB* patterns, and 'Punk Rock' employs mostly *ABBA* patterns, while 'Grunge' makes particular little use of the latter.

Figure 5 presents plots of the most relevant part-of-speech features. Adverbs seem to help discriminating 'Hip-Hop' with low and 'Pop' and 'R&B' with higher values over the other classes. 'R&B' further can be well discriminated due to the infrequent usage of articles in the lyrics. Modals, on the other hand, are rarely used in 'Hip-Hop'.

Finally, the most interesting features from the text statistics type are illustrated in Figure 6. 'Reggae', 'Punk Rock', 'Metal', and, to some extent, also 'Hip-Hop' seem to use very expressive language, which manifests in the higher percentage of exclamation marks appearing in the lyrics. 'Hip-Hop' and 'Folk' in general seem to have more creative lyrics, indicated by the higher percentage of unique words used as compared to other genres, which may have more repetitive lyrics. 'Words per Minute' appears to be a very good feature to distinguish 'Hip-Hop' as the genre with the fastest sung (or spoken) lyrics from music styles such as 'Grunge', 'Metal' and 'Slow Rock'. The latter frequently have longer instrumental phases, especially longer lead-ins and fade-outs, and the pace of singing is adapted towards the general slower tempo of the (guitar) music. Comparing this feature with the well-known 'Beats per Minute' descriptor, it can be noted that the high tempo of 'Hip-Hop' lyrics coincides with the high number of beats per minute. 'Reggae' on the other hand has an even higher number of beats, and even though there are several pieces with fast lyrics, it is also characterised by longer instrumental passages, as well as words accentuated longer.

4.3 Experimental Results

After describing our experimental setup, we then discuss in detail the performance of the different audio and lyrics-only feature sets, and their combinations. We evaluate the impact of manually cleansing the lyrics, and specifically the performance of the newly added genre of children's music.

4.3.1 Setup

For each of the databases, we extract the audio and lyrics feature sets described in Section 3. We then build several combinations of these different feature sets, both separately within the audio and lyrics modalities, as well as combinations of audio and lyrics feature sets. This results in several dozens of different feature set combinations, out of which the most interesting ones are presented here. Most combina-

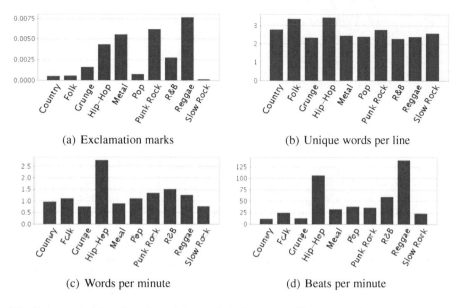

(a) Exclamation marks   (b) Unique words per line

(c) Words per minute   (d) Beats per minute

Fig. 6 Average values for selected text statistic features and beats-per-minute

tions with audio features are done with the SSD, as those are the best performing audio feature set.

For all our experiments, we employed the WEKA machine learning toolkit[6], and unless otherwise noted used the default settings for the classifiers and tests. We utilised mainly k-Nearest-Neighbour, Naïve Bayes and Support Vector Machines. We performed the experiments based on a ten-fold cross-validation, which is further averaged over five repeated runs. All results given in this sections are micro-averaged classification accuracies. i.e. they are calculated giving equal weight to each document. Statistical significance testing is performed per column, using a paired t-test with an α value of 0.05. In the following tables, plus signs (+) denote a significant improvement, whereas minus signs (−) denote significant degradation. The best results for each group of features are indicated by bold print.

4.3.2   Small Database – Collection 600

Table 4 shows the results for genre classification experiments performed on the small collection using only audio-based feature sets. The columns show the results for three different types of machine learning algorithms, with different parameter settings: k-NN with $k = 4$ and $k = 5$ and employing Euclidean distance, Support Vector Machine with linear (SVM/lin), polynomial (quadratic, SVM/pol), and radial basis function (SVM/rbf) kernels, and a Naïve Bayes (NB) classifier. All six algorithm variations were applied to the six single feature sets, as well as nine dif-

---

[6] http://www.cs.waikato.ac.nz/ml/weka/

Table 4 Classification accuracies and results of significance testing for various combinations of audio features for the 600 song collection (*collection_600_cleansed*). Statistically significant improvement or degradation over datasets (column-wise) is indicated by (+) or (−), respectively

Feature set	Dim.	4-NN	5-NN	NB	SVM/pol	SVM/lin	SVM/rbf
Chroma	48	18.33 -	18.50 -	18.77 -	19.60 -	22.53 -	14.63 -
MFCC	52	26.43 -	27.43 -	23.37 -	29.63 -	29.80 -	18.70 -
Marsyas	68	28.63 -	30.33 -	25.70 -	31.63 -	30.53 -	21.43 -
RP	1440	32.27 -	31.77 -	37.60 -	46.30 -	48.47 -	**44.20**
RH	60	29.73 -	29.03 -	31.13 -	36.03 -	36.47 -	28.97 -
SSD (base-line)	168	**48.97**	**49.57**	**44.57**	**56.63**	**59.37**	**44.20**
SSD / Chroma	216	50.70	**51.90**	42.37	59.30	59.17	43.13
SSD / Mars.	236	48.70	49.17	44.20	58.27	59.83	46.13
SSD / Mars. / Chroma	284	47.53	49.10	43.30	58.30	59.33	45.57
SSD / Mars. / Chroma / RH	344	47.73	48.60	42.67	59.67	60.90	46.97
SSD / Mars. / RH	296	49.50	49.63	43.90	**59.93**	**61.10**	47.67
SSD / MFCC	220	**51.23**	51.07	**44.73**	58.93	59.77	45.83
SSD / RH	228	49.37	49.80	43.17	58.57	60.37	46.83
SSD / RP	1608	41.77 -	39.87 -	41.77	57.73	60.23	52.87 +
SSD / RP / RH	1668	41.63 -	40.27 -	41.40	57.50	60.43	**53.30 +**

ferent combinations thereof. Significance testing is performed per column, using the SSD features as the base line.

Generally, the highest classification results, sometimes by far better, are achieved with the SVM, which is thus the most interesting classifier for a more in-depth analysis. For the single audio feature sets, the Statistical Spectrum Descriptors (SSD) achieves the highest accuracy (59.37%) of all, followed by Rhythm Patterns (RP) with an accuracy of 48.47%, both with the SVM with linear kernel. SSD clearly outperforms all the other feature sets with statistical significance, except for the SVM classifier with the RBF kernel, which achieves the exact same result on both SSD and RP. Apart from the Rhythm Patterns on the different SVM kernel variations, SSD features outperform the other sets by factors of 1.6 to 3.0.

Regarding the combinations of different audio feature sets, it was possible to increase the SSD baseline with some of the combinations with other feature sets, on one classifier with the Chroma features, on two in combination with MFCCs and by adding Marsyas and Rhythm Histograms (RH), and once by combing the SSD with the RH and RP. For most of those combinations, however, no change in performance that would be of statistical significance could be obtained; the only notable exception are the combination of SSD with RP only and both the RP and RH, yielding a significant degradation on the *k*-NN classifiers and a significant improvement utilising the SVM with RBF kernels. For the former, this most likely can be attributed to the generally poor performance of *k*-NN on high-dimensional feature sets, while for the latter, the baseline on the SSD-only features with the RBF kernel is very low compared to the other classifiers, even lower than *k*-NN.

Table 5 Classification accuracies and results of significance testing for various combinations of lyrics features for the 600 song collection (*collection_600_cleansed*). Statistically significant improvement or degradation over datasets (column-wise) is indicated by (+) or (−), respectively

Feature set	Dim.	3-NN	4-NN	5-NN	NB	SVM/lin
Rhyme	6	13.93	13.20	13.37	15.00	13.77
POS	9	16.13	18.23	17.57	19.63	20.03
Text-stat (base line)	23	21.00	21.20	22.00	21.70	29.73
Text-stat / POS	32	**25.87** +	**25.17**	24.77	22.80	31.27
Text-stat / Rhyme	29	23.73	23.13	23.60	22.87	31.03
Text-stat / POS / Rhyme	38	22.90	24.47	**26.07**	24.20 +	30.63
Chroma (base-line)	48	17.87	18.33	18.50	18.77	22.53
Chroma / Text-stat	71	**23.07** +	**23.93** +	**23.87** +	22.33 +	32.87 +
Chroma / Text-stat / POS	80	21.43	21.00	21.57	22.53 +	34.87 +
Chroma / Text-stat / POS / Rhyme	86	21.47	20.83	21.53	**23.27** +	**35.07** +
Chroma / Text-stat / Rhyme	77	21.80	22.47	22.83	23.53 +	33.43 +
MFCC (base-line)	52	24.50	26.43	27.43	23.37	29.80
MFCC / Text-stat	75	27.83	**31.50** +	31.47	29.57 +	38.43 +
MFCC / Text-stat / POS	84	29.07 +	31.17 +	32.07	30.13 +	38.27 +
MFCC / Text-stat / POS / Rhyme	90	28.77	30.90	**32.50** +	31.33 +	**39.63** +
MFCC / Text-stat / Rhyme	81	**29.53** +	30.87	31.40	29.90 +	38.50 +
MFCC / POS / Rhyme	67	23.37	26.20	28.10	26.53 +	34.53 +
Marsyas (base-line)	68	26.00	28.63	30.33	25.70	30.53
Mars. / Text-stat	91	29.23	30.60	32.90	30.50 +	37.83 +
Mars. / Text-stat / POS	100	29.57	**33.27** +	32.47	31.03 +	37.50 +
Mars. / Text-stat / POS / Rhyme	106	**30.30**	32.53	**34.10**	31.83 +	**39.37** +
Mars. / Text-stat / Rhyme	97	28.97	32.37	33.73	30.90 +	39.00 +
SSD (base-line)	168	48.60	48.97	49.57	44.57	59.37
SSD / Text-stat	191	51.20	**53.07** +	**53.30** +	46.80	**64.53** +
SSD / Text-stat / POS	200	**51.97** +	51.00	51.70	46.73	64.07 +
SSD / Text-stat / POS / Rhyme	206	50.63	51.90	53.00	47.37 +	62.90 +
SSD / Text-stat / Rhyme	197	50.17	52.30 +	52.93	**47.57** +	63.93 +

Regarding the individual performance of the different classifiers, for k-NN it can be noted that there is no clear pattern on a better performance of a single classifier, even though the 5-NN seems to perform slightly better on most feature sets. For the SVMs, except for two out of the 15 feature sets, the linear kernel always got the highest results; especially the RBF kernel-based SVMs performed significantly worse. Thus, for the following experiments, we employ only the linear kernel.

After these initial experiments, we chose the highest result achievable with audio-only features, the SSD features, as the baseline we want to improve on. The SSDs show in general very good performance on our databases, with the achieved almost 60% clearly outperforming the minimal baseline of 10% on a database of ten equally-sized classes. Thus, they are as such a challenging baseline.

In Table 5, we detail the results of the different lyrics features, and their combination with the audio-only feature sets on the small, cleansed database (*collection_600_cleansed*), that is with automatic lyric fetching and manual checking of the retrieved lyrics. The experiments were again performed with six different classifiers, in contrary to those in Table 4 we employ also a 3-NN instead of the RBF kernel SVM, to give more details on the behaviour of the different k values. Indeed, 3-NN is the best performing of the k-NN family on a number of low-dimensional feature sets.

For the lyrics-only features, the rhyme features yield the lowest accuracies, while the Text-Statistics feature achieve a 29.73% accuracy, using a linear SVM. This result is remarkable, as it significantly outperforms the Chroma features, and is nearly achieving the results of MFCCs (0.07% short) and Marsyas (0.8% difference), coming at a very low dimensionality of only 23 features, while they are fast in computation. All combinations of the Text-Statistics features with the Part-of-Speech and / or Rhyme features achieve better results than Text-Statistics features alone.

When combining the different audio-features with the lyrics-based feature sets, it can be noted that in any combination, we achieve higher results than with the lyrics features alone. Especially for SVM, those improvements are always statistically significant when we include the Text-statistics features, which is also the case for all but two combinations when applying Naïve Bayes classification. For the k-NN, there is almost always one combination of features that leads to significant improvement. The combination of MFCCs is the only one where we can achieve *significant* improvement with adding just the Rhyme and POS features on SVM and NB, not using the Text-statistics features.

Compared to the baseline results achieved with SSDs, all four combinations of SSDs with the text statistic features yield higher performance, and at least one (and even all four in the case of SVMs) are statistically significant. The highest accuracy values are obtained for an SSD and text-statistic feature combination (64.53%), which is 5.15%-points higher than the SSD-only value. It is interesting to note that adding part-of-speech and rhyme features does not help to improve on this result on SVMs, while it does on Naïve Bayes and 3-NN.

4.3.3   Small Database with Children's Music – Collection 660

Table 6 illustrates the results of adding the additional genre of 'Children's music' to the small collection, thus forming a database of 660 songs. First, it can be noted that, when compared to the results on the smaller collection, with SVM classification and linear kernel, the audio-only feature sets had mostly improved classification results (except Chroma and RH). The improvements range from 0.5% for RP to 2% for the SSD, which thus now achieves 61.36%. They are remarkable, as the classification task per-se has become a bit harder, with a minimal baseline of now 9.09%. The improvements thus already indicate that the new genre can be well captured by audio-only features. Again, combinations with the rhyme and style features can improve the results significantly in many combinations.

On this database, we also present results from the bag-of-words features. In fact, we show a number of bag-of-words feature sets of different feature dimensions, which were obtained using different parameters for the document frequency thresholding based feature selection. Using this feature set alone, with a still moderate dimensionality of 653 topic terms, the best results are at around 33% for both SVM and Naïve Bayes. Notably, k-NN has rather poor performance, and further degrades with higher dimensionality. Also for the other classifiers it has to be noted that with a rising dimensionality, the accuracy starts to degrade again. Interestingly, both SVM and Naïve Bayes on BOW with 653 features can outperform the audio-only features Marsyas, MFCC and Chroma, most of it statistically significant, except for SVMs on the Marsays feature set. Rhythm histograms are outperformed on the Naïve Bayes classifier, while Rhythm Patterns and SSD are significantly outperforming any of the bag-of-words features.

Also, it can be observed that adding the bag-of-words features can significantly improve the results obtained with the Marsyas features, even over the best combination of Marsyas with the rhyme and style features. Finally, adding bag-of-words to this aforementioned combination leads to a further improvement of more than 5%-points with SVM, thus totally more than 15%-points difference to the Marsyas-only features. Similar effects can be achieved for the other audio-only feature sets.

Regarding SSD features, the combination with the rhyme and style features again yields significant improvement on all classifiers. Combining them with the bag-of-words features can still yield better results than the SSD-only features, however, it leads to an improvement over the best combination with the rhyme and style features only on the Naïve Bayes classifier.

Finally, we want to examine the classification performance for each individual genre; for this, we train SVMs with a linear kernel on the SSD and the combination of SSD and Text-statistics feature set, which achieved the highest results. Table 7 gives the confusion matrix and the precision and recall values per class (in percent) for both feature sets, SSD on the left side, and SSD combined with Text-statistics on the right hand side.

With the audio features, high precision values can be achieved for the Children's music, R&B, Reggae, Punk Rock and Folk music, while Country, Slow Rock and especially Grunge perform poor.

When adding the Text-statistics features to the SSD features, eight out of eleven classes achieve a higher precision (of up to 25%), while the other three classes degrade in performance only by one percent; two out of those, namely Folk and R&B, however, gain 7% and 8%, resp, in recall. Overall, the average precision, as well as the recall and the F-measure[7], thus rise from values around 61% to approximately 67%. The biggest increase in precision is achieved for Hip-Hop, which improves from 63% to 88%; much of this increase is likely to be attributed to the 'words-per-minute' feature. Other genres that improve greatly in precision are Reggae and

---

[7] The F-Measure or F-score is a commonly used measure including both precision and recall. In our case, we specifically employ the F_1-measure, calculated as $\frac{2 \times precision \times recall}{precision + recall}$.

Table 6 Classification accuracies and results of significance testing for various combinations of lyrics features for the 660 song collection (*collection_660*). Statistically significant improvement or degradation over the resp. audio-features-only baseline (column-wise) is indicated by $(+)$ or $(-)$, respectively

Feature set	Dim.	3-NN	4-NN	5-NN	NB	SVM/lin
Chroma	48	15.94	18.21	19.03	17.94	22.06
MFCC	52	25.39	26.55	27.67	24.27	30.06
Mars.	68	28.33	29.94	30.21	27.33	31.88
RH	60	27.82	29.27	28.76	30.55	36.42
RP	1440	28.12	30.67	30.55	37.06	48.70
SSD	168	**49.18**	**50.15**	**51.97**	**44.21**	**61.36**
BOW_{59}	59	**15.24**	**15.85**	**15.06**	20.64	26.18
BOW_{150}	150	10.97	10.24	9.42	24.97	29.52
BOW_{194}	194	9.64	9.55	9.18	28.58	32.73
BOW_{653}	653	11.21	10.03	10.03	**32.58**	**33.52**
BOW_{1797}	1797	10.47	11.20	10.87	30.90	31.23
Mars. / Text-stat / POS	100	**32.55** +	**34.27** +	35.27 +	33.03 +	39.94 +
Mars. / Text-stat / Rhyme	97	30.73	32.67	34.39 +	33.06 +	41.33 +
Mars. / BOW_{248}	316	26.06	25.73	26.42	28.91	41.67 +
Mars. / BOW_{653}	721	17.00 -	19.18 -	21.64 -	33.27 +	42.61 +
Mars. / BOW_{194} / Text-stat	285	31.33	33.30	**35.48** +	32.36 +	44.52 +
Mars. / BOW_{653} / Text-stat	744	24.06 -	26.09 -	27.45	**34.45** +	**46.61** +
SSD / Text-stat	191	**53.42** +	54.06 +	**55.06** +	47.15 +	**66.27** +
SSD / Text-stat / Rhyme	197	**53.42** +	**54.12** +	54.91 +	**48.39** +	65.55 +
SSD / BOW_{14}	182	48.03	50.85	50.76	47.06 +	58.88
SSD / BOW_{573}	741	45.30	47.30	46.70 -	38.45 -	63.21
SSD / BOW_{385} / Text-stat	576	50.67	53.76	54.88	38.00 -	65.30 +
SSD / BOW_{10} / Text-stat / POS	210	49.33	51.36	53.97	48.24 +	62.09
SSD / BOW_{248} / Text-stat / POS / Rhyme	454	50.85	53.88	53.06	37.39 -	65.70 +

Pop, even though the latter still has a low absolute precision. Of special interest is also the genre of Children's music. We noted before that children's music has some specific characteristics. This manifests e.g. in a focus on vocals, and a very limited set of instruments used, mainly guitars and pianos. Therefore, this genre can be well identified with audio-based feature sets, and indeed has a high recall of 52 out of 60, or 87%, and a high precision of 78% already with the SSD features. However, even in such cases, combining the audio features with lyrics-based features can improve the performance, in this specific case by raising the recall to 92%, and the precision to 86%, when combined with the Text-statistics features. The number of songs wrongly assigned into this genre greatly reduces, from 15 to only 8 songs.

4.3.4 Large Database

To confirm our findings from the small database, we further performed experiments on the large collections (*collection_3000, collection_3120*). We again compare the results of the single audio and lyrics feature sets, and the combinations thereof.

Table 7 Confusion matrix on the *collection_660*: SSD (left) vs. SSD combined with Text-statistics (right). Precision and recall are measured per class.

			classified as								genre				classified as							
a	b	c	d	e	f	g	h	i	j	k		a	b	c	d	e	f	g	h	i	j	k
34	3	0	0	2	8	0	0	2	10	1	a = Country	38	5	0	0	2	6	0	0	1	8	0
9	39	0	1	1	4	0	0	0	5	1	b = Folk	6	43	1	0	0	1	2	0	0	7	0
0	2	47	0	1	4	1	0	1	4	0	c = Grunge	1	2	49	1	1	2	1	0	0	3	0
0	2	0	39	0	3	1	6	8	0	1	d = Hip-Hop	0	0	0	50	0	0	1	6	2	0	1
2	3	3	0	34	4	10	0	0	4	0	e = Metal	0	6	2	0	36	4	8	0	0	4	0
10	3	9	4	4	11	3	2	1	11	2	f = Pop	9	3	8	0	4	16	3	4	0	12	1
5	2	5	0	10	2	36	0	0	0	0	g = Punk Rock	4	2	6	0	9	1	37	0	0	1	0
2	0	0	10	0	3	0	40	2	1	2	h = R&B	3	0	0	3	0	3	0	45	5	1	0
0	1	0	7	0	1	0	2	45	0	4	i = Reggae	2	0	0	2	0	1	0	3	47	0	5
8	1	8	1	3	5	1	1	1	27	4	j = Slow Rock	7	1	5	1	5	9	0	2	2	26	2
1	0	0	0	0	1	0	1	3	2	52	k = Children's	0	1	0	0	0	0	0	0	3	1	55
47	69	65	63	62	23	69	76	71	42	77	Precision	54	68	69	88	63	37	71	75	78	41	86
57	65	78	65	57	18	6	67	75	45	87	Recall	63	72	82	83	6	27	62	75	78	43	92

As there is not much difference in the variations of the k-NN algorithms, we now only present the results of the best-performing of the tested versions, 5-NN. For the SVMs, we again used a linear kernel.

The three centre columns in Table 8 give an overview of the accuracies of the different feature sets. For the audio-only features, we can observe an increased accuracy for most of the features set and classifier combinations, as compared to the smaller collection. In the case of the best-performing SSDs on SVMs, the increase is of 7%-points to 66.35%. Similar patterns can be observed for the lyrics-based features, even though the flagship Text-statistics feature set achieves a 1% lower result on the SVM.

Also for the combination of the audio feature sets with the lyrics based features, a generally higher accuracy than on the smaller database can be noted, with total gains of 12.18% (Chroma), 8.71% (MFCCs), and 7.43% (Marsyas). The improvement over the SSD when combining them with the lyrics features is not as high as on the smaller collection – the accuracy raised to 68.91%, constituting an improvement of 2.55%-points, which is statistically significant. In general, it seems that the influence of part-of-speech and rhyme features is higher in this database, as they are more often part of the highest-performing feature set combination than in the smaller collection.

The right columns in Table 8 finally show a summary of the results on the large database, extended by adding about 110 songs from the children's music genre. The audio-only features generally perform a bit worse, between 0.1 and 0.8% when using SVMs, a bit more on some of the other classifiers. The same holds true for the rhyme and style features, though in their combinations among themselves, for some classifiers, the results are slightly, at most about 0.4%, higher than without the children's music genre. Similarly, most of the combinations of audio and lyrics features perform slightly better on this database.

Table 8 Classification accuracies and results of significance testing for the large collections. Statistically significant improvement or degradation over datasets (column-wise) is indicated by $(+)$ or $(-)$, respectively

Feature set	Dim.	Collection_3000			Collection_3120		
		NB	SVM	5-NN	NB	SVM	5-NN
Chroma	48	18.96	24.01	21.24	16.57	23.18	20.35
MFCC	52	27.64	33.57	31.27	26.55	32.66	31.21
Mars.	68	30.20	37.65	34.14	29.23	36.97	33.93
RP	1440	34.44	55.65	41.10	34.27	55.73	39.90
RH	60	29.26	35.05	34.44	29.03	34.17	33.86
SSD	168	**42.04**	**66.35**	**61.85**	**39.29**	**65.84**	**60.79**
Rhyme	6	16.68	16.11	16.97	16.57	16.09	17.70
POS	9	**23.67**	23.94	21.14	**23.60**	23.22	20.46
Text-stat	23	17.27	**28.70**	**24.78**	17.09	**28.16**	**24.79**
POS / Rhyme	15	**23.20**+	24.43-	21.66-	**23.27**+	24.55-	21.41-
Text-stat / POS	32	18.84+	31.23+	**25.91**	19.18+	31.41+	**25.72**
Text-stat / POS / Rhyme	38	20.15+	**31.24**+	25.10	20.88+	**31.64**+	25.47
Chroma / Text-stat	71	21.00+	32.61+	26.00+	20.64+	32.34+	25.97+
Chroma / Text-stat / POS	80	21.97+	35.95+	27.19+	22.10+	35.94+	26.44+
Chroma / Text-stat / POS / Rhyme	86	**22.31**+	**36.19**+	**27.35**+	**22.97**+	**36.54**+	**27.20**+
Chroma / Text-stat / Rhyme	77	21.53+	33.45+	26.06+	21.89+	33.45+	25.83+
Chroma / POS / Rhyme	63	21.48+	30.12+	25.19+	21.57+	30.01+	24.47+
MFCC / Text-stat	75	24.86-	40.48+	33.98+	24.35	40.40+	33.69+
MFCC / Text-stat / POS	84	26.03	41.71+	**35.50**+	26.21	41.76+	**34.94**+
MFCC / Text-stat / POS / Rhyme	90	26.83	**42.28**+	34.07+	27.55	**42.35**+	33.67+
MFCC / POS	61	**30.22**+	36.84+	32.52	29.42+	36.07+	31.80
MFCC / POS / Rhyme	67	30.13+	37.15+	32.10	**30.40**+	37.29+	31.44
Mars. / Text-stat	91	27.08-	43.44+	35.77+	27.13	43.44+	35.92+
Mars. / Text-stat / POS	100	28.33	44.98+	**37.17**+	28.50	44.91+	**36.32**+
Mars. / Text-stat / POS / Rhyme	106	29.70	**45.08**+	35.71	30.22	**45.38**+	36.11+
Mars. / POS / Rhyme	83	**32.67**+	41.54+	34.14	**32.91**+	41.82+	33.68
SSD / Text-stat	191	43.70+	68.57+	62.41	42.14+	68.38+	61.80
SSD / Text-stat / POS	200	44.29+	**68.91**+	**62.77**	42.86+	**68.94**+	61.44
SSD / Text-stat / POS / Rhyme	206	**44.51**+	68.35+	62.36	**43.44**+	68.36+	61.35
SSD / Text-stat / Rhyme	197	44.10+	68.00+	62.02	42.75+	68.01+	**61.81**

5 Beyond Audio and Lyrics

Much of today's research in Music Information Retrieval is driven by audio-only genres, and classification of pieces of music therein. However, user studies have revealed that this narrow focus poses certain problems. For example, *semantic genres* such as Christmas songs or love-songs, cannot be adequately captured by audio features, as they might comprise musical genres – Christmas songs can actually be classical music, pop songs, or punk rock. Christian Rock is a genre that can virtually only be detected via the song texts. Similarly, pop music is a genre that is generally difficult to grasp with only acoustical features, as the common property of pop music is maybe more in the orientation towards being commercial music, rather than in musical characteristics. Thus, it is important to incorporate additional modalities as sources for features describing music. Such sources can e.g. be the song lyrics, album covers, social web data, etc.

In this paper, we thus presented a set of rhyme and style features for automatic lyrics processing, namely features to capture characteristics such as rhyme, parts-of-speech, and text statistics of song lyrics. We further combined these new feature sets with the standard bag-of-words features and well-known feature sets for acoustic analysis of digital audio tracks. To show the positive effects of feature combination on classification accuracies in musical genre classification, we performed experiments on two test collections. A smaller collection, consisting of 600 songs was manually edited and contains high quality unabridged lyrics. We then extended this database by adding songs from the children's music genre, which are already well distinguishable on the audio features, and thus posed an interesting challenge on whether there could be further performance gains with this new dataset. We further compiled a larger test collection, comprising more than 3000 songs, which was again analysed in two flavours, with and without the children's music. Using only automatically fetched lyrics, we achieved similar results in genre classification.

The most notable results reported in this paper are statistically significant improvements in musical genre classification. We outperformed both audio features alone as well as their combination with simple bag-of-words features. We conclude that combination of feature sets is beneficial in two ways: a) possible reduction in dimensionality, and b) statistically significant improvements in classification accuracies. Future work hence is motivated by the promising results presented in this paper. Noteworthy future research areas in terms of machine learning are on more sophisticated ways of feature combination via ensemble classifiers, which pay special attention to the unique properties of single modalities and the different characteristics of certain genres in specific parts of the feature space. Additionally, a more comprehensive investigation of feature selection techniques and the impact of individual/global feature selection might further improve results.

Another topic for future research is the continued expansion of modalities and types of feature representations to be used for music analysis. A 'glass-ceiling' of achievable performance in regards to music information retrieval based on naïve timbral audio features only is discussed in [2]. It is further suggested that more high-level musical features are needed to overcome this limitation. While improved audio feature sets have been designed to address this issue, it is certainly of interest to look beyond the audio-only domain.

Steps in this direction have been discussed in this paper. Yet, we need to expand way beyond this scope. Album covers, for example, are carefully designed for specific target groups. Searching for music in a record shop is facilitated by browsing through album covers. There, album covers can, and have to, reveal very quickly the musical content of the album, and are thus used as strong visual clues [7]. Due to well-developed image recognition abilities of humans, this task can be performed very efficiently, much faster than listening to excerpts of the songs. Also, [4] suggests that 'an essential part of human psychology is the ability to identify music, text, images or other information based on associations provided by contextual information of different media'. It further suggests that a well-chosen cover of a book can reveal it's contents, or that lyrics of a familiar song can remind one of the song's melody.

However, capturing the semantic meaning of album covers is a challenging task, requiring advanced pattern recognition and image retrieval methods. Concepts in the covers are more difficult to grasp than by simple colour histograms (even though for some genres, such as Gothic with a focus on dark/black colours, this feature might be a suitable candidate). More than that, it seems necessary to employ algorithms to detect the fonts used, face recognition to detect whether or not the singer or band feature on the cover, what scenery is depicted to e.g. indicate folk music, or which objects, instruments, etc. are present, down to understanding the sentiment and emotions of cover images.

This breath of information extends way beyond a cover, the song itself, or its recording. It encompasses cultural aspects and community feelings as expressed by subculture language, clothes and other aspects of social groupings.

Music may seem to be mono modal, audio only at first glance. Yet, it is inherently multimodal, living from, playing with and serving information on a multitude of layers. It needs to be appreciated and covered in all its multimodal complexities if we want to fully explore its richness and do justice to its versatility.

References

1. Allamanche, E., Herre, J., Hellmuth, O., Fröba, B., Kastner, T., Cremer, M.: Content-based identification of audio material using MPEG-7 low level description. In: Proceedings of the International Symposium on Music Information Retrieval (ISMIR), Bloomington, IN, USA, October 15-17, pp. 197–204 (2001)
2. Aucouturier, J.-J., Pachet, F.: Improving timbre similarity: How high is the sky? Journal of Negative Results in Speech and Audio Sciences 1(1) (2004)
3. Baumann, S., Pohle, T., Vembu, S.: Towards a socio-cultural compatibility of MIR systems. In: Proceedings of the 5th International Conference of Music Information Retrieval (ISMIR 2004), Barcelona, Spain, October 10-14, pp. 460–465 (2004)
4. Brochu, E., de Freitas, N., Bao, K.: The sound of an album cover: Probabilistic multimedia and IR. In: Proceedings of the 9th International Workshop on Artificial Intelligence and Statistics, Key West, FL, USA, January 3-6 (2003)
5. Cavnar, W.B., Trenkle, J.M.: N-gram-based text categorization. In: Proceedings of the 3rd Annual Symposium on Document Analysis and Information Retrieval (SDAIR 1994), Las Vegas, USA, pp. 161–175 (1994)
6. Crysandt, H., Wellhausen, J.: Music classification with MPEG-7. In: Proceedings of SPIE-IS&T Electronic Imaging, Santa Clara (CA), USA, January 2003. Storage and Retrieval for Media Databases, vol. 5021, pp. 307–404. The International Society for Optical Engineering (2003)
7. Cunningham, S.J., Reeves, N., Britland, M.: An ethnographic study of music information seeking: implications for the design of a music digital library. In: Proceedings of the 3rd ACM/IEEE-CS joint conference on Digital libraries (JCDL 2003), Washington, DC, USA, pp. 5–16. IEEE Computer Society, Los Alamitos (2003)

8. Stephen Downie, J.: Music Information Retrieval. In: Annual Review of Information Science and Technology, vol. 37, pp. 295–340. Information Today, Medford (2003)
9. Elovitz, H.S., Johnson, R., McHugh, A., Shore, J.E.: Letter-to-sound rules for automatic translation of English text to phonetics. IEEE Transactions on Acoustics, Speech and Signal Processing 24(6), 446–459 (1976)
10. Foote, J.: An overview of audio information retrieval. Multimedia Systems 7(1), 2–10 (1999)
11. Goto, M.: A chorus section detection method for musical audio signals and its application to a music listening station. IEEE Transactions on Audio, Speech & Language Processing 14(5), 1783–1794 (2006)
12. Iskandar, D., Wang, Y., Kan, M.-Y., Li, H.: Syllabic level automatic synchronization of music signals and text lyrics. In: Proceedings of the ACM 14th International Conference on Multimedia (MM 2006), New York, NY, USA, pp. 659–662(2006)
13. Knees, P., Schedl, M., Pohle, T., Widmer, G.: An Innovative Three-Dimensional User Interface for Exploring Music Collections Enriched with Meta-Information from the Web. In: Proceedings of the ACM 14th International Conference on Multimedia (MM2006), Santa Barbara, California, USA, October 23-26, pp. 17–24 (2006)
14. Knees, P., Schedl, M., Widmer, G.: Multiple lyrics alignment: Automatic retrieval of song lyrics. In: Proceedings of 6th International Conference on Music Information Retrieval (ISMIR 2005), London, UK, September 11-15, pp. 564–569 (2005)
15. Laurier, C., Grivolla, J., Herrera, P.: Multimodal music mood classification using audio and lyrics. In: Proceedings of the Seventh International Conference on Machine Learning and Applications (ICMLA 2008), San Diego, CA, USA, December 11-13, pp. 688–693 (2008)
16. Li, T., Ogihara, M., Li, Q.: A comparative study on content-based music genre classification. In: Proceedings of the International ACM Conference on Research and Development in Information Retrieval (SIGIR), Toronto, Canada, pp. 282–289 (2003)
17. Lidy, T., Rauber, A.: Evaluation of feature extractors and psycho-acoustic transformations for music genre classification. In: Proceedings of the 6th International Conference on Music Information Retrieval (ISMIR 2005), London, UK, September 11-15, pp. 34–41 (2005)
18. Lidy, T., Rauber, A., Pertusa, A., Inesta, J.M.: Improving genre classification by combination of audio and symbolic descriptors using a transcription system. In: Proceedings of the International Conference on Music Information Retrieval (ISMIR), Vienna, Austria, September 23-27, pp. 61–66 (2007)
19. Logan, B.: Mel frequency cepstral coefficients for music modeling. In: Proceedings of the International Symposium on Music Information Retrieval (ISMIR), Plymouth, Massachusetts, USA, October 23-25 (2000)
20. Logan, B., Kositsky, A., Moreno, P.: Semantic analysis of song lyrics. In: Proceedings of the IEEE International Conference on Multimedia and Expo (ICME 2004), Taipei, Taiwan, June 27-30, pp. 827–830 (2004)
21. Logan, B., Salomon, A.: A music similarity function based on signal analysis. In: Proceedings of the IEEE International Conference on Multimedia and Expo (ICME), Tokyo, Japan (August 2001)
22. Mahedero, J.P.G., Martínez, Á., Cano, P., Koppenberger, M., Gouyon, F.: Natural language processing of lyrics. In: Proceedings of the ACM 13th International Conference on Multimedia (MM 2005), New York, NY, USA, pp. 475–478 (2005)
23. Mayer, R., Neumayer, R., Rauber, A.: Combination of audio and lyrics features for genre classification in digital audio collections. In: Proceedings of the ACM Multimedia 2008, October 27-31, pp. 159–168. ACM, New York (2008)

24. Mayer, R., Neumayer, R., Rauber, A.: Rhyme and style features for musical genre classification by song lyrics. In: Proceedings of the 9th International Conference on Music Information Retrieval (ISMIR 2008), Philadelphia, PA, USA, September 14-18 (2008)
25. Neumayer, R., Rauber, A.: Integration of text and audio features for genre classification in music information retrieval. In: Proceedings of the 29th European Conference on Information Retrieval (ECIR 2007), Rome, Italy, April 2-5, pp. 724–727 (2007)
26. Neumayer, R., Rauber, A.: Multi-modal music information retrieval - visualisation and evaluation of clusterings by both audio and lyrics. In: Proceedings of the 8th Conference Recherche d'Information Assistée par Ordinateur (RIAO 2007), Pittsburgh, PA, USA, May 29th - June 1 (2007)
27. Orio, N.: Music retrieval: A tutorial and review. Foundations and Trends in Information Retrieval 1(1), 1–90 (2006)
28. Pampalk, E., Flexer, A., Widmer, G.: Hierarchical organization and description of music collections at the artist level. In: Proceedings of the 9th European Conference on Research and Advanced Technology for Digital Libraries (ECDL 2005), pp. 37–48 (2005)
29. Pampalk, E., Rauber, A., Merkl, D.: Content-based Organization and Visualization of Music Archives. In: Proceedings of the ACM 10th International Conference on Multimedia (MM 2002), Juan les Pins, France, December 1-6, pp. 570–579 (2002)
30. Ross Quinlan, J.: Induction of decision trees. Machine Learning 1(1), 81–106 (1986)
31. Rabiner, L., Juang, B.-H.: Fundamentals of Speech Recognition. Prentice-Hall, Englewood Cliffs (1993)
32. Rauber, A., Pampalk, E., Merkl, D.: Using psycho-acoustic models and self-organizing maps to create a hierarchical structuring of music by musical styles. In: Proceedings of the 3rd International Symposium on Music Information Retrieval (ISMIR 2002), Paris, France, October 13-17, pp. 71–80 (2002)
33. Rauber, A., Pampalk, E., Merkl, D.: The SOM-enhanced JukeBox: Organization and visualization of music collections based on perceptual models. Journal of New Music Research 32(2), 193–210 (2003)
34. Salton, G.: Automatic text processing – The Transformation, Analysis, and Retrieval of Information by Computer. Addison-Wesley Longman Publishing Co., Inc., Amsterdam (1989)
35. Shepard, R.N.: Circularity in judgments of relative pitch. The Journal of the Acoustical Society of America 36(12), 2346–2353 (1964)
36. Tzanetakis, G., Cook, P.: Marsyas: A framework for audio analysis. Organized Sound 4(30), 169–175 (2000)
37. Tzanetakis, G., Cook, P.: Sound analysis using MPEG compressed audio. In: Proceedings of the International Conference on Audio, Speech and Signal Processing (ICASSP), Istanbul, Turkey (2000)
38. Tzanetakis, G., Cook, P.: Musical genre classification of audio signals. IEEE Transactions on Speech and Audio Processing 10(5), 293–302 (2002)
39. Xiong, Z., Radhakrishnan, R., Divakaran, A., Huang, T.S.: Comparing MFCC and MPEG-7 audio features for feature extraction, maximum likelihood HMM and entropic prior HMM for sports audio classification. In: Proceedings of the International Conference on Multimedia and Expo, ICME (2003)
40. Yang, D., Lee, W.: Disambiguating music emotion using software agents. In: Proceedings of the International Conference on Music Information Retrieval (ISMIR), Barcelona, Spain (October 2004)
41. Zhu, Y., Chen, K., Sun, Q.: Multimodal content-based structure analysis of karaoke music. In: Proceedings of the ACM 13th International Conference on Multimedia (MM 2005), Singapore, pp. 638–647 (2005)

Melodic Grouping in Music Information Retrieval: New Methods and Applications

Marcus T. Pearce, Daniel Müllensiefen, and Geraint A. Wiggins

Abstract. We introduce the MIR task of segmenting melodies into phrases, summarise the musicological and psychological background to the task and review existing computational methods before presenting a new model, IDyOM, for melodic segmentation based on statistical learning and information-dynamic analysis. The performance of the model is compared to several existing algorithms in predicting the annotated phrase boundaries in a large corpus of folk music. The results indicate that four algorithms produce acceptable results: one of these is the IDyOM model which performs much better than naive statistical models and approaches the performance of the best-performing rule-based models. Further slight performance improvement can be obtained by combining the output of the four algorithms in a hybrid model, although the performance of this model is moderate at best, leaving a great deal of room for improvement on this task.

1 Introduction

The segmentation of music into meaningful units is a fundamental (pre-)processing step for many MIR applications including melodic feature computation, melody indexing, and retrieval of melodic excerpts. Here, we focus on the grouping of musical elements into contiguous segments that occur sequentially in time or, to put it another way, the identification of boundaries between the final element of one

Marcus T. Pearce
Department of Computing, Goldsmiths, University of London, SE14 6NW, UK
e-mail: m.pearce@gold.ac.uk

Daniel Müllensiefen
Department of Computing, Goldsmiths, University of London, SE14 6NW, UK
e-mail: d.mullensiefen@gold.ac.uk

Geraint A. Wiggins
Department of Computing, Goldsmiths, University of London, SE14 6NW, UK
e-mail: g.wiggins@gold.ac.uk

Z.W. Raś and A.A. Wieczorkowska (Eds.): Adv. in Music Inform. Retrieval, SCI 274, pp. 365–389.
springerlink.com   © Springer-Verlag Berlin Heidelberg 2010

segment and the first element of the subsequent one. This way of structuring a musical surface is usually referred to as *grouping* (Lerdahl & Jackendoff, 1983) or *segmentation* (Cambouropoulos, 2006) and is distinguished from the grouping of musical elements that occur simultaneously in time, a process usually referred to as *streaming* (Bregman, 1990). In musical terms, the kinds of groups we shall consider might correspond with motifs, phrases, sections and other aspects of musical form, so the scope is rather general. Just as speech is perceptually segmented into phonemes, and then words which subsequently provide the building blocks for the perception of phrases and complete utterances (Brent, 1999b; Jusczyk, 1997), motifs or phrases in music are identified by listeners, stored in memory and made available for inclusion in higher-level structural groups (Lerdahl & Jackendoff, 1983; Peretz, 1989; Tan et al., 1981). The low-level organisation of the musical surface into groups allows the use of these primitive perceptual units in more complex structural processing and may alleviate demands on memory.

We restrict ourselves primarily to research on symbolic representations of musical structure that take discrete events (individual musical notes in this work) as their musical surface (Jackendoff, 1987). Working at this level of abstraction, the task is to gather events (represented in metrical time as they might be in a musical score) into sequential groups. Research on segmentation from sub-symbolic or acoustic representations of music is not discussed as it generally operates either at the level of larger sections of music differing in instrumentation (e.g., Abdallah et al., 2006) or at the lower level of separating a continuous audio stream into individual note events (e.g., Gjerdingen, 1999; Todd, 1994). Furthermore, the present work emphasises melody (although not exclusively) reflecting the predominant trends in theoretical and computational treatments of perceived grouping structure in music.

Grouping structure is generally agreed to be logically independent of metrical structure (Lerdahl & Jackendoff, 1983) and some evidence for a separation between the psychological processing of the two kinds of structure has been found in cognitive neuropsychological (Liegeoise-Chauvel et al., 1998; Peretz, 1990) and neuroimaging research (Brochard et al., 2000). In practice, however, metrical and grouping structure are often intimately related and both are likely to serve as inputs to the processing of more complex musical structures (Lerdahl & Jackendoff, 1983). Nonetheless, most theoretical, empirical and computational research has considered the perception of grouping structure independently of metrical structure (Stoffer, 1985, and Temperley, 2001, being notable exceptions).

Melodic segmentation is a key task in the storage and retrieval of musical information. The melodic phrase is often considered one of the most important basic units of musical content (Lerdahl & Jackendoff, 1983) and many large electronic corpora of music are structured or organised by phrases, for example, the Dictionary of Musical Themes by Barlow & Morgenstern (1949), the Essen Folksong Collection (EFSC, Schaffrath, 1995) or the RISM collection (RISM-ZENTRALREDAKTION, RISM-ZENTRALREDAKTION). At the same time, melodic grouping is thought to be an important part of the perceptual processing of music (Deliège, 1987; Frankland & Cohen, 2004; Peretz, 1989). It is also fundamental to the phrasing of a melody when sung or played: melodic segmentation is a task that musicians and musical listeners perform regularly in their everyday musical practice.

Several algorithms have been proposed for the automated segmentation of melodies. These algorithms differ in their modelling approach (supervised learning, unsupervised learning, music-theoretic rules), and in the type of information they use (global or local). In this chapter, we review these approaches before introducing a new statistical model of melodic segmentation and comparing its performance to several existing algorithms on a melody segmentation task. The motivation for this model comparison is two-fold: first, we are interested in the performance differences between different types of model; and second, we aim to build a hybrid model that achieves superior performance by combining boundary predictions from different models.

2 Background

The segmentation of melodies is a cognitive process performed by the minds and brains of listeners based on their musical and auditory dispositions and experience. Therefore, an MIR system must segment melodies in a musically and psychologically informed way if it is to be successful. Before reviewing computational models of melodic segmentation and their use in MIR, we consider it appropriate to survey the musicological and psychological literature that has informed the development of these models.

2.1 Music-Theoretic Approaches

2.1.1 A Generative Theory of Tonal Music

Melodic grouping has traditionally been modelled through the identification of local discontinuities or changes between events in terms of temporal proximity, pitch, duration and dynamics (Cambouropoulos, 2001; Lerdahl & Jackendoff, 1983; Temperley, 2001). Perhaps the best known examples are the Grouping Preference Rules (GPRs) of the Generative Theory of Tonal Music (GTTM, Lerdahl & Jackendoff, 1983). The most widely studied of these GPRs predict that phrase boundaries will be perceived between two melodic events whose temporal proximity is less than that of the immediately neighbouring events due to a slur, a rest (GPR 2a) or a relatively long inter-onset interval or IOI (GPR 2b) or when the transition between two events involves a greater change in register (GPR 3a), dynamics (GPR 3b), articulation (GPR 3c) or duration (GPR 3d) than the immediately neighbouring transitions. Another rule, GPR 6, predicts that grouping boundaries are perceived in accordance with musical parallelism (e.g., at parallel points in a metrical hierarchy or after a repeated motif). The GPRs were directly inspired by the principles of proximity (GPR 2) and similarity (GPR 3) developed to account for figural grouping in visual perception by the Gestalt school of psychology (e.g., Koffka, 1935).

2.1.2   The Implication-Realisation Theory

Narmour (1990, 1992) presents the *Implication-Realisation* (IR) theory of music cognition which, like GTTM, is intended to be general (although the initial presentation was restricted to melody). However, while GTTM operates statically on an entire piece of music, the IR theory emphasises the dynamic processes involved in perceiving music as it occurs in time. The theory posits two distinct perceptual systems: the *bottom-up* system is held to be hard-wired, innate and universal while the *top-down system* is held to be learnt through musical experience. The two systems may conflict and, in any given situation, one may over-ride the implications generated by the other.

In the bottom-up system, sequences of melodic intervals vary in the degree of *closure* that they convey. An interval which is unclosed (i.e., one that generates expectations for a subsequent interval) is said to be an *implicative interval* and generates expectations for the following interval, termed the *realised interval*. The expectations generated by implicative intervals for realised intervals are described by Narmour (1990) in terms of several principles of continuation which are, again, influenced by the Gestalt principles of proximity, similarity, and good continuation. Strong closure, however, signifies the termination of ongoing melodic structure (i.e., a boundary) and the melodic groups formed either side of the boundary thus created can share different amounts of structure depending on the degree of closure conveyed. Furthermore, structural notes marked by strong closure at one level can *transform* to a higher level, itself amenable to analysis as a musical surface in its own right, thus allowing for the emergence of hierarchical levels of structural description of a melody.

2.2   *Psychological Studies*

Early studies of musical segmentation (Gregory, 1978; Sloboda & Gregory, 1980; Stoffer, 1985) provided basic evidence that listeners perceptually organise melodies into structural groups using a click localisation paradigm adapted from research on perceived phrase structure in spoken language (Fodor & Bever, 1965; Ladefoged & Broadbent, 1960). More recently, two kinds of experimental task have been used to study perceptual grouping in music.

The first is a short-term memory recognition paradigm introduced by Dowling (1973), based on studies of phrase perception in language (Bower, 1970; Waugh & Norman, 1965). In a typical experiment listeners are first presented with a musical stimulus containing one or more hypothesised boundaries before being presented with a short excerpt (the probe) and asked to indicate whether it appeared in the stimulus. The critical probes either border on or straddle a hypothesised boundary and it is expected that due to perceptual grouping, the former will be recalled more accurately or efficiently than the latter. Dowling's original experiment demonstrated that silence contributes to the perception of melodic segment boundaries. Using the same paradigm, Tan et al. (1981) demonstrated the influence of harmonic closure

Table 1 The quantification by Frankland & Cohen (2004) of GTTM's grouping preference rules which identify boundaries between notes based on their properties (n) including local proximity to other notes (GPR 2) or the extent to which they reflect local changes in pitch or duration (GPR 3). $\perp$ indicates that the result is undefined.

GPR	Description	n	Boundary Strength
2a	Rest		absolute length of rest (semibreve = 1.0)
2b	Attack-point	length	$\begin{cases} 1.0 - \frac{n_1+n_3}{2\times n_2} & \text{if } n_2 > n_3 \wedge n_2 > n_1 \\ \perp & \text{otherwise} \end{cases}$
3a	Register change	pitch height	$\begin{cases} 1.0 - \frac{|n_1-n_2|+|n_3-n_4|}{2\times|n_2-n_3|} & \begin{array}{l} \text{if } n_2 \neq n_3 \wedge \\ |n_2-n_3| > |n_1-n_2| \wedge \\ |n_2-n_3| > |n_3-n_4| \end{array} \\ \perp & \text{otherwise} \end{cases}$
3d	Length change	length	$1.0 - \begin{cases} n_1/n_3 & \text{if } n_3 \geq n_1 \\ n_3/n_1 & \text{if } n_3 < n_1 \end{cases}$

(e.g., a cadence to the tonic chord) with an effect of musical training such that musicians were more sensitive to this parameter than non-musicians.

In the second paradigm, subjects provide explicit judgements of boundary locations while listening to the musical stimulus. The indicated boundaries are subsequently analysed to discover what principles guide perceptual segmentation. Using this approach with short musical excerpts, Deliège (1987) found that musicians and (to a lesser extent) non-musicians identify segment boundaries in accordance with the GPRs of GTTM (Lerdahl & Jackendoff, 1983) especially those relating to rests or long notes and changes in timbre or dynamics. These factors have also been found to be important in large-scale segmentation by musically-trained listeners of piano works composed by Stockhausen and Mozart (Clarke & Krumhansl, 1990). Frankland & Cohen (2004) collected explicit boundary judgements from participants listening to six melodies (nursery rhymes and classical themes) and compared these to the boundaries predicted by quantitative implementations of GPRs 2a, 2b, 3a and 3d (see Table 1). The results indicated that GPR 2b (Attack-point) produced consistently strong correlations with the empirical boundary profiles, while GPR 2a (Rest) also received support in the one case where it applied. No empirical support was found for GPRs 3a (Register Change) and 3d (Length change).

Given the differences between these two experimental paradigms, it is not certain that they probe the same cognitive systems. Peretz (1989) addressed this question by comparing both methods on one set of stimuli (French folk melodies). The judgement paradigm (online, explicit) showed that musicians and non-musicians responded significantly more often in accordance with GPR 3d (Length change) than they did with GPR 3a (Register Change). However, the recognition-memory paradigm (offline, implicit) showed no effect of boundary type for either group of participants. To test the possibility that this discrepancy is due to a loss of information in the offline probe-recognition task, Peretz carried out a third experiment in which participants listened to a probe followed by the melody and were asked to indicate as quickly and accurately as possible whether the probe occurred in the melody. As predicted, the results demonstrated an influence of GPR 3d, but not 3a, on boundary perception. In contrast to these results, however, Frankland & Cohen (2004) found no major difference between the results of their explicit judgement task and a retrospective recognition-memory task using the same materials.

Many questions remain open and further empirical study is necessary to fully understand perceptual grouping. Nonetheless, psychological research has guided the development of computational models of melodic segmentation, which can be applied to practical tasks in MIR.

2.3 Computational Models

Tenney & Polansky (1980) were perhaps the first to propose formal models of melodic segmentation based on Gestalt-like rules, which became the dominant paradigm in the years to come. In this section, we review three models developed within this tradition: quantified versions of the GPRs from GTTM (Frankland & Cohen, 2004); the Local Boundary Detection Model (Cambouropoulos, 2001); and Grouper (Temperley, 2001). We also summarise previous studies that have evaluated the comparative performance of some of these models of melodic segmentation. Recently, there has been increasing interest in using machine learning to build models that learn about grouping structure, in either a supervised or unsupervised manner, through exposure to large bodies of data (Bod, 2001; Brent, 1999a; Ferrand et al., 2003; Saffran et al., 1999). The model we present follows this tradition and we include some related work in our review. In another direction, some researchers have combined Gestalt-like rules with higher-level principles based on parallelism and music structure (Ahlbäck, 2004; Cambouropoulos, 2006) in models which are mentioned for the sake of completeness but not reviewed in detail.

2.3.1 Grouping Preference Rules

Inspired by the GTTM, Frankland & Cohen (2004) quantified GPRs 2a, 2b, 3a and 3d as shown in Table 1. Since a slur is a property of the IOI while a rest is an absence of sound following a note, they argued that these two components of GPR 2a should be separated and, in fact, only quantified the rest aspect. Since GPRs 2a (Rest), 2b (Attack-point) and 3d (Length change) concern perceived duration, they were based on linearly scaled time in accordance with psychoacoustic research (Allan, 1979). Finally, a natural result of the individual quantifications is that they can be combined using multiple regression (a multivariate extension to linear correlation, Howell, 2002) to quantify the implication contained in GPR 4 (Intensification) that co-occurrences of two or more aspects of GPRs 2 and 3 lead to stronger boundaries.

2.3.2 The Local Boundary Detection Model

Cambouropoulos (2001) proposes a model related to the quantified GPRs in which boundaries are associated with any local change in interval magnitudes. The

Local Boundary Detection Model (LBDM) consists of a *change* rule, which assigns boundary strengths in proportion to the degree of change between consecutive intervals, and a *proximity* rule, which scales the boundary strength according to the size of the intervals involved. The LBDM operates over several independent parametric melodic profiles $P_k = [x_1, x_2, \ldots, x_n]$ where $k \in \{$pitch, ioi, rest$\}$, $x_i > 0, i \in \{1, 2, \ldots, n\}$ and the boundary strength at interval x_i (a pitch interval in semitones, inter-onset interval, or offset-to-onset interval) is given by:

$$s_i = x_i \times (r_{i-1,i} + r_{i,i+1}) \tag{1}$$

where the degree of change between two successive intervals:

$$r_{i,i+1} = \begin{cases} \frac{|x_i - x_{i+1}|}{x_i + x_{i+1}} & \text{if } x_i + x_{i+1} \neq 0 \wedge x_i, x_{i+1} \geq 0 \\ 0 & \text{if } x_i = x_{i+1} = 0. \end{cases} \tag{2}$$

For each parameter k, the boundary strength profile $S_k = [s_1, s_2, \ldots, s_n]$ is calculated and normalised in the range $[0, 1]$. A weighted sum of the boundary strength profiles is computed using weights derived by trial and error (.25 for *pitch* and *rest*, and .5 for *ioi*), and boundaries are predicted where the combined profile exceeds a threshold which may be set to any reasonable value (Cambouropoulos used a value such that 25% of notes fell on boundaries).

Cambouropoulos (2001) found that the LBDM obtained a recall of 63-74% of the boundaries marked on a score by a musician (depending on the threshold and weights used) although precision was lower at 55%. In further experiments, it was demonstrated that notes falling before predicted boundaries were more often lengthened than shortened in pianists' performances of Mozart piano sonatas and a Chopin étude. This was also true of the penultimate notes in the predicted groups.

More recently, Cambouropoulos (2006) proposed a complementary model which identifies instances of melodic repetition (or parallelism) and computes a pattern segmentation profile. While repetitions of melodic patterns are likely to contribute to the perception of grouping (see GPR 6 above), this model is not yet a fully developed model of melodic segmentation as it operates at a "local level (i.e. within a time window rather than [on] a whole piece)" (Emilios Cambouropoulos, personal email communication, 09/2007).

2.3.3 Grouper

Temperley (2001) introduces a model called *Grouper* which accepts as input a melody, in which each note is represented by its onset time, off time, chromatic pitch and level in a metrical hierarchy (which may be computed using a beat-tracking algorithm or computed from the time signature and bar lines if these are available), and returns a single, exhaustive partitioning of the melody into non-overlapping groups. The model operates through the application of three *Phrase Structure Preference Rules* (PSPRs):

PSPR 1 (Gap Rule):   prefer to locate phrase boundaries at (a) large IOIs and (b) large offset-to-onset intervals (OOI); PSPR 1 is calculated as the sum of the IOI and OOI divided by the mean IOI of all previous notes;

PSPR 2 (Phrase Length Rule):   prefer phrases with about 10 notes, achieved by penalising predicted phrases by $|(\log_2 N) - \log_2 10|$ where N is the number of notes in the predicted phrase – the preferred phrase length is chosen *ad hoc* (see Temperley, 2001, p. 74), to suit the corpus of music being studied (in this case Temperley's sample of the EFSC) and therefore may not be general;

PSPR 3 (Metrical Parallelism Rule):   prefer to begin successive groups at parallel points in the metrical hierarchy (e.g., both on the first beat of the bar).

The first rule is another example of the Gestalt principle of temporal proximity (cf. GPR 2 above) while the third is related to GPR 6; the second was determined through an empirical investigation of the typical phrase lengths in a collection of folk songs. The best analysis of a given piece is computed offline using a dynamic programming approach where candidate phrases are evaluated according to a weighted combination of the three rules. The weights were determined through trial and error. Unlike the other models, this procedure results in binary segmentation judgements rather than continuous boundary strengths. By way of evaluation, Temperley used Grouper to predict the phrase boundaries marked in 65 melodies from the EFSC, a collection of several thousand folk songs with phrase boundaries annotated by expert musicologists, achieving a recall of .76 and a precision of .74.

2.3.4   Data Oriented Parsing

Bod (2001) argues for a supervised learning approach to modelling melodic grouping structure as an alternative to the rule-based approach. He examined three grammar induction algorithms originally developed for automated language parsing in computational linguistics: first, the treebank grammar learning technique which reads all possible context free rewrite rules from the training set and assigns each a probability proportional to its relative frequency in the training set (Manning & Schütze, 1999); second, the Markov grammar technique which assigns probabilities to context free rules by decomposing the rule and its probability by a Markov process, allowing the model to estimate the probability of rules that have not occurred in the training set (Collins, 1999); and third, a Markov grammar augmented with a Data Oriented Parsing (DOP, Bod, 1998) method for conditioning the probability of a rule over the rule occurring higher in the parse tree. A best-first parsing algorithm based on Viterbi optimisation (Rabiner, 1989) was used to generate the most probable parse for each melody in the test set given each of the three models. Bod (2001) evaluated the performance of these three algorithms in predicting the phrase boundaries in the EFSC using F1 scores (Witten & Frank, 1999). The results demonstrated that the treebank technique yielded moderately high precision but very low recall ($F1 = .07$), the Markov grammar yielded slightly lower precision but much higher recall ($F1 = .71$) while the Markov-DOP technique yielded the highest precision and recall ($F1 = .81$). A qualitative examination of the folk song data revealed several cases (15% of the phrase boundaries in the test set) where the annotated phrase boundary cannot be accounted for by Gestalt principles but is predicted by the Markov-DOP parser.

2.3.5 Transition Probabilities and Pointwise Mutual Information

In research on language acquisition, it has been shown that infants and adults reliably identify grouping boundaries in sequences of synthetic syllables on the basis of statistical cues (Saffran et al., 1996). In these experiments participants are exposed to long, isochronous sequences of syllables where the only reliable cue to boundaries between groups of syllables consist of higher transition probabilities within than between groups. A *transition (or digram) probability* (TP) is the conditional probability of an element e_i at index $i \in \{2, \ldots, j\}$ in a sequence e_1^j of length j given the preceding element e_{i-1}:

$$p(e_i|e_{i-1}) = \frac{count(e_{i-1}^i)}{count(e_{i-1})}. \tag{3}$$

where e_m^n is the subsequence of e between indices m and n, e_m is the element at index m of the sequence e and $count(x)$ is the number of times that x appears in a training corpus. Further research using the same experimental paradigm has demonstrated that infants and adults use the implicitly learnt statistical properties of pitch (Saffran et al., 1999), pitch interval (Saffran & Griepentrog, 2001) and scale degree (Saffran, 2003) sequences to identify segment boundaries on the basis of higher digram probabilities within than between groups.

In a comparison of computational methods for word identification in unsegmented speech, Brent (1999a) quantified these ideas in a model that puts a word boundary between phonemes whenever the transition probability at e_i is lower than at both e_{i-1} and e_{i+1}. Brent also introduced a related model that replaces digram probabilities with *pointwise mutual information* (PMI), $I(e_i, e_{i-1})$, which measures how much the occurrence of one event reduces the model's uncertainty about the co-occurrence of another event (Manning & Schütze, 1999) and is defined as:

$$I(e_i, e_{i-1}) = \log_2 \frac{p(e_{i-1}^i)}{p(e_i)p(e_{i-1})}. \tag{4}$$

While digram probabilities are asymmetrical with respect to the order of the two events, pointwise mutual information is symmetrical in this respect.[1] Brent (1999a) found that the pointwise mutual information model outperformed the transition probability model in predicting word boundaries in phonemic transcripts of

---

[1] Manning & Schütze (1999) note that pointwise mutual information is biased in favour of low-frequency events inasmuch as, all other things being equal, I will be higher for digrams composed of low-frequency events than for those composed of high-frequency events. In statistical language modelling, pointwise mutual information is sometimes redefined as $count(xy)I(x,y)$ to compensate for this bias.

phonemically-encoded infant-directed speech from the CHILDES collection (MacWhinney & Snow, 1985).

Brent (1999a) implemented these models such that a boundary was placed whenever the statistic (TP or PMI) was higher at one phonetic location than in the immediately neighbouring locations. By contrast, here we construct a boundary strength profile P at each note position i for each statistic $S = \{TP, PMI\}$ such that:

$$P_i = \begin{cases} \frac{2S_i}{S_{i-1}+S_{i+1}} & \text{if } S_i > S_{i-1} \wedge S_i > S_{i+1} \\ 0 & \text{otherwise.} \end{cases} \tag{5}$$

2.3.6 Model Comparisons

The models reviewed above differ along several different dimensions. For example, the GPRs, LBDM and Grouper use rules derived from expert musical knowledge while DOP and TP/PMI rely on learning from musical examples. Looking in more detail, DOP uses supervised training while TP/PMI uses unsupervised induction of statistical regularities. Along another dimension, the GPRs, LBDM and TP/PMI predict phrase boundaries locally while Grouper and DOP attempt to find the best segmentation of an entire melody.

Most of these models were evaluated to some extent by their authors and, in some cases, compared quantitatively to other models. Bod (2001), for example, compared the performance of his data-oriented parsing with other closely related methods (Markov and treebank grammars). In addition, however, a handful of studies has empirically compared the performance of different melodic segmentation models. These studies differ in the models compared, the type of ground truth data used and the evaluation metrics applied. Melucci & Orio (2002), for example, collected the boundary indications of 17 expert musicians and experienced music scholars on melodic excerpts from 20 works by Bach, Mozart, Beethoven and Chopin. Having combined the boundary indications into a ground truth, they evaluated the performance of the LBDM against three models that inserted boundaries after a fixed (8 and 15) or random (in the range of 10 and 20) numbers of notes. Melucci & Orio report false positives, false negatives and a measure of disagreement which show that the LBDM outperforms the other models.

Melucci & Orio noticed a certain amount of disagreement between the segmentation markings of their participants. However, as they did not observe clear distinctions between participants when their responses were scaled by MDS and subjected to a cluster analysis, they aggregated all participants' boundary markings to binary judgements using a probabilistic procedure.

Bruderer (2008) evaluated a broader range of models in a study of the grouping structure of melodic excerpts from six Western pop songs. The ground truth segmentation was obtained from 21 adults with different degrees of musical training; the boundary indications were summed within consecutive time windows to yield a quasi-continuous boundary strength profile for each melody. Bruderer examined the performance of three models: Grouper, LBDM and the summed GPRs (GPR 2a, 2b, 3a and 3d) quantified by Frankland & Cohen (2004). The output of each model was convolved with a 2.4s Gaussian window to produce a boundary strength profile that was then correlated with the ground truth. Bruderer reports that the LBDM achieved the best and the GPRs the worst performance.

In another study, Thom et al. (2002) compared the predictions of the LBDM and Grouper with segmentations at the phrase and subphrase level provided (using a pen on a minimal score while listening to a MIDI file) by 19 musical experts for 10 melodies in a range of styles. In a first experiment, Thom et al. examined the average F1 scores between experts for each melody, obtaining values ranging between .14 and .82 for phrase judgements and .35 and .8 for subphrase judgements. The higher consistencies tended to be associated with melodies whose phrase structure was emphasised by rests. In a second experiment, the performance of each model on each melody was estimated by averaging the F1 scores over the 19 experts. Model parameters were optimised for each individual melody. The results indicated that Grouper tended to outperform the LBDM. Large IOIs were an important factor in the success of both models. In a third experiment, the predictions of each model were compared with the transcribed boundaries in several datasets from the EFSC. The model parameters were optimised over each dataset and the results again indicated that Grouper (with mean F1 between .6 and .7) outperformed the LBDM (mean F1 between .49 and .56). Finally, in order to examine the stability of the two models, each was used to predict the expert boundary profiles using parameters optimised over the EFSC. The performance of both algorithms was impaired, most notably for the subphrase judgements of the experts.

To summarise, the few existing comparative studies suggest that more complex models such as Grouper and LBDM outperform the individual GPR rules even when the latter are combined in an additive manner (Bruderer, 2008). Whether Grouper or LBDM exhibits a superior performance seems to depend on the data set and experimental task. Finally, most of these comparative studies used ground truth segmentations derived from manual annotations by human judges. However, only a limited number of melodies can be tested in this way (ranging from 6 in the case of Bruderer, 2008 to 20 by Melucci & Orio, 2002). Apart from Thom et al. (2002, Experiment D), there has been no thorough comparative evaluation over a large corpus of melodies annotated with phrase boundaries. However, that study did not include the GPRs and to date, no published study has directly compared these rule-based models with learning-based models (as we do here).

2.4 A New Segmentation Model

2.4.1 The IDyOM Model

As we have seen, most existing models of melodic grouping consist of collections of symbolic rules that describe the musical features corresponding to perceived groups. Such models have to be adjusted by hand using detailed *a priori* knowledge of a musical style. Therefore, these models are not only domain-specific, pertaining only to music, but also potentially style specific, pertaining only to Western tonal music or even a certain genre.

We present a new model of melodic grouping (the Information Dynamics Of Music, or IDyOM, model) which, unlike the GPRs, the LBDM and Grouper, uses unsupervised learning from experience rather than expert-coded symbolic rules. The model differs from DOP in that it uses unsupervised, rather than supervised, learning which makes it more useful for identifying grouping boundaries in corpora where phrase boundaries are not explicitly marked. The IDyOM model takes the same overall approach and inspiration from experimental psychology (Saffran, 2003; Saffran & Griepentrog, 2001; Saffran et al., 1999) as the TP/PMI models (see §2.3.5). In contrast to these models, however, IDyOM uses a range of strategies to improve the accuracy of its conditional probability estimates. Before describing these aspects of the model, we first review related research in musicology, cognitive linguistics and machine learning that further motivates a statistical approach to segmentation.

From a musicological perspective, it has been proposed that perceptual groups are associated with points of closure where the ongoing cognitive process of expectation is disrupted either because the context fails to stimulate strong expectations for any particular continuation or because the actual continuation is unexpected (Meyer, 1957; Narmour, 1990, see §2.1.2). These proposals may be given precise definitions in an information-theoretic framework (MacKay, 2003; Manning & Schütze, 1999) which we define by reference to a model of sequences, e_i, composed of symbols drawn from an alphabet $\mathcal{E}$. The model estimates the conditional probability of an element at index i in the sequence given the preceding elements in the sequence: $p(e_i|e_1^{i-1})$. Given such a model, the degree to which an event appearing in a given context in a melody is unexpected can be defined as the *information content* (MacKay, 2003), $h(e_i|e_1^{i-1})$, of the event given the context:

$$h(e_i|e_1^{i-1}) = \log_2 \frac{1}{p(e_i|e_1^{i-1})}. \tag{6}$$

The information content can be interpreted as the contextual unexpectedness or surprisal associated with an event. The contextual uncertainty of the model's expectations in a given melodic context can be defined as the *entropy* (or average information content) of the predictive context itself:

$$H(e_1^{i-1}) = \sum_{e \in \mathcal{E}} p(e_i|e_1^{i-1})h(e_i|e_1^{i-1}). \tag{7}$$

We hypothesise that boundaries are perceived before events for which the unexpectedness of the outcome (h) and the uncertainty of the prediction (H) are high. These correspond to two ways in which the prior context can fail to inform the model's sequential predictions leading to the perception of a discontinuity in the sequence. Segmenting at these points leads to cognitive representations of the sequence (in this case a melody) that maximise likelihood and simplicity (cf. Chater, 1996, 1999). In the current work, we focus on the information content (h), leaving the role of entropy (H) for future work.

There is evidence that related information-theoretic quantities are important in cognitive processing of language. For example, it has recently been demonstrated that the difficulty of processing words is related both to their information content

(Levy, 2008) and the induced changes in entropy over possible grammatical contin-
uations (Hale, 2006). Furthermore, in machine learning and computational linguistics, algorithms based on the idea of segmenting before unexpected events can identify word boundaries in infant-directed speech with some success (Brent, 1999a).
Similar strategies for identifying word boundaries have been implemented using recurrent neural networks (Elman, 1990). Recently, Cohen et al. (2007) proposed a
general method for segmenting sequences based on two principles: first, so as to
maximise the probability of events to the left and right of the boundary; and second,
so as to maximise the entropy of the conditional distribution across the boundary.
This algorithm was able to successfully identify word boundaries in text from four
languages as well as episode boundaries in the activities of a mobile robot.

The digram models used by TP and PMI are specific examples of a larger class
of models called n-gram models (Manning & Schütze, 1999). An n-gram is a sequence of n symbols consisting of a *context* of $n - 1$ symbols followed by a single
symbol *prediction*. A digram, for example, is a sequence of two symbols ($n = 2$)
with a single symbol context and a single symbol prediction. An n-gram model is
simply a collection of n-grams each of which is associated with a frequency count.
The quantity $n - 1$ is known as the *order* of the model and represents the number
of symbols making up the sequential context within which the prediction occurs.
During the *training* of the statistical model, these counts are acquired through an
analysis of some corpus of sequences (the training set) in the target domain. When
the trained model is exposed to an unseen sequence drawn from the target domain,
it uses the frequency counts associated with n-grams to estimate a probability distribution governing the identity of the next symbol in the sequence given the $n - 1$
preceding symbols. Therefore, an assumption made in n-gram modelling is that the
probability of the next event depends only on the previous $n - 1$ events:

$$p(e_i|e_1^{i-1}) \approx p(e_i|e_{(i-n)+1}^{i-1})$$

However, n-gram models suffer from several problems, both in general and specifically when applied to music. The TP and PMI models are conceptually simple but,
as models of musical structure, they have at least two major shortcomings. The first
is general: probabilities are estimated purely on the basis of digram (first order)
statistics collected from some existing corpus. The second problem is representational and specific to music: in estimating the probability of a note, only its pitch
(and that of its predecessor) are taken into consideration - the timing of the note is
ignored. In the IDyOM model, we address these shortcomings as described below.

Regarding the first problem, that of probability estimation, IDyOM uses several
methods drawn from the literature on text compression (Bell et al., 1990; Bunton,
1997) and statistical language modelling (Manning & Schütze, 1999) to improve
the prediction performance of the model. The following is a brief description of
the principal methods used; technical details can be found elsewhere (Conklin &
Witten, 1995; Pearce et al., 2005; Pearce & Wiggins, 2004).

Since the model is based on n-grams, one obvious improvement would be to increase the model order (i.e., n). However, while low-order models fail to provide an

adequate account of the structural influence of the context, increasing the order can prevent the model from capturing much of the statistical regularity present in the training set (an extreme case occurring when the model encounters an n-gram that does not appear in the training set and returns an estimated probability of zero). To address this problem (and maximise the benefits of both low- and high-order models) the IDyOM model maintains frequency counts during training for n-grams of all possible values of n in any given context. This results in a large number of n-grams; the time and space complexity of both storage and retrieval are rendered tractable through the use of suffix trees augmented with frequency counts (Bunton, 1997; Larsson, 1996; Ukkonen, 1995). During prediction, distributions are estimated using a weighted sum of all models below an order bound that varies depending on the context (Cleary & Teahan, 1997; Pearce & Wiggins, 2004). This bound is determined in each predictive context using simple heuristics designed to minimise uncertainty (Cleary & Teahan, 1997). The combination is designed such that higher-order predictions, which are more specific to the context, receive greater weighting than lower-order predictions, which are more general (Witten & Bell, 1991).

Another problem with many n-gram models is that a static (pre-trained) model will fail to make use of local statistical structure in the music it is currently analysing. To address this problem, IDyOM includes two kinds of model: first, a static *long-term* model that learns from the entire training set before being exposed to the test data; and second, a *short-term* model that is constructed dynamically and incrementally throughout each individual melody to which it is exposed (Conklin & Witten, 1995; Pearce & Wiggins, 2004). The distributions returned by these models are combined using an entropy-weighted multiplicative combination scheme corresponding to a weighted geometric mean (Pearce et al., 2005) in which greater weights are assigned to models whose predictions are associated with lower entropy (or uncertainty) at that point in the melody.

A final issue regards the fact that music is an inherently multi-dimensional phenomenon. Musical events have many perceived attributes including pitch, onset time (the start point of the event), duration, timbre and so on. In addition, *sequences* of these attributes may have multiple relevant emergent dimensions. For example, pitch interval, pitch class, scale degree, pitch contour (rising, falling or unison) and many other derived features are important in the perception and analysis of pitch structure. To accommodate these properties of music into the model, we use a multiple viewpoint approach to music representation (Conklin & Witten, 1995). The modelling process begins by choosing a set of basic properties of musical events (e.g., pitch, onset, duration, loudness etc) that we are interested in predicting. As these basic features are treated as independent attributes, their probabilities are computed separately and the probability of a note is simply the product of the probabilities of its attributes. Each basic feature (e.g., pitch) may then be predicted by any number of models for different derived features (e.g., pitch interval, scale degree) whose distributions are combined using the same entropy-weighted scheme (Pearce et al., 2005).

The use of long- and short-term models, incorporating models of derived features, the entropy-based weighting method and the use of a multiplicative (as opposed to a weighted linear or additive) combination scheme all improve the performance of IDyOM in predicting the pitches of unseen melodies; technical details of the model and its evaluation can be found elsewhere (Conklin & Witten, 1995; Pearce et al., 2005; Pearce & Wiggins, 2004). The goal in the current work, however, is to test its performance in retrieving segmentation boundaries in large corpora of melodies. Here, we use the model to predict the pitch, IOI and OOI associated with melodic events, multiplying the probabilities of these attributes together to yield the overall probability of the event. For simplicity, we use no derived features. We then focus on the unexpectedness of events (information content, h) using this as a boundary strength profile from which we compute boundary locations, as described in §2.4.2.

2.4.2 Peak Picking

To convert the boundary strength profile produced by IDyOM into a concrete segmentation, we devised a simple method that achieves this using three principles. First, given a vector S of boundary strengths for each note in a melody, the note following a boundary should have a greater or equal boundary strength than the note following it: $S_n \geq S_{n+1}$. Second, the note following a boundary should have a greater boundary strength than the note preceding it: $S_n > S_{n-1}$. Third, the note following a boundary should have a high boundary strength relative to the local context. We implement this principle by requiring the boundary strength to be k standard deviations greater than the mean boundary strength computed in a linearly weighted window from the beginning of the piece to the preceding event:

$$S_n > k\sqrt{\frac{\sum_{i=1}^{n-1}(w_i S_i - \overline{S}_{w,1...n-1})^2}{\sum_{1}^{n-1} w_i} + \frac{\sum_{i=1}^{n-1} w_i S_i}{\sum_{1}^{n-1} w_i}}. \qquad (8)$$

where w_i are the weights associated with the linear decay (triangular window) and the parameter k is allowed to vary depending on the nature of the boundary strength profile.

3 Method

3.1 The Ground Truth Data

The IDyOM model was tested against existing segmentation models on a subset of the EFSC, database Erk, containing 1705 Germanic folk melodies encoded in symbolic form with annotated phrase boundaries which were inserted during the encoding process by folk song experts. The dataset contains 78,995 note events at an average of about 46 events per melody and overall about 12% of notes fall before boundaries (a boundary occurs between two notes). There is only one hierarchical level of phrasing and the phrase structure exhaustively subsumes all the events in a melody.

3.2   The Models

The models included in the comparison are as follows:

Grouper: as implemented by Temperley (2001);[2]
LBDM: as specified by Cambouropoulos (2001) with $k = 0.5$;
IDyOM: as specified in §2.4.1 with $k = 2$;
GPR2a: as quantified by Frankland & Cohen (2004) with $k = 0.5$;
GPR2b: as quantified by Frankland & Cohen (2004) with $k = 0.5$;
GPR3a: as quantified by Frankland & Cohen (2004) with $k = 0.5$;
GPR3d: as quantified by Frankland & Cohen (2004) with $k = 2.5$;
TP: as defined in §2.3.5 with $k = 0.5$;
PMI: as defined in §2.3.5 with $k = 0.5$;
Always: every note falls on a boundary;
Never.   no note falls on a boundary.

The Always model predicts a boundary for every note while the Never model never predicts a boundary for any note. Grouper outputs binary boundary predictions. These models, therefore, do not use the peak-picking and are not associated with a value of k. The output of every other model was processed by Simple Picker using a value of k chosen from the set $\{0.5, 1, 1.5, 2, 2.5, 3, 3.5, 4\}$ so as to maximise F1 (and secondarily Recall in the case of ties).

The DOP method (Bod, 2001) is not included due to the complexity of its implementation and lack of any third party software that is straightforwardly applicable to musical data.

The IDyOM, TP and PMI models were trained and evaluated on melodies taken from the Erk dataset. In order to demonstrate generalisation, we adopted a cross-validation strategy in which the dataset is divided into k disjoint subsets of approximately equal size. The model is trained k times, each time leaving out a different subset to be used for testing. A value of $k = 10$ was used which has been found to produce a good balance between the bias associated with small values of k and the high variance associated with large values of k (Kohavi, 1995).

3.3   Making Model Outputs Comparable

The outputs of the algorithms tested vary considerably. While Grouper marks each note with a binary indicator (1 = boundary, 0 = no boundary), the other models output a positive real number for each note which can be interpreted as a boundary strength. In contrast to Bruderer (2008) we chose to make all segmentation algorithms comparable by picking binary boundary indications from the boundary strength profiles.

To do so, we applied the peak-picking procedure described in S2.4.2 to the boundary profiles of all models (except Grouper which produces binary boundary judgements) and chose a value of k to optimise the performance of each model

---

[2] Adapted for use with Melconv 2 by Klaus Frieler.

Fig. 1 An example showing the binary vectors representing the segmentation of a melody

individually. In practice, the optimal value of k varies between algorithms depending on the nature of the boundary strength profiles they produce.

In addition, we modified the output of all models to predict an implicit phrase boundary on the last note of a melody.

3.4 Evaluation Measures

It is common to represent a segmentation of a melody using a binary vector with one element for each event in the melody indicating, for each event, whether or not that event falls on a grouping boundary. An example is shown in Figure 1.

Given this formulation, we can state the problem of comparing the segmentation of a model with the ground truth segmentation in terms of computing the similarity or distance between two binary vectors. Many methods exist for comparing binary vectors. For example, version 14 of the commercial statistical software package SPSS provides 27 different measures for determining the similarity or distance between binary variables. Additional measures have been proposed in the areas of data mining and psychological measurement. The appropriate measure to use depends on the desired comparison and the nature of the data (Sokolova & Lapalme, 2007). Here we introduce and compare five methods that are widely used in psychology, computer science and biology.

These methods enable us to compute the similarity between phenomenal data encoded as a binary vector, the *ground truth*, and the output of a model of the process generating that data, the *prediction*, encoded in the same way.

All methods start with the 2 x 2 table shown in Table 2 which summarises the co-occurrences of binary events between the ground truth and the prediction. The ground truth positives (P) and negatives (N), respectively, are the numbers of positions where the ground truth vector contains 1 and 0. The predicted positives (p) and negatives (n) indicate numbers of positions where the prediction vector contains 1 and 0 respectively. The *true positives (TP)* is the number of positions where both ground truth and prediction vectors indicate 1 while the *true negatives (TN)* is the number of positions where both vectors contain 0. *False positives (FP)* and *false negatives (FN)* are the numbers of locations where the ground truth and prediction vectors differ. In the former case, the prediction contains 1 where the ground truth contains 0, and *vice versa* for the latter.

Table 2 A summary of the outcomes of comparing prediction and ground truth binary data

		Ground Truth	
		P	N
Prediction	p	TP	FP
	n	FN	TN

One of the most intuitive measures for comparing binary vectors is *accuracy*, defined as the number of times the prediction vector and ground truth vector agree as a proportion of the total number of entries in the vector:

$$accuracy = \frac{TP + TN}{P + N} \tag{9}$$

However, this measure of *accuracy* can be misleading when the ground truth data is skewed. For example, if the proportion of negative cases in the ground truth is .8, a model that always gives a negative answer will achieve an accuracy of 80%. The following measures take into account the proportion of positive and negative instances in the ground truth data which means that the values are comparable across the distributions occurring in different datasets.

Psychologists are often interested in the agreement between human raters or judges when they assess the same items and Kappa (κ) has become one of the most frequently used measures for assessing inter-rater agreement. It is conceptually related to the accuracy measure but takes the distribution of the two binary classes into account and thus resembles the well-known χ^2 distribution. The variant known as *Fleiss' κ* (Fleiss, 1971) is formulated for multiple-class ratings and multiple raters. Reducing κ to binary markings from only two sources (raters) and using the notation introduced above, κ is defined as the difference between the proportions of actual agreement ($Pr = accuracy$) and expected agreement (Pr_e):

$$\kappa = \frac{Pr - Pr_e}{1 - Pr_e} \tag{10}$$

where:

$$Pr = \frac{TP + TN}{P + N}, \quad Pr_e = Pr_1^2 + Pr_0^2, \tag{11}$$

$$Pr_1 = \frac{P + p}{2 \cdot (P + N)}, \quad Pr_0 = \frac{N + n}{2 \cdot (P + N)}. \tag{12}$$

$$\tag{13}$$

Another measure, d' (Green & Swets, 1966), was developed in psychophysics and is often used to measure human ability to detect a particular cue in a signal or distinguish two stimuli differing along some dimension. It has been also widely used to analyse experimental data in other areas of cognitive psychology such as memory. It is defined as:

$$d' = z(\frac{TP}{TP+FN}) - z(\frac{FP}{FP+TN}) \qquad (14)$$

where $z()$ is the cumulative distribution function of the normal probability distribution.

In modern data mining, the following three measures are standard methods for evaluating query-based systems for document retrieval (Witten & Frank, 1999). *Precision* reflects the true positives as a proportion of the positive output of the prediction while *Recall* reflects the true positives as a proportion of the positive data in the ground truth. *F1* is the harmonic mean of the two.

$$Precision = \frac{TP}{TP+FP},$$
$$Recall = \frac{TP}{TP+FN},$$
$$F1 = \frac{2 \cdot precision \cdot recall}{precision + recall}.$$

4 Results

Before comparing the performance of the models, it is instructive to consider the problem of how to evaluate quantitatively the degree of correspondence between two segmentations of a melody. To do so, we compute the Pearson correlation coefficients between the different evaluation measures described in §3.4 for each pairwise comparison between each models output for each melody in the dataset. The results are shown in Table 3.

Table 3 Correlations between evaluation measures over models and melodies

	Accuracy	Precision	Recall	F1	d'	κ
Accuracy	1					
Precision	0.56	1				
Recall	-0.31	0.08	1			
F1	0.45	0.69	0.63	1		
d'	0.52	0.48	0.64	0.91	1	
κ	0.86	0.70	0.17	0.83	0.84	1

Precision and Recall each only take into consideration one kind of error (i.e., FP or FN) and show low or moderate correlations with the other measures (and very low correlations with each other as expected). Here, however, we want a measure that takes into account both kinds of error. κ, F1 and d' all correlate very highly with each other because they all reflect TP in relation to FP and FN. Although κ is also influenced by TN, the proportion of true negatives is constrained given a fixed number of data points (i.e. if we know TP, FP, and FN and the total number of notes

Table 4 The model comparison results in order of mean F1 scores. See text for details of the Hybrid model.

Model	Precision	Recall	F1
Hybrid	0.87	0.56	0.66
Grouper	0.71	0.62	0.66
LBDM	0.70	0.60	0.63
IDyOM	0.76	0.50	0.58
GPR2a	0.99	0.45	0.58
GPR2b	0.47	0.42	0.39
GPR3a	0.29	0.46	0.35
GPR3d	0.66	0.22	0.31
PMI	0.16	0.32	0.21
TP	0.17	0.19	0.17
Always	0.13	1.00	0.22
Never	0.00	0.00	0.00

then TN is fixed; we have 3 degrees of freedom and not 4 for pairs of vectors of the same length). Accuracy exhibits only small correlations with these three measures (except κ to which it is closely related) and is not appropriate here due to the unequal proportions of positive and negative values in the data (see §3.4). The results of the correlational analysis suggest that we could have used any one of d', F1 or κ for evaluating our models against the ground truth. Following common practice in data mining and information retrieval, we use *F1* to compare model performance.

The results of the model comparison are shown in Table 4. The four models achieving mean F1 values of over 0.5 (Grouper, LBDM, GPR2a, IDyOM) were chosen for further analysis. Sign tests between the F1 scores on each melody indicate that all differences between these models are significant at an alpha level of 0.01, with the exception of that between GPR2a and LBDM. In order to see whether further performance improvement could be achieved by a combined model, we constructed a logistic regression model including Grouper, LBDM, IDyOM and GPR2a as predictors. Backwards stepwise elimination using the Bayes Information Criterion (BIC) failed to remove any of the predictors from the overall model (Venables & Ripley, 2002). The performance of the resulting model is shown in the top row of Table 4. Sign tests demonstrated that the Hybrid model achieved better F1 scores on significantly more melodies than each of the other models (including Grouper, in spite of the fact that the average performance, shown in Table 4, was the same). Compared to Grouper and LBDM, the hybrid model has slightly worse recall but much better precision; compared to IDYOM, the hybrid model has better precision and recall; while compared to GPR2a, the lower precision achieved by the hybrid model is balanced by it's better recall.

5 Discussion

We would like to highlight four results of this evaluation study. First, we were sur-
prised by the strong performance of one of the GTTM preference rules, GPR2a.
This points to the conclusion that rests, perhaps above all other melodic parameters,
have a large influence on boundaries for this set of melodies. Consequently, all of the
high-performing rule-based models (Grouper, LBDM, GPR2a) make use of a rest or
temporal gap rule while IDyOM includes rests in its probability estimation. Future
research should undertake a more detailed qualitative comparison of the kinds of
musical context in which each model succeeds or fails to predict boundaries. This
suggests that future research should focus on boundaries not indicated explicitly by
rests.

Second, it is interesting to compare the results to those reported in other studies.
In general, the performance of Grouper and LBDM are comparable to their performance on a different subset of the EFSC reported by Thom et al. (2002). The
performance of Grouper is somewhat lower than that reported by Temperley (2001)
on 65 melodies from the EFSC. The performance of all models is lower than that of
the supervised learning model reported by Bod (2001).

Third, the hybrid model which combines Grouper, LBDM, GPR2a and IDyOM
generated better performance values than any of its components. The fact that the
$F1$ value seems to be only slightly better than Grouper is due to the fact that logistic
regression optimises the log-likelihood function for whether or not a note is a boundary given the boundary indications of the predictor variables (models). It therefore
uses information about positive boundary indications (P) and negative boundary indications (N) to an equal degree, in contrast to $F1$. This suggests options, in future
research, for assigning different weights to P and N instances or including the raw
boundary profiles of LBDM and IDyOM (i.e., without peak-picking) in the logistic regression procedure. Another possibility is to use boosting (combining multiple
weak learners to create a single strong learner, Schapire, 2003) to combine the different models which may lead to better performance enhancements than logistic
regression.

Finally, it is interesting to note that an unsupervised learning model (IDyOM) that
makes no use of music-theoretic rules about melodic phrases performed as well as
it does. It not only performs much better than simple statistical segmenters (the TP
and PMI models) but also approaches the performance of sophisticated rule-based
models. In fact, IDyOM's precision is better than LBDM and Grouper although it's
Recall is worse (this is a common tradeoff in MIR). In comparison to supervised
learning methods such as DOP, IDyOM does not require pre-segmented data as a
training corpus. This may not be an issue for folk-song data where we have large
corpora with annotated phrase boundaries but is a significant factor for other musical styles such as pop. IDyOM learns regularities in the melodic data it is trained
on and outputs probabilities of note events which are ultimately used to derive an
information content (unexpectedness) for each note event in a melody. In turn, this
information-theoretic quantity (in comparison to that of previous notes) is used to
decide whether or not the note falls on a boundary.

These findings have been corroborated by a recent study comparing computational models of melodic segmentation to perceived segmentations indicated by human listeners for 10 popular melodies (de Nooijer et al., 2008). The results showed that IDyOM's segmentations did not differ significantly from those of the listeners and, furthermore, that the segmentations of IDyOM, LBDM and Grouper did not differ.

We argue that the present results provide preliminary evidence that the notion of expectedness is strongly related to boundary detection in melodies. In future research, we hope to achieve better performance by tailoring IDyOM specifically for segmentation including a metrically-based (i.e., we represent whatever is happening in each metrical time slice) rather than an event-based representation of time, optimising the derived features that it uses to make event predictions and using other information-theoretic measures such as entropy or predictive information (Abdallah & Plumbley, 2009).

Acknowledgements. This research was supported by EPSRC via grant numbers GR/S82220/01 and EP/D038855/1.

References

Abdallah, S., Plumbley, M.: Information dynamics: Patterns of expectation and surprise in the perception of music. Connection Science 21(2-3), 89–117 (2009)

Abdallah, S., Sandler, M., Rhodes, C., Casey, M.: Using duration models to reduce fragmentation in audio segmentation. Machine Learning 65(2-3), 485–515 (2006)

Ahlbäck, S.: Melody beyond notes: A study of melody cognition. Doctoral dissertation, Göteborg University, Göteborg, Sweden (2004)

Allan, L.G.: The perception of time. Perception and Psychophysics 26(5), 340–354 (1979)

Barlow, H., Morgenstern, S.: A dictionary of musical themes. Ernest Benn (1949)

Bell, T.C., Cleary, J.G., Witten, I.H.: Text Compression. Prentice Hall, Englewood Cliffs (1990)

Bod, R.: Beyond Grammar: An experience-based theory of language. CSLI Publications, Standford (1998)

Bod, R.: Memory-based models of melodic analysis: Challenging the Gestalt principles. Journal of New Music Research 30(3), 27–37 (2001)

Bower, G.: Organizational factors in memory. Cognitive Psychology 1, 18–46 (1970)

Bregman, A.S.: Auditory Scene Analysis: The perceptual organization of sound. MIT Press, Cambridge (1990)

Brent, M.R.: An efficient, probabilistically sound algorithm for segmentation and word discovery. Machine Learning 34(1-3), 71–105 (1999a)

Brent, M.R.: Speech segmentation and word discovery: A computational perspective. Trends in Cognitive Science 3, 294–301 (1999b)

Brochard, R., Dufour, A., Drake, C., Scheiber, C.: Functional brain imaging of rhythm perception. In: Woods, C., Luck, G., Brochard, R., Seddon, F., Sloboda, J.A. (eds.) Proceedings of the Sixth International Conference of Music Perception and Cognition. University of Keele, Keele (2000)

Bruderer, M.J.: Perception and Modeling of Segment Boundaries in Popular Music. Doctoral dissertation, J.F. Schouten School for User-System Interaction Research, Technische Universiteit Eindhoven, Nederlands (2008)

Bunton, S.: Semantically motivated improvements for PPM variants. The Computer Journal 40(2/3), 76–93 (1997)

Cambouropoulos, E.: The local boundary detection model (LBDM) and its application in the study of expressive timing. In: Proceedings of the International Computer Music Conference, ICMA, San Francisco, pp. 17–22 (2001)

Cambouropoulos, E.: Musical parallelism and melodic segmentation: A computational approach. Music Perception 23(3), 249–269 (2006)

Chater, N.: Reconciling simplicity and likelihood principles in perceptual organisation. Psychological Review 103(3), 566–581 (1996)

Chater, N.: The search for simplicity: A fundamental cognitive principle? The Quarterly Journal of Experimental Psychology 52A(2), 273–302 (1999)

Clarke, E.F., Krumhansl, K.L.: Perceiving musical time. Music Perception 7(3), 213–252 (1990)

Cleary, J.G., Teahan, W.J.: Unbounded length contexts for PPM. The Computer Journal 40(2/3), 67–75 (1997)

Cohen, P.R., Adams, N., Heeringa, B.: Voting experts: An unsupervised algorithm for segmenting sequences. Intelligent Data Analysis 11(6), 607–625 (2007)

Collins, M.: Head-Driven Statistical Models for Natural Language Parsing. Doctoral dissertation, Department of Computer and Information Science, University of Pennsylvania, USA (1999)

Conklin, D., Witten, I.H.: Multiple viewpoint systems for music prediction. Journal of New Music Research 24(1), 51–73 (1995)

de Nooijer, J., Wiering, F., Volk, A., Tabachneck-Schijf, H.J.M.: An experimental comparison of human and automatic music segmentation. In: Miyazaki, K., Adachi, M., Hiraga, Y., Nakajima, Y., Tsuzaki, M. (eds.) Proceedings of the 10th International Conference on Music Perception and Cognition, pp. 399–407. Causal Productions, Adelaide (2008)

Deliège, I.: Grouping conditions in listening to music: An approach to Lerdahl and Jackendoff's grouping preference rules. Music Perception 4(4), 325–360 (1987)

Dowling, W.J.: Rhythmic groups and subjective chunks in memory for melodies. Perception and Psychophysics 14(1), 37–40 (1973)

Elman, J.L.: Finding structure in time. Cognitive Science 14, 179–211 (1990)

Ferrand, M., Nelson, P., Wiggins, G.: Memory and melodic density: a model for melody segmentation. In: Bernardini, N.G.F., Giosmin, N. (eds.) Proceedings of the XIV Colloquium on Musical Informatics, Firenze, Italy, pp. 95–98 (2003)

Fleiss, J.L.: Measuring nominal scale agreement among many raters. Psychological Bulletin 76(5), 378–382 (1971)

Fodor, J.A., Bever, T.G.: The psychological reality of linguistic segments. Journal of Verbal Learning and Verbal Behavior 4, 414–420 (1965)

Frankland, B.W., Cohen, A.J.: Parsing of melody: Quantification and testing of the local grouping rules of Lerdahl and Jackendoff's A Generative Theory of Tonal Music. Music Perception 21(4), 499–543 (2004)

Gjerdingen, R.O.: Apparent motion in music? In: Griffith, N., Todd, P.M. (eds.) Musical Networks: Parallel Distributed Perception and Performance, pp. 141–173. MIT Press/Bradford Books, Cambridge (1999)

Green, D., Swets, J.: Signal Detection Theory and Psychophysics. Wiley, New York (1966)

Gregory, A.H.: Perception of clicks in music. Perception and Psychophysics 24(2), 171–174 (1978)

Hale, J.: Uncertainty about the rest of the sentence. Cognitive Science 30(4), 643–672 (2006)

Howell, D.C.: Statistical methods for pscyhology. Duxbury, Pacific Grove (2002)

Jackendoff, R.: Consciousness and the Computational Mind. MIT Press, Cambridge (1987)

Jusczyk, P.W.: The Discovery of Spoken Language. MIT Press, Cambridge (1997)

Koffka, K.: Principles of Gestalt Psychology. Harcourt, Brace and World, New York (1935)

Kohavi, R.: Wrappers for Performance Enhancement and Oblivious Decision Graphs. Doctoral dissertation, Department of Computer Science, Stanford University, USA (1995)

Ladefoged, P., Broadbent, D.E.: Perception of sequences in auditory events. Journal of Experimental Psychology 12, 162–170 (1960)

Larsson, N.J.: Extended application of suffix trees to data compression. In: Storer, J.A., Cohn, M. (eds.) Proceedings of the IEEE Data Compression Conference, pp. 190–199. IEEE Computer Society Press, Washington (1996)

Lerdahl, F., Jackendoff, R.: A Generative Theory of Tonal Music. MIT Press, Cambridge (1983)

Levy, R.: Expectation-based syntactic comprehension. Cognition 16(3), 1126–1177 (2008)

Liegeoise-Chauvel, C., Peretz, I., Babai, M., Laguitton, V., Chauvel, P.: Contribution of different cortical areas in the temporal lobes to music processing. Brain 121(10), 1853–1867 (1998)

MacKay, D.J.C.: Information Theory, Inference, and Learning Algorithms. Cambridge University Press, Cambridge (2003)

MacWhinney, B., Snow, C.: The child language data exchange system. Journal of Child Language 12, 271–296 (1985)

Manning, C.D., Schütze, H.: Foundations of Statistical Natural Language Processing. MIT Press, Cambridge (1999)

Melucci, M., Orio, N.: A comparison of manual and automatic melody segmentation. In: Fingerhut, M. (ed.) Proceedings of the Third International Conference on Music Information Retrieval, pp. 7–14. IRCAM, Paris (2002)

Meyer, L.B.: Meaning in music and information theory. Journal of Aesthetics and Art Criticism 15(4), 412–424 (1957)

Narmour, E.: The Analysis and Cognition of Basic Melodic Structures: The Implication-realisation Model. University of Chicago Press, Chicago (1990)

Narmour, E.: The Analysis and Cognition of Melodic Complexity: The Implication-realisation Model. University of Chicago Press, Chicago (1992)

Pearce, M.T., Conklin, D., Wiggins, G.A.: Methods for combining statistical models of music. In: Wiil, U.K. (ed.) Computer Music Modelling and Retrieval, pp. 295–312. Springer, Berlin (2005)

Pearce, M.T., Wiggins, G.A.: Improved methods for statistical modelling of monophonic music. Journal of New Music Research 33(4), 367–385 (2004)

Peretz, I.: Clustering in music: An appraisal of task factors. International Journal of Psychology 24(2), 157–178 (1989)

Peretz, I.: Processing of local and global musical information by unilateral brain-damaged patients. Brain 113(4), 1185–1205 (1990)

Rabiner, L.R.: A tutorial on Hidden Markov Models and selected applications in speech recognition. Proceedings of the IEEE 77(2), 257–285 (1989)

RISM-ZENTRALREDAKTION. Répertoire international des sources musicales (rism)

Saffran, J.R.: Absolute pitch in infancy and adulthood: The role of tonal structure. Developmental Science 6(1), 37–49 (2003)

Saffran, J.R., Aslin, R.N., Newport, E.L.: Statistical learning by 8-month old infants. Science 274, 1926–1928 (1996)

Saffran, J.R., Griepentrog, G.J.: Absolute pitch in infant auditory learning: Evidence for developmental reorganization. Developmental Psychology 37(1), 74–85 (2001)

Saffran, J.R., Johnson, E.K., Aslin, R.N., Newport, E.L.: Statistical learning of tone sequences by human infants and adults. Cognition 70(1), 27–52 (1999)

Schaffrath, H.: The Essen folksong collection. In: Huron, D. (ed.) Database containing 6,255 folksong transcriptions in the Kern format and a 34-page research guide [computer database]. CCARH, Menlo Park (1995)

Schapire, R.E.: The boosting approach to machine learning: An overview. In: Denison, D.D., Hansen, M.H., Holmes, C., Mallick, B., Yu, B. (eds.) Nonlinear Estimation and Classification. Springer, Berlin (2003)

Sloboda, J.A., Gregory, A.H.: The psychological reality of musical segments. Canadian Journal of Psychology 34(3), 274–280 (1980)

Sokolova, M., Lapalme, G.: Performance measures in classification of human communications. In: Kobti, Z., Wu, D. (eds.) Canadian AI 2007. LNCS (LNAI), vol. 4509, pp. 159–170. Springer, Heidelberg (2007)

Stoffer, T.H.: Representation of phrase structure in the perception of music. Music Perception 3(2), 191–220 (1985)

Tan, N., Aiello, R., Bever, T.G.: Harmonic structure as a determinant of melodic organization. Memory and Cognition 9(5), 533–539 (1981)

Temperley, D.: The Cognition of Basic Musical Structures. MIT Press, Cambridge (2001)

Tenney, J., Polansky, L.: Temporal Gestalt perception in music. Contemporary Music Review 24(2), 205–241 (1980)

Thom, B., Spevak, C., Höthker, K.: Melodic segmentation: Evaluating the performance of algorithms and musical experts. In: Proceedings of the International Computer Music Conference, pp. 65–72. ICMA, San Francisco (2002)

Todd, N.P.M.: The auditory "primal sketch": A multiscale model of rhythmic grouping. Journal of New Music Research 23(1), 25–70 (1994)

Ukkonen, E.: On-line construction of suffix trees. Algorithmica 14(3), 249–260 (1995)

Venables, W.N., Ripley, B.D.: Modern Applied Statistics with S. Springer, New York (2002)

Waugh, N., Norman, D.A.: Primary memory. Psychological Review 72, 89–104 (1965)

Witten, I.H., Bell, T.C.: The zero-frequency problem: Estimating the probabilities of novel events in adaptive text compression. IEEE Transactions on Information Theory 37(4), 1085–1094 (1991)

Witten, I.H., Frank, E. (eds.): Data mining: Practical machine learning tools and techniques with Java implementations. Morgan Kaufmann, San Francisco (1999)

Automatic Musical Genre Classification and Artificial Immune Recognition System

Shyamala Doraisamy and Shahram Golzari

Abstract. Artificial Immune Recognition System (AIRS) has been shown to be an effective classifier for several machine learning problems. In this study, AIRS is investigated as a classifier for musical genres from differing cultures. Musical data of two cultures were used – Traditional Malay Music (TMM) and Latin Music (LM). The performance of AIRS for the classification of these genres was compared with performances using several commonly used classifiers. The best classification accuracy for TMM was obtained using AIRS and was comparable, almost similar, to the performance obtained with the popular classifiers. However, the performance of AIRS for LM genre classification was shown to be not promising.

1 Introduction

Interest on music information retrieval systems for the storage, retrieval and classification of large collections of digital musical files has grown in recent years. Metadata such as filename, author, file size, date and genres are commonly used to classify and retrieve these documents. Such manual classification is highly labor-intensive and costly both in terms of time and money [1]. An automatic classification system that is able to analyze and extract implicit knowledge of the musical files is therefore highly sought. One approach to automated musical classification that is currently being widely studied is classification based on musical genres.

Musical genres are labels created and used by humans for categorizing and describing music [2]. Examples of musical genres include Pop, Rock, Hip-hop and Classical. Several systems for automated genre classification and retrieval of musical files have been researched and developed [2, 3]. However, most of these studies

Shyamala Doraisamy
Faculty of Comp. Sc. & IT, University Putra Malaysia
e-mail: shyamala@fsktm.upm.edu.my

Shahram Golzari
Faculty of Comp. Sc. & IT, University Putra Malaysia
e-mail: golzari@ieee.com

Z.W. Raś and A.A. Wieczorkowska (Eds.): Adv. in Music Inform. Retrieval, SCI 274, pp. 391–403.
springerlink.com

were conducted using only genres such as those given earlier in this paragraph. These genres in general have originated from North America and Europe. In this study we focus on two sets of musical genres from differing cultures Traditional Malaysian Music (TMM) and Latin Music (LM). Music from different cultures has different influences and instrumentalisation.

Studies by Norowi et. al. [4] and Silla et. al. [5] have investigated automated musical genre classification of TMM and LM respectively. The study by Norowi et. al. [4] showed the significance of beat features for TMM genre classification in comparison to classifying the genre set of – Blues, Classical, Jazz, Pop and Rock. This genre set of music was referred to as Western music in the scope of their study. We will continue to refer to Western music as music from North America and Europe.

As for feature selection with LM genre classification, the study by Silla et. al. [5] showed that the use of multiple features from the three main groups of features, i.e. Timbral Texture, Beat Related and Pitch Related, was useful for Latin music genre classification. The study also showed that using the middle segment of a musical piece enables better classification. With TMM however, using the first 30 seconds was shown to be better for classification. One of the motivations in the work by Silla et. al. [5] work was to analyze Latin music audio signals, which present a great variation in time. This characteristic also opposes the main characteristic of TMM which is very repetitive and its main beats clearly audible with gong hits. These differences were used as a basis for selecting these two datasets in this study investigating the performance of Artificial Immune Recognition System (AIRS) as a classifier for automated musical genre classification.

Artificial Immune System (AIS) is a computational method inspired by the biological immune system. It is progressing slowly and steadily as a new branch of computational intelligence and soft computing [6, 7]. One AIS-based algorithm is AIRS. AIRS is a supervised immune-inspired classification system capable of assigning data items unseen during training to one of any number of classes based on previous training experience. AIRS is probably the first and best known AIS for classification, first introduced in the study by Watkins [8].

In this study, the automated genre classification approach consists of three phases. i.e., feature extraction, feature selection preprocessing and then classification with AIRS (in comparison to using just the first two phases with most automated genre classification systems). Feature extraction was not performed in this study. We continue to use the appropriate features that had been identified as efficient for the classification of the respective genres from these previous studies. This is due to differences in data pre-processing strategies needed for efficient classification. Based on the strategies used for the two genre sets as shown in the studies by Norowi et. al. [4] and Silla et.al. [5] respectively, we continue to use the appropriately pre-processed data-sets obtained from these studies. More discussion on these data sets is presented in later sections. Feature selection has been shown to be useful with TMM genre classification in an earlier study [9]. In this study, we included this feature selection phase for further pre-processing of the obtained LM dataset from the study by Silla et.al. [5]. The aim and focus of this study is to investigate the feasibility of using AIRS for automated musical genre classification. The performance of AIRS is tested with regard to classification accuracy. This performance is also compared with the obtained accuracies by popularly used classifiers.

The remainder of this paper is organized as follows: Section 2 and 3 give the briefly description about TMM and LM respectively. Section 4 describes about feature selection method. AIRS is explained in Section 5. In Section 6, we explain the experiments and discuss the results and lastly in Section 7, we conclude the paper.

2 Traditional Malaysian Music

TMM encompasses traditional music of both the Malay (largely from West Malaysia) and native communities of Sabah and Sarawak (East Malaysian states on the Island of Borneo). We use this definition based on the scope defined by Nasaruddin [10], which categorises TMM into six broad categories: shadow puppet music, dance theatre, music with Indonesian influence, percussion music known as *nobat*, syncreatic Malaysian music and music from Sabah and Sarawak. The work by Nasaruddin [10] also attempts to notate the microtonal instruments using standard Western notation. However, it is just an approximation and does not reflect actual pitches.

TMM is mainly derivative, influenced by the initial overall Indian and Middle Eastern music during the trade era and later from colonial powers of countries and nations such as Thailand, Indonesia, Portugal and Britain who introduced their own culture including dance and music. A thorough overview on the origin and history of TMM can be found in [11]. The taxonomy of TMM depends on the nature of the theatre forms they serve and their instrumentations. Categorization of TMM genres has also been studied extensively by Ang [12]. These genres are usually disseminated non-commercially, usually performed by persons who are not highly trained musical specialists, undergoes change arising from creative impulses and exists in many forms. The musical ensembles usually include drums known as *gendangs* that are used to provide constant rhythmic beat of the songs and gongs to mark the end of a temporal cycle at specific part of the song [13].

One common attribute that is shared by most TMM genres is that they are generally repetitive in nature and exist in *gongan*-like cycle. *Gongan* is defined as a temporal cycle marked internally at specific points by specific gongs and at the end by the lowest-pitched gong of an ensemble [11]. It is an important structural function as it divides the musical pieces into temporal sections. Once every measure has been played, musicians continue playing in a looping motion by repeating the cycle from the beginning again until one of the lead percussionists signals the end of the song by varying their rhythms noticeably. In general, TMM does not have a chorus that plays differently than other parts of the songs, which is the usual occurrence in western music. Its repetitiveness and constant rhythms are two aspects that are taken into account to facilitate classification by genre.

Norowi et. al. [4] studied the effects of various factors and audio feature set combinations towards the classification of TMM genres. Results from experiments conducted in several phases in [4] show that factors such as dataset size, track length and location together with various combinations of audio feature sets comprising Short

Time Fourier Transform (STFT), Mel-Frequency Cepstral Coefficients (MFCCs) and Beat Features affect classification. A detailed discussion of the data collection and treatment of TMM files obtained in both digital analogue forms is available in Noris et. al. [4]. The track-length of 30 seconds from the beginning of the recording was concluded to be efficient for the classification. This could be due to the repetitive nature of the music with little variation through the composition. As for the classification, only the J48 classifier was used in the study which achieved 66.3% classification accuracy for TMM genres [4]. The features were extracted from the music files through MARSYAS-0.2.2; a free framework that enables the evaluation of computer audition applications. MARSYAS is a semi-automatic music classification system that is developed as an alternative solution for the existing audio tools that are incapable of handling the increasing amount of computer data [2]. It enables the three feature sets for representing the timbral texture, rhythmic content and pitch content of the music signals and uses trained statistical pattern recognition classifiers for evaluation.

Table 1 Overall number of musical instances for each TMM genre

NO	Genre	Class Label	Number of Instances
1	Dikir Barat	A	31
2	Etnik Sabah	B	12
3	Gamelan	C	23
4	Ghazal	D	19
5	Inang	E	10
6	Joget	F	15
7	Keroncong	G	43
8	Tumbuk Kalang	H	13
9	Wayang Kulit	I	17
10	Zapin	J	10

Ten TMM genres were involved in the study. The breakdown for each genre and its number of musical files are listed in Table 1. We continue to use these extracted features in this study as it was shown in the previous study to be efficient in the recognition of TMM genres.

3   Latin Music

The rhythmic structure of Latin musical genres in general is syncopated with high temporal variations, i.e change of speed, in comparison to TMM. A pre-processed collection of 3000 samples of Latin music was used in this study. As discussed in Section 1 , this collection was used in a prior study by Silla et. al. [5] evaluating their automated musical genre classification system for LM. We continue to use this pre-processed dataset with ten different genres of Latin music comprising Tango,

Table 2 Overall number of musical instances for each Latin genre

NO	Genre	Class Label	Number of Instances
1	Tango	A	300
2	Salsa	B	300
3	Forro	C	300
4	Axe	D	300
5	Bachata	E	300
6	Bolero	F	300
7	Merengue	G	300
8	Gaucha	H	300
9	Sertaneja	I	300
10	Pagode	J	300

Salsa, Forro, Axe, Bachata, Bolero, Merengue, Gaucha, Sertaneja and Pagode. The number of instances and class labels assigned is shown in Table 2. Rhythms of genres such as Rock and Classic from Western music, and TMM in this case, in general are more constant than Latin rhythms. In the dataset description of Silla et. al. [5], most of the music samples used were said to start slow (sometimes as slow as a Bolero) in the introduction and after a while they "explode" (at the time when all instruments come into play). This variation in speed differs to the strict tempo of TMM marked by cyclic gongs.

Multiple features were extracted using MARSYAS, the similar framework used by Norowi et. al. [4] for TMM. The breakdown of the combination of this multiple feature set of timbral, beat-related and pitch-related features extracted for the LM data set comprises are as follows: - timbral texture: nine FFT and ten MFCC; beat: six; pitch: five. As discussed in Section 1 , the data was further pre-processed with a feature selection phase. This is discussed in the following section.

4 Feature Selection

Feature selection is the process of removing features from the data set that are irrelevant with respect to the task that is to be performed. Feature selection can be extremely useful in reducing the high dimensionality of features to be processed by the classifier, reducing execution time and improving predictive accuracy (inclusion of irrelevant features can introduce noise into the data, thus obscuring relevant features important for accurate classification). It is worth noting that even though some machine learning algorithms perform some degree of feature selection themselves (such as classification trees), feature space reduction can be useful even for these algorithms. Reducing the dimensionality of the data reduces the size of the hypothesis space and thus results in faster execution time.

Feature selection techniques can be split into two categories filter methods and wrapper methods. Filter methods determine whether features are predictive using heuristics based on characteristics of the data. Wrapper methods make use of the classification algorithm that will ultimately be applied to the data in order to evaluate the predictive power of features. Wrapper methods generally result in better performance than filter methods because the feature selection process is optimized for the classification algorithm to be used. However, they are generally far too expensive to be used if the number of features is large because each feature set considered must be evaluated with the trained classifier. For this reason, wrapper methods will not be considered in this study. Filter methods are much faster than wrapper methods and therefore are better suited to high dimensional data sets.

Preliminary experiments were conducted for these datasets to compare the performance of feature subset selection methods such as: Correlation based methods, Principal Component Analysis (PCA), Information gain, Gain ratio and Chi square. Based on preliminary experiments, gain ratio measure achieved the best results for both data sets. This is due to the data characteristics.

Gain ratio is an extension of information gain. Information gain is an entropy based measure for feature selection. Entropy is a commonly used measure in information technology. For set D of data, entropy is defined as equation (1).

$$Entropy(D) = \sum_{i=1}^{c} -p_i \log_2 p_i \tag{1}$$

p_i is the proportion of D belonging to class i, and c is the number of classes. Entropy is a measure of impurity of training data and information gain is a measure of the effectiveness of a feature in classifying of the training data.

The information gain of feature F in D, is defined as equation (2).

$$Gain(D, F) = Entropy(D) - \sum_{v \in Values(F)} \frac{|D_v|}{|D|} Entropy(D_v) \tag{2}$$

Values of F is the set of all possible values for feature F. D_v is the subset of D for which feature F has value v. In information gain feature selection, features have effectiveness based on their gain in descending order i.e. the feature F with the highest gain is chosen as the first feature.

The information gain measure is biased towards tests with many values. That is, it prefers to select features having a large number of values. For example, if instances contains a feature such as id (identification number), information gain selects the id number feature as the first choice. However, id number is not very effective or useful towards classification.

Gain ratio measure is an extension of information gain to solve this drawback. It applies a kind of normalization to information gain using a "split information" value defined as equation (3).

$$SplitInfo(D, F) = \sum_{v \in Values(F)} -\frac{|D_v|}{|D|} \log_2 \frac{|D_v|}{|D|} \tag{3}$$

For each given value, it considers the number of instances having that value in relation to the total number of instances in D. The gain ratio is defined as equation (4).

$$GainRatio(D, F) = \frac{Gain(D, F)}{SplitInfo(D, F)} \tag{4}$$

The feature with the maximum gain ratio is selected as the appropriate feature. More detailed explanation about feature selection and Gain ratio can be found in [14, 15].

5 AIRS

Artificial Immune Recognition System (AIRS) was investigated by Watkins [8]. AIRS can be applied to classification problems, which is a very common real world data mining task. Most other artificial immune system (AIS) research concerns unsupervised learning and clustering. The only other attempt to use immune systems for supervised learning is the work of Carter [16]. The AIRS design refers to many immune system metaphors including resource competition, clonal selection, affinity maturation, memory cell retention. It also includes the resource limited artificial immune system concept investigated by [17]. In this algorithm, the feature vectors presented for training and test are named as antigens while the system units are called B cells. Similar B cells are represented with Artificial Recognition Balls (ARBs) and these ARBs compete with each other for a fixed resource number. This provides ARBs with higher affinities to the training antigen to improve the training process. The memory cells formed after the presentation of all training antigens, would be used to classify test antigens.

AIRS has four stages. The first is performed once at the beginning of the process (normalization and initialization), and other stages constitute a loop and are performed for each antigen in the training set: ARB generation, Competition for resources and nomination of candidate memory cell, promotion of candidate memory cell into memory pool. The mechanism to develop a candidate memory cell is as follows [8, 18]:

1. A training antigen is presented to all the memory cells belonging to the same class as the antigen. The memory cell most stimulated by the antigen is cloned. The memory cell and all the just generated clones are put into the ARB pool. The number of clones generated depends on the affinity between the memory cell and antigen, and affinity in turn is determined by Euclidean distance between the feature vectors of the memory cell and the training antigen. The smaller the Euclidean distance, the higher the affinity, the more is the number of clones allowed.

2. Next, the training antigen is presented to all the ARBs in the ARB pool. All the ARBs are appropriately rewarded based on affinity between the ARB and the antigen as follows: An ARB of the same class as the antigen is rewarded highly for high affinity with the antigen. On the other hand, an out of class ARB is rewarded highly for a low value of affinity measure. The rewards are in the form of number of resources. After all the ARBs have been rewarded, the sum of all the resources in the system typically exceeds the maximum number allowed for the system. The excess number of resources held by ARBs are removed in order starting from the ARB of lowest affinity and moving higher until the number of

resources held does not exceed the number of resources allowed for the system. Those ARBs, which are not left with any resources, are removed from the ARB pool. The remaining ARBs are tested for their affinities towards the training anti-gen. If for any class of ARB the total affinity over all instances of that class does not meet a user defined stimulation threshold, then the ARBs of that class are mutated and their clones are placed back in the ARB pool. Step 2 is repeated until the affinity for all classes meet the stimulation threshold.

3. After ARBs of all classes have met the stimulation threshold, the best ARB of the same class as the antigen is chosen as a candidate memory cell. If its affinity for the training antigen is greater than that of the original memory cell selected for cloning at step 1, then the candidate memory cell is placed in the memory cell pool. If in addition to this the difference in affinity of these two memory cells is smaller than a user defined threshold, the original memory cell is removed from the pool.

These steps are repeated for each training antigen. After completion of training the test data are presented only to the memory cell pool, which is responsible for actual classification. The class of a test antigen is determined by majority voting among the k most stimulated memory cells, where k is a user defined parameter.

Several studies have been done to evaluate the performance of AIRS [18, 19, 20, 21, 22]. The results show that AIRS is comparable (almost similar) with famous and powerful classifiers.

6 Experiments and Results

As discussed in Sections 3 and 4, feature selection was performed on the dataset ob-tained from Silla et. al. [5]. The Gain Ratio feature subset evaluation with best first search strategy was used to reduce the dimensional of features. In the best first strat-egy, the best features are inserted into the feature subset to achieve highest accuracy. The features are evaluated based on effectiveness measures such as information gain and gain ratio. With TMM, the dataset from the study by Norowi et. al. [4] discussed in Section 2 with 63 features and 193 instances was used.

Some experiments were carried out in order to determine how AIRS performed TMM and LM genre classification in comparison to some other famous classifiers. One advantage of AIRS is that it is not necessary to know the appropriate settings and parameters for the classifier. The most important element of the classifier is its ability of self-determination [22] the AIRS is able to determine the suitable final structure of the system on its own. In general, the various parameters of the algo-rithm has minimal effect on the performance of system. However, based on some ex-perimentation, appropriate AIRS algorithm parameters were determined and shown in Table 3 .

These parameters have been discussed in the algorithm description presented in Section 5. In addition, the Affinity Threshold Scalar (ATS) parameter value shown in Table 3, is a value between 0 and 1 that provides a cut-off value for memory cell replacement in the AIRS training routine.

Table 3 Algorithm Parameters

NO	Used Parameter	Value
1	Clonal rate	10
2	Mutation rate	0.1
3	ATS	0.2
4	Stimulation threshold	0.99
5	Resources	150
6	Hyper mutation rate	2.00
7	K value in KNN classifier	4

To evaluate the performance of AIRS for musical genre classification, the following classifiers listed below were selected. This list includes a wide range of paradigms. The available programmes in the WEKA [15] data mining workbench and the default parameters were used for each algorithm.

- Bagging
- Bayesian Network
- Cart
- Conjunctive rule learner (Conj-Rules)
- Decision Stump
- Decision Table
- IB1
- J48 (an implementation of C4.5)
- Kstar
- Logistic
- LogitBoost
- Multi-layer neural network with back propagation (MLP)
- Naïve Bayesian
- Nbtree
- PART (a decision list [23])
- RBF Network
- SMO (a support vector machine implementation [24])

A 10-fold cross validation approach was used to estimate the predictive accuracy of the algorithms. In this approach, data instances are randomly assigned to one of 10 approximately equal size subsets. At each iteration, all but one of these sets are merged to form the training set while the classification accuracy of the algorithm is measured on the remaining subset. This process is repeated 10 times, choosing a different subset as the test set each time until all data instances have been used 9 times for training and once for testing. The final predictive accuracy is computed by dividing the number of correctly classified instances to the number of tested instances. This approach was used in all experiments to control the validity of experiments.

6.1 TMM Results

The accuracies achieved by classifiers for TMM are shown in Table 4. AIRS, together
Logistic and SMO obtained highest classification accuracy of 86%. Table 5 shows the
confusion matrix obtained by using AIRS for TMM genre classification. Results show
the class B has the worst behavior among classes and only 50% of this class instances
are classified correctly. More exploration on the data collection and feature extraction
for this class would need to be done as future work to achieve more accuracy.

Table 4 TMM Genre Classification Accuracies

NO	Method	Accuracy (%)
1	Conj-Rules	31.60
2	Decision Stump	33.68
3	Decision Table	52.85
4	CART	61.67
5	PART	68.39
6	J48	73.06
7	Nbtree	75.13
8	Bagging	76.68
9	Naïve Bayesian	77.72
10	RBF	80.31
11	Bayesian Network	80.83
12	Kstar	80.83
13	LogitBoost	81.35
14	MLP	84.47
15	IB1	84.97
16	Logistic	86.01
17	SMO	86.01
18	AIRS	86.01

Table 5 Confusion Matrix for TMM Genre classification

	A	B	C	D	E	F	G	H	I	J
A	28	0	0	0	0	2	0	0	0	1
B	1	6	1	0	1	0	0	0	3	0
C	0	1	22	0	0	0	0	0	0	0
D	0	0	2	17	0	0	0	0	0	0
E	0	0	0	0	8	0	0	0	0	2
F	0	0	0	0	1	12	0	0	0	2
G	0	1	1	1	0	0	40	0	0	0
H	0	0	0	0	1	0	0	11	0	1
I	0	0	0	0	1	0	2	0	14	0
J	0	0	0	0	2	0	0	0	0	8

6.2   LM Results

Table 6 shows the results for LM genre classification. The accuracy achieved by AIRS is not as good as TMM. However, the performance is better than the rule-based, tree-based and Bayesian classifiers. The confusion matrix for LM classification using AIRS and the additional feature selection face is shown in Table 7. Results show that class I achieves less than 50% accuracy.

Table 6 Latin Genre Classification Accuracies

NO	Method	Accuracy (%)
1	Conj-Rules	18.77
2	Decision Stump	18.83
3	Decision Table	47.80
4	CART	60.00
5	PART	62.23
6	J48	58.50
7	Nbtree	56.27
8	Bagging	69.10
9	Naïve Bayesian	58.53
10	RBF	63.73
11	Bayesian Network	61.63
12	Kstar	69.37
13	LogitBoost	63.67
14	MLP	71.70
15	IB1	74.80
16	Logistic	79.37
17	SMO	75.17
18	AIRS	67.47

Table 7 Confusion Matrix for LM Genre classification

	A	B	C	D	E	F	G	H	I	J
A	293	1	0	0	0	5	0	1	0	0
B	0	209	17	17	4	12	5	17	14	5
C	0	13	172	23	7	16	1	22	20	26
D	0	14	18	186	2	2	4	27	35	12
E	0	16	8	3	264	3	3	1	0	2
F	16	13	18	4	2	186	0	12	31	18
G	0	18	9	19	11	0	234	8	0	1
H	2	15	45	44	2	9	0	164	13	6
I	0	13	37	42	0	40	0	20	128	20
J	0	21	30	17	1	13	1	10	9	180

As for the performances of the classifiers between both the musical genre sets, the Logistic classifier performed best for both cases. The Logistic Classifier in WEKA uses logistic regression with a ridge estimator to classify instances. Further reading on ridge estimators can be found in Houwelingen et. al. [25]. More tests would be needed in comparing musical genres from various cultures to confirm the usefulness of the Logistic classifier as a musical genre classifier in addition to investigating methods to improve the performance of AIRS as a musical genre classifier. A more in-depth analysis of the misclassification of certain genres, such as Etnik Sabah of TMM and Sertaneja of LM in particular, from each of the musical genre sets would also be needed.

7 Conclusions

AIRS is the most important classifier among the AIS based classifiers. In this study AIRS was used to classify TMM and LM genres. Experiments were conducted to test the feasibility of AIRS as a musical genre classifier. According to experimental results tested on two sets of musical genres from differing cultures, AIRS showed a considerably high performance with regard to the classification accuracy for TMM genres. The performance was close to other popular classifiers such as Logistic and SMO. However, this performance was not achieved by AIRS for the Latin musical genres. AIRS does not show promising performance for LM genres in comparison to some popular classifiers. More in-depth study and comparison with musical genre sets from a larger number of cultures would need to be investigated as future work.

References

1. Dannenberg, R., Foote, J., Tzanetakis, G., Weare, C.: Panel: New Directions in Music Information Retrieval. In: Int. Computer Music Conf., Int. Computer Music Association, pp. 52–59 (2001)
2. Tzanetakis, G., Cook, P.: Musical Genre Classification of Audio Signals. IEEE Transactions on Speech and Audio Processing 10, 293–302 (2002)
3. Wold, E., Blum, T., Keislar, D., Wheaton, J.: Content-based Classification, Search, and Retrieval of Audio. IEEE Multimedia 3, 27–36 (1996)
4. Norowi, N.M., Doraisamy, S., Rahmat, R.W.: Traditional Malaysian musical genres classification based on the analysis of beat feature in audio. J. Information Technology in Asia (2007)
5. Silla, J., Carlos, N., Kaestner, C.A.A., Koerich, A.L.: Automatic Music Genre Classification Using Ensemble of Classifiers. In: IEEE Int. Conf. of Systems, Man and Cybernetics (SMC), pp. 1687–1692 (2007)
6. de Castro, L.N., Timmis, J.: Artificial Immune Systems as a novel Soft Computing Paradigm. J. Soft Computing 7, 526–544 (2003)
7. de Castro, L.N., Timmis, J.: Artificial Immune Systems: A New Computational Intelligence Approach. Springer, Heidelberg (2002)
8. Watkins, A.: A Resource Limited Artificial Immune Classifier, M.S. thesis, Department of Computer Science, Mississippi State University, USA (2001)

9. Golzari, S., Doraisamy, S., Sulaiman, M.N., Udzir, N.I.: A Comprehensive Study in Benchmarking Feature Selection and Classification Approaches for Traditional Malay Music Genre Classification. In: Proc. of the 2008 Int. Conf. on Data Mining (DMIN), pp. 71–77 (2008)
10. Nasaruddin, M.G.: Muzik Tradisional Malaysia. Dewan bahasa dan pustaka, Kuala Lumpur, Malaysia (2003)
11. Matusky, P.: Malaysian Shadow Play and Music: Continuity of an Oral Tradition. Oxford University Press, Oxford (1993)
12. Ang, M.: Layered Architectural Model for Music: Malaysian Music on the World Wide Web. Ph.D. dissertation, UPM (1998)
13. Becker, J.: The Percussive Patterns in the Music of Mainland Southeast Asia. J. Ethnomusicology 2, 173–191 (1968)
14. Hall, M.A., Smith, L.A.: Practical feature subset selection for machine learning. In: Proc. of the 21st Australian Computer Science Conf., pp. 181–191 (1998)
15. Witten, H., Frank, E.: Mining: Practical Machine Learning Tools and Techniques, 2nd edn. Morgan Kaufmann, San Francisco (2005)
16. Carter, J.H.: The immune systems as a model for pattern recognition and classification. J. American Medical Informatics Association 7, 28–41 (2000)
17. Timmis, J., Neal, M.: A Resource Limited Artificial Immune System. Knowledge Based Systems 14, 121–130 (2001)
18. Marwah, G., Boggess, L.: Artificial Immune Systems for Classification: Some Issues. In: Proc. of the 1st Int. Conf. on Artificial Immune Systems (ICARIS), pp. 149–153 (2002)
19. Watkins, A., Boggess, L.: A New Classifier Based on Resource Limited Artificial Immune Systems. In: Congress on Evolutionary Computation, Part of the World Congress on Computational Intelligence, pp. 1546–1551 (2002)
20. Watkins, A., Timmis, J.: Artificial Immune Recognition System (AIRS): Revisions and Refinements. In: Proc. of the 1st Int. Conf. on Artificial Immune Systems (ICARIS), pp. 173–181 (2002)
21. Watkins, A.: Exploiting Immunological Metaphors in the Development of Serial, Parallel, and Distributed Learning Algorithms. Ph.D dissertation, Computer Science, University of Kent, England (2005)
22. Watkins, A., Timmis, J., Boggess, L.: Artificial Immune Recognition System (AIRS): An Immune-Inspired Supervised Learning Algorithm. Genetic Programming and Evolvable Machines 5, 291–317 (2004)
23. Frank, E., Witten, I.H.: Generating Accurate Rule Sets without Global Optimization. In: 15th Int. Conf. on Machine Learning (1998)
24. Keerthi, S.S., Shevade, S.K., Bhattacharyya, C., Murthy, K.R.K.: Improvements to Platt's SMO Algorithm for SVM Classifier Design. Neural Computation 13, 637–649 (2001)
25. van Houwelingen, J.C., le Cessie, S.: Ridge Estimators in Logistic Regression. J. Royal Statistical Society - Series C: Applied Statistics 41, 191–201 (1992)

Author Index

Glossary

Audio Engineering – acoustic and electrical technology applied to audible sound signal

Augmented Fifth – see: interval

Beat – pulse, the regular rhythmic pattern (template) in a piece of music

Chord – a combination of three or more notes sounding simultaneously

Chromatic Scale – musical scale with twelve pitches, each a semitone apart.

Consonance – a chord or a set of sounds sounding in concordance

Diatonic Scale – seven note musical scale, e.g. major scale, minor scale

Digital Audio – sound signal represented in digital form

Dissonance – a chord or a set of sounds sounding in discord

Envelope – a curve connecting the peaks of a graph of sound wave; can be also calculated for amplitude spectrum

Fifth – see: interval

Formant – a region of concentration of energy in amplitude spectrum, consisting of a number of harmonic partials

Fourth – see: interval

Frequency – number of cycles of a repeating event per unit of time

Fundamental Frequency (of a periodic signal) – the inverse of the period length

Fundamental – the lowest frequency in a harmonic series

Harmonic (of a wave) – component frequency of the signal that is an integer multiple of the fundamental frequency

Harmony – describes simultaneous sounding of two or more notes (chords) and their arrangement in a succession

Hertz (Hz) – 1 herz (Hz) is a frequency equal to one cycle per second

Interval – relationship between two pitches
- Unison (perfect unison) – both pitches are the same; an example of consonance
- Minor second – the pitches are semitone apart
- Major second – the pitches are two semitones apart
- Minor third - a musical interval of three semitones
- Major third – an interval of fourth semitones
- Fourth (perfect fourth) - the notes are five semitones apart; can be augmented (by a semitone) or diminished (also by a semitone)
- Tritone, or augmented fourth, or diminished fifth – the interval of six semitones; an example of dissonance
- Fifth (perfect fifth) - musical interval between a note and the note seven semitones above it; can be augmented (by a semitone) or diminished (also by a semitone)
- Minor sixth – an interval between pitches which are eight semitones apart
- Major sixth - an interval between pitches which are nine semitones apart
- Minor seventh – between pitches which are ten semitones apart
- Major seventh – between pitches which are eleven semitones apart
- Octave (perfect octave) - between one musical pitch and another with half or double its frequency, i.e. twelve semitones apart

Key – a family of diatonic tones; the tone to which a scale is referred. There are 24 major or minor diatonic scales
Key – the part of an instrument (e.g. piano) which is used to play it

Major Scale – diatonic scale made up of seven distinct notes (plus an eighth which duplicates the first an octave higher). The key note is called tonic. The notes in the scale are in the following steps: tonic, major second above the tonic, major third above the tonic, fourth above the tonic, fifth above the tonic, major sixth above the tonic, major seventh above the tonic, and octave

Melody – a succesion of notes

Metadata – data about the data

Meter - the rhythmic element measured by division into parts of equal time value; describes beat

MIDI – Musical Instrument Digital Interface, protocol for controlling digital audio devices. MIDI data represent such events as key number (representing pitch), note on, note off (pressing and releasing the key of a keyboard), voice number (timbre selected), and so on

Minor Scale - diatonic scale, with a third scale degree at an interval of a minor third above the tonic, and various versions of higher steps (sixth and seventh); see Major Scale for comparison

Music Genre - category that identifies music and distinguishes from other types of music, e.g. popular music, blues, country, jazz, rock, and so on

Musical Scale – a group of musical notes arranged in ascending and descending order. Examples: major scale, minor scale, chromatic scale

Octave - see: interval

Offset - the ending of a musical note or other sound, in which the amplitude decreases

Onset - the beginning of a musical note or other sound, in which the amplitude rises

Pitch - represents the perceived fundamental frequency of a sound

Pitch Tracking - estimation of pitch of note events in a melody or a piece of music

Polyphonic Music - music arranged in parts for several voices or instruments

Rhythm - basic temporal element of music, arrangement of notes into regular patterns according to their relative duration and relative accentuation

Second – see: interval

Semitone – the smallest interval used in Western music. Example: the distance between two neighboring keys of the keyboard is a semitone.

Signal - physical (e.g. electrical) varying quantity that carries information. Example: audio signal, conveying information on changes of amplitude of sound wave in time

Sound – vibrations of frequency withing the range of approximately 16 Hz – 20 kHz, transmitted through the air or other medium, and producing the sensation of hearing

Spectrum – the result of analysis that can be performed for sound data using e.g. Fourier transform, transforming time domain into frequency domain (for complex-valued functions), and showing magnitudes (amplitude spectrum) and phases (phase spectrum) of sound components, represented here in sinusoidal form

Temperament - system of tuning in music. Example: equal temperament – the ratio of frequency of any adjacent notes in this system is constant and equal to $2^{1/12}$, since an octave in divided into 12 parts (semitones) with equal step in logarithmic scale.

Tempo - speed of a given piece of music

Third – see: interval

Timbre - the quality of a musical sound that distinguishes sounds, even if they are of the same pitch or loudness

Transient - a short-duration part of signal that represents transitory phase of a musical sound. Example: onset (the beginning of the sound), offset (the end of the sound)

Triad – a group of three notes in a chord, consisting of a given tone, a third (minor or major), and a fifth (perfect, augmented, or diminished)

Unison – see: interval

Voicing – ordering of notes in a chord